Topics in
Current Physics

30

Topics in Current Physics

Founded by Helmut K. V. Lotsch

Applications of the
Monte Carlo Method
in Statistical Physics

Edited by K. Binder

With Contributions by
A. Baumgärtner   K. Binder   J. P. Hansen
M. H. Kalos   K. W. Kehr   D. P. Landau   D. Levesque
H. Müller-Krumbhaar   C. Rebbi   Y. Saito
K. E. Schmidt   D. Stauffer   J. J. Weis

Second Edition

With 90 Figures

Springer-Verlag Berlin Heidelberg New York
London Paris Tokyo

Professor Dr. Kurt Binder

Fachbereich Physik, Johannes-Gutenberg-Universität, Postfach 39 80,
D-6500 Mainz 1, Fed. Rep. of Germany

ISBN 3-540-17650-0 2. Auflage Springer-Verlag Berlin Heidelberg New York Tokyo
ISBN 0-387-17650-0 2nd edition Springer-Verlag New York Heidelberg Berlin Tokyo

ISBN 3-540-12764-X 1. Auflage Springer-Verlag Berlin Heidelberg New York
ISBN 0-387-12764-X 1st edition Springer-Verlag New York Heidelberg Berlin

Library of Congress Cataloging-in-Publication Data. Applications of the Monte Carlo method in statistical physics. (Topics in current physics ; 36) Includes bibliographies and index. 1. Monte Carlo method. 2. Statistical physics. I. Binder, K. (Kurt), 1944–. II. Baumgärtner, A. (Artur). III. Series. QC174.85.M64A67 1987 530.1′3 87-4882

Offset printing and bookbinding: Konrad Triltsch, Graphischer Betrieb, Würzburg.
2153/3150-543210

Preface to the Second Edition

Only three years have passed since this volume first appeared. The fact that a second edition is already necessary is characteristic of the great interest in this rapidly expanding field. Together with a companion volume ("Monte Carlo Methods in Statistical Physics", Topics in Current Physics, Vol. 7), the second edition of which was published in 1986, and an introductory monograph to computer simulation for the newcomer (D.W. Heermann "Introduction to Computer Simulation Methods in Theoretical Physics", Springer, Berlin, Heidelberg 1986), this book gives an up-to-date survey of numerous applications of the Monte Carlo method in various branches of statistical physics, including fields such a polymer physics and lattice gauge theory.

To update this book, a new chapter "Recent Developments" was written as a joint effort of all the authors of the various chapters in this book. This new chapter contains about 300 recent references and gives a condensed description of many aspects of this new research. With respect to the old chapters, typographical errors have been corrected and fuller references have been given wherever appropriate, but otherwise the layout and contents of the book have been left unchanged. It is hoped that this paperback edition will be useful to a wide range of scientists and students in experimental and theoretical condensed-matter physics and related disciplines (physical chemistry, materials science, applied mathematics, elementary particle physics, etc.).

Once again it is a pleasure for the editor to thank his coauthors who contributed to this book for their valuable and fruitful cooperation, which has been absolutely essential for the success of this effort. He also wishes to thank numerous colleagues for their helpful comments and advice, and for generously supplying preprints and reprints.

Mainz, March 1987   *Kurt Binder*

Preface to the First Edition

Monte Carlo computer simulations are now a standard tool in scientific fields such as condensed-matter physics, including surface-physics and applied-physics problems (metallurgy, diffusion, and segregation, etc.), chemical physics, including studies of solutions, chemical reactions, polymer statistics, etc., and field theory. With the increasing ability of this method to deal with quantum-mechanical problems such as quantum spin systems or many—fermion problems, it will become useful for other questions in the fields of elementary-particle and nuclear physics as well.

The large number of recent publications dealing either with applications or further development of some aspects of this method is a clear indication that the scientific community has realized the power and versatility of Monte Carlo simulations, as well as of related simulation techniques such as "molecular dynamics" and "Langevin dynamics," which are only briefly mentioned in the present book. With the increasing availability of recent very-high-speed general-purpose computers, many problems become tractable which have so far escaped satisfactory treatment due to practical limitations (too small systems had to be chosen, or too short averaging times had to be used). While this approach is admittedly rather expensive, two cheaper alternatives have become available, too: (i) array or vector processors specifically suited for wide classes of simulation purposes; (ii) special purpose processors, which are built for a more specific class of problems or, in the extreme case, for the simulation of one single model system. In this way one hopes to answer subtle theoretical questions which cannot be settled even with the most advanced and successful analytical tools such as renormalization group theory, nor by experiment. In this way, Monte Carlo simulations make a considerable impact on the further development of scientific computing, including hardware.

On the other hand, simulation methods are not only useful for research, but it is increasingly realized that they can play a role in student education in advanced university courses and seminars, too. First, it is advantageous to learn and practice this relatively simple and widely useful tool; second, many general concepts of statistical physics (subsystems, statistical fluctuations, probability distributions) can be literally and explicitly demonstrated.

The present book complements a previous volume with the same title[*]. It starts with a rather simple introduction, which should help the beginner become a success-

---

[*]See Volume 7 listed on inside back cover.

ful practitioner in the field, and then emphasizes either fields which have not been treated at all in the previous volume (diffusion phenomena, polymers, lattice gauge theory) or subjects which have seen dramatic activity in the last five years (water and other molecular or ionic fluids; two-dimensional melting; phase diagrams and multicritical phenomena; fermion problems; spin glasses and percolation). It contains about one thousand references and thus should be a very useful guide to the original literature. Together with the first volume, it forms a thorough overview of all applications of Monte Carlo simulations in "statistical physics" taken in a very wide sense (including ordinary solid-state physics, field and elementary particle theory, quantum-mechanical problems, metallurgy, surface physics). Since no other literature with a similar scope exists as yet, this undertaking was possible only through very constructive collaboration and fruitful interaction among the leading experts in the various fields who have contributed to the present book. It is a great pleasure for me to acknowledge their very valuable efforts in compiling all the material described here, and to thank them for stimulating and pleasant collaboration.

Jülich, November 1983   *Kurt Binder*

Contents

List of Contributors

Baumgärtner, Artur

   IBM Research Laboratory, San José, CA 95139, USA

   Permanent address: Institut für Festkörperforschung, Kernforschungsanlage
Jülich, D-5170 Jülich, Fed. Rep. of Germany

Binder, Kurt

   Institut für Festkörperforschung, Kernforschungsanlage Jülich,
D-5170 Jülich, Fed. Rep. of Germany

Hansen, Jean-Pierre

   Université Pierre et Marie Curie, Lab. de Physique Théoretique des Liquides,
4, Place Jussieu, F-75239 Paris, Cedex 05, France

Kalos, Malvin H.

   Courant Institute, New York University, 251 Mercer Street, New York, NY 10012, USA

Kehr, Klaus W.

   Institut für Festkörperforschung, Kernforschungsanlage Jülich,
D-5170 Jülich, Fed. Rep. of Germany

Landau, David P.

   Department of Physics and Astronomy, The University of Georgia,
Athens, GA 30602, USA

Levesque, Dominique

   Laboratoire de Physique Théorique et Hautes Energies, Université de Paris Sud,
F-91405 Orsay, France

Müller-Krumbhaar, Heiner

   Institut für Festkörperforschung, Kernforschungsanlage Jülich,
D-5170 Jülich, Fed. Rep. of Germany

Rebbi, Claudio

   Department of Physics, Brookhaven National Laboratory, Upton,
Long Island, NY 11973, USA

Saito, Yukio

   Institut für Festkörperforschung, Kernforschungsanlage Jülich,
D-5170 Jülich, Fed. Rep. of Germany

   Permanent address: Department of Physics, KEIO University,
3-14-1 Hiyoshi, Kohoku-ku, Yokohama 223, Japan

Schmidt, Kevin E.

Theoretical Division, Los Alamos National Laboratory, Los Alamos, NM 87545, USA

Permanent address: Courant Institute, New York University, 251 Mercer Street, New York, NY 10012, USA

Stauffer, Dietrich

Institut für Theoretische Physik, Universität zu Köln, Zülpicherstraße 77, D-5000 Köln 41, Fed. Rep. of Germany

Weis, Jean-Jacques

Laboratoire de Physique Théorique et Hautes Energies, Université de Paris Sud, F-91405 Orsay, France

1. A Simple Introduction to Monte Carlo Simulation and Some Specialized Topics

K. Binder and D. Stauffer

With 6 Figures

The aim of this introductory chapter is twofold.

i) It should help someone who never made a Monte Carlo simulation but knows how to program a computer to start with a Monte Carlo study of simple systems. For this purpose we include simple explicit examples of such programs written in FORTRAN language. We start by outlining how one gets random numbers (Sect.1.1.1) and sketch the percolation problem as an example of their immediate use (Sect.1.1.2). The Ising system is presented as the simplest case of a system with thermal fluctuations (Sect.1.1.3). The discreteness of variables in both cases is a simplifying feature, which enables computer time and/or computer memory to be saved by special "tricks" such as the multispin coding technique (App.1.A). Section 1.1.4 describes how systems with continuous variables (like Heisenberg spins) are treated, followed by an introductory discussion of the various pitfalls which may hamper such simulations (Sect.1.1.5). With this background, it should be possible to start doing some Monte Carlo work; readers interested in the more formal and theoretical aspects of this technique should consult Chap.1 of [1.1].

ii) The second aim is to review some technical aspects of simulations of "classical" systems where recent progress has been achieved (Sect.1.2), and thus update the corresponding sections of [1.1]. This section should be useful particularly if the reader has already had some experience with Monte Carlo techniques. Again the emphasis will be on problems illustrated with examples taken from the authors own research: information drawn from the study of subsystems and distribution functions (Sect.1.2.1), details on the thermodynamic integration method for obtaining free energy and entropy (Sect.1.2.2), sampling of intensive variables (like the chemical potential of a lattice gas simulated at a given density), Sect.1.1.3, surface tension between coexisting phases (Sect.1.2.4), the problems of identifying a first-order transition (Sect.1.2.5), and techniques relying on the use of the fluctuation-dissipation theorem relating correlation functions and response functions (Sect. 1.2.6). We emphasize that there are many other technical points where important progress was achieved, which will be treated elsewhere in this book; e.g., when simulating a fluid where atoms are randomly moved from their old position r to a new position r' it is advantageous to move preferentially in the direction of the force acting on the atom ([1.2], for more details, see Chap.2); in simulating polymer adsorption it is advantageous to use "biased simple sampling" where polymer configur-

1

ations with many contacts to the wall are generated more often ([1.3], for more details, see Chap.5); particular technical problems also occur when one is sampling both thermal and compositional disorder (Chap.8). Last but not least, a real breakthrough has been achieved with the further development of *Monte Carlo renormalization group techniques* for the study of critical phenomena; this progress was recently reviewed in [1.4], and thus here only the additional most recent work is described briefly in Chap.3.

1.1 A First Guide to Monte Carlo Sampling

1.1.1 Random Numbers

The name Monte Carlo method arises from the fact that this method uses random numbers similar to those coming out of roulette games. But producing up to 10^{10} such random numbers in Monte Carlo in this way would consume a lot of time and money; thus computers are used instead. For understanding the principle of "random number generators" implemented for the software of digital computers, we recall that each "word" of computer memories consists of m bits for one integer. One of these bits indicates the sign, the others give the magnitude. Thus $2^{m-1}-1$ is the largest integer the computer can handle. Typically $m = 32$ or 60, but some machines also have $m = 16$ or $m = 48$. Thus one may produce a "pseudorandom" integer I_1 between 0 and 2^{m-1} from an initial integer I_0 by multiplying I_0 with a suitable number like 65539, which in general would produce an integer distinctly exceeding 2^{m-1}, and by putting the product back into the interval from 0 to 2^{m-1} by a modulo operation

$$I_1 = 65549*I_0 \text{ modulo } 2^{m-1} \quad . \tag{1.1}$$

This modulo operation can be done on many but not all computers by simply omitting the leading bits of the product, which is done automatically when too large integers are multiplied. But one has to ensure that the result is not negative. Thus on a 32-bit computer the following FORTRAN statements produce from a random integer (stored as variable I) another integer (stored at the end again under the name I) and a real number X between zero and unity:

    I = I*65539

    IF(I.LT.0) I = I + 2147483647 + 1

    X = I*0.4656612E - 9 .   (1.2)

By repeating this procedure one produces another integer I and real number X, and this can be repeated again and again; the distribution of real numbers is *approximately homogeneous* and *approximately random* between zero and one (as a rule of thumb, the larger m is the better the approximation). The resulting series of "random" numbers is completely reproducible, however, if one always starts with the same value for I, which repeats itself after a period of at most 2^{m-2}. The initial value of I

2

must be a positive *odd* integer. If it is not very large, one needs several iterations of this "random number generator" (1.2) to get numbers X which are not very
small. Thus at the beginning of the program one may use a short loop which "warms
up" this generator. It is also safer to put $I = 2*I + 1$ after I is set initially, so
that one never gets erroneously an even I. Computer time can be saved if this generator is not written as a separate subroutine but inserted in the main program
wherever it is needed. Also the last line of (1.2) where integers are converted to
real can sometimes be avoided. For example, if during the program one must compare
millions of times random numbers X to a fixed probability P [or a finite set of
fixed probabilities P(J)], it is better to define at the beginning an integer probability $I_p = 2^{m-1}*P$, and later one simply compares the I's from (1.2) to this I_p
without ever calculating the X's.

The above generator (1.2) is widespread under various names usually beginning
with RAND..., but has disadvantages. The reader can easily test it (or any other
random number generator) by trying to fill up a simple cubic lattice of L^3 sites
$n(k_1, k_2, k_3)$, where the k_i run from 1 to L. Initially all elements n are empty
$(n = 0)$. Then one point is selected randomly by calculating its three coordinates
k_1, k_2 and k_3 consecutively from three consecutive random numbers X through $k_1 = 1$
$+ X*L$, etc. If the randomly selected lattice site is still empty it is then
filled $(n = 1)$; otherwise it is left filled. This procedure is repeated again and
again, about $t*L^3$ times, with t of order 10. Theoretically the fraction of empty
lattice sites should decay as exp(-t). With (1.2), one gets good results for L = 10,
but already at L = 15 some deviations are seen. Very drastically, at L = 20 more
than 2000 sites remain empty even if theoretically less than one should be left
empty. This failure is due to strong (triplet) correlations between the consecutive
pseudorandom numbers generated by (1.2).

This problem can be avoided by "mixing" two different random number generators
[1.5]. The function RANF listed in (1.3) determines randomly an integer K between
1 and 256, extracts a random number RANF from a table of 256 random numbers RN, and
puts in its place a freshly calculated number RN(K):

```
FUNCTION RANF(I)
DIMENSION RN(256)
COMMON  RN,J
I=I*32771
IF(I.LT.0) I=I+2147483647+1
K=1+I/8388608
RANF=RN(K)
J=J*65539
IF(J.LT.0) J=J+2147483647+1
RN(K)=I*0.4656612E-9
RETURN
END  .
```

(1.3)

3

At the beginning of the program I and J have to be set, and the array RN has to be filled with 256 random numbers. In this way, the quality of the pseudorandom numbers on 32-bit computers is distinctly improved, but procedure (1.3) is distinctly slower than (1.2), and some problems still remain. Of course, one again saves some time by inserting the statements of (1.3) into the main program instead of calling them through a function. A random number generator of about the speed of (1.2) and at least the quality of (1.3) has recently been developed [1.6]; we do not list its statements here since they involve using "machine language."

For computers using about 60 bits per word the very fast random generators RANF supplied with their FORTRAN compiler seem to be good enough according to all practical experience with statistical physics problems. One thus uses it by simply calling it as one calls the exponential function.

The development of "special-purpose" processors working in parallel [1.7] has become a very powerful tool for simulations. The development of efficient random number generators for such machines is still an active field of research [1.8]. We are not going to discuss this problem here, nor shall we discuss the problem of doing Monte Carlo simulations on commercially available vector machines in an optimal way, but refer the interested reader to [1.9]. More discussion on the generation of random numbers in general is found in [1.10].

1.1.2 An Example of "Simple Sampling": The Percolation Problem

What can we do with these random numbers in the field of statistical physics? Most of the Monte Carlo work described in this book refers to thermal (or quantum) fluctuations to be sampled. This application will be introduced in the following section. Even simpler is the purely geometric problem of the so-called percolation transition: one considers an (infinite) lattice where each site is randomly occupied with probability p and empty with probability 1-p; neighboring occupied sites are said to form "clusters" [1.11]. Some questions concerning this problem are: how many clusters $n_\ell(p)$ containing ℓ occupied sites exist in the lattice? At which probability p_c does an infinite cluster form for the first time, that "percolates" from one boundary of the lattice to the opposite one?

A Monte Carlo procedure which generates sample configurations of such a partially filled lattice then simply consists of FORTRAN statements like

```
    DO   1    K1=1,L
    DO   1    K2=1,L
    DO   1    K3=1,L
    N(K1,K2,K3)=0
  1 IF(RANF(I).LT.P) N(K1,K2,K3)=1   .
```

(1.4)

One sweep through the lattice already determines the whole system; there is no need to wait until some "equilibrium" is established, in contrast to thermal or quantum systems. By using rather sophisticated programs (see [1.12] or Chap.8 for a program

4

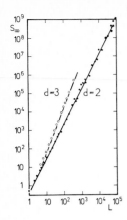

Fig.1.1. Size S_∞ of largest cluster in an $L \times L$ triangular lattice right at the percolation treshold ($p_C = 1/2$) and in a $L \times L \times L$ simple cubic lattice (at $p_C \cong 0.311$ [1.13])

listing) the computer groups the occupied sites into clusters. Since each place is randomly occupied or empty, independent of what happens at neighboring sites, one needs to store only the current plane of a three-dimensional lattice (or the current row in two dimensions) for this analysis. Thus rather large lattices have been successfully investigated. Figure 1.1 shows the size S_∞ of the largest cluster for the two-dimensional $L \times L$ triangular lattice and the case where exactly half the lattice sites are filled [1.13]. Except for small L the data follow a straight line on this log-log plot, which corresponds to a power law $S_\infty \propto L^{91/48}$. This so-called critical exponent or fractal dimension 91/48 for the largest cluster agrees with the actual expectations [1.12-14]. This example, which to our knowledge includes one of the largest systems ever simulated ($95\,000 \times 95\,000$), shows that the size dependence of some quantities may contain useful information. A theoretical basis for such an analysis is provided by the so-called finite-size scaling theory (see [1.1,15-17] and Sects.1.1.5 and 2.1). More information on the percolation problem is found in Chap.8.

1.1.3 An Example of "Importance Sampling": The Ising Model

We now consider problems of statistical thermodynamics, where the state of the system (among other variables) is characterized by some temperature $T > 0$. Microscopically, the state is described by ascribing certain values to the microscopic degrees of freedom (atom positions, spin directions) of the system. It is not just a single configuration of these variables that contributes to a thermal average at nonzero temperature, but rather all possible configurations are weighted according to a probability proportional to the Boltzmann factor $\exp(-E/k_B T)$, E being the system's internal energy in the respective configuration.

A particularly simple system is the spin 1/2 Ising magnet. It can be used as a model of many cooperative systems, where each of the many identical units can be in one of two states only. We call this degree of freedom "spin up" and "spin down", but in spite of this magnetic language we do not really deal with the truly quantum-mechanical aspects of these spins. The magnetization M is the difference between the

5

number of up- and down-spins divided by the total number N of spins. If $S_i = +1$ for up-spins and $S_i = -1$ for down-spins, then $M = \sum_i S_i / N$. Two parallel (antiparallel) spins i and k have energy $-(+)J_{ik}$ which depends on the distance between site i and site k (usually the spins are thought to be located on the sites of a regular lattice). In a magnetic field H pointing upwards, each spin has the additional energy $\pm H$ (plus for down- and minus for up-spins; we choose the units of H such that the magnetic moment per spin is unity). In this example, the total energy is hence

$$E = - \sum_{j<k} J_{ik} S_i S_k - H \sum_i S_i \; . \tag{1.5}$$

In a "simple sampling" procedure for the simulation of thermal fluctuations in such a model system, one would randomly generate configurations of the degrees of freedom, the set of spin directions $\{S_i\}$, calculate for each state $\{S_i\}$ the energy E, and give the former a weight proportional to $\exp(-E/k_BT)$ in calculating any average properties, such as the average magnetization $\langle M \rangle_T = \sum_\nu M_\nu \exp(-E_\nu/k_BT)/\sum_\nu \exp(-E_\nu/k_BT)$, where the index ν labels the generated states. Note that from a single simulation of a set of states ($\{M_\nu\}$) one can obtain the desired averages ($\langle M \rangle_T$) in principle at any temperature.

In practice, this method works for rather small sizes N only (for useful applications in polymer science, see Chap.5), because it samples the configurations of the system *uniformly*. On the other hand, thermodynamic fluctuation theory [1.18] tells us that the distribution function of any macroscopic variable will be sharply peaked around its average value, e.g., for the energy E itself, the distribution will be peaked at $\langle E \rangle_T$, which is proportional to N, while the width of the distribution will be proportional to $\sqrt{N}$. On a scale of energy per spin, the width of the distribution shrinks to zero as $1/\sqrt{N}$ for $N \rightarrow \infty$. Thus at any temperature a rather narrow region of the configuration space of the system contributes significantly to the averages; a very small fraction of the generated states would actually lie in this important region of configuration space. In fact, for (1.5) it is obvious that randomly generated spin configurations will have energies distributed according to a Gaussian distribution around zero energy, rather than around $\langle E \rangle_T$, because parallel and antiparallel pairs of spins will occur on the average equally often irrespective of their distance. Thus, the method above is very inefficient for large N.

This problem is avoided by the *Metropolis* "importance sampling" method [1.19]. This is an algorithm which generates states already with a probability proportional to the Boltzmann factor itself, i.e., the states are distributed according to a Gaussian distribution around the appropriate average value. In the Ising model example, this procedure (see [1.1,19], for a justification) works as follows: one starts with some initial spin configuration, and then repeats again and again the following six steps which simulate the real thermal fluctuations:
1) Select one spin to be considered for flipping ($S_i \rightarrow -S_i$).
2) Compute energy change ΔE connected with that flip.

3) Calculate the "transition probability" $W = \exp(\Delta E/k_B T)$ for that flip.

4) Calculate a random number x between zero and unity.

5) If $x < W$ flip the spin, otherwise do not flip.

6) Analyze the resulting configuration as desired, store its properties to calculate the necessary averages, etc.

It is obvious from this list how in principle the Monte Carlo method works on systems other than Ising models: one only has to replace the words "spin" and "flip(-ping)" by the appropriate words for that system.

Several comments should now be made.

i) By Steps (3-5), the spin flip *on the average* is executed with probability W (if W is <1; for $W > 1$ the flip would always be executed; often W is replaced by a properly normalized transition probability $W' = \exp(-\Delta E/k_B T)/[1 + \exp(-\Delta E/k_B T)])$. By choosing W instead of W' "equilibrium" is approached mostly even somewhat quicker.

ii) Since subsequent states differ only by a single spin flip, and hence their physical properties are very strongly correlated, it often saves time to perform at least parts of Step (6) not after every (attempted) flip, but only after much larger "time" intervals. We define one Monte Carlo step (MCS) per spin by taking the above (5 or 6) steps once for every lattice site, if the spins are selected consecutively. If we select them at random, the MCS is defined by requiring that on the average each spin is selected once. It then often makes sense to perform Step 6 only once after every MCS/spin or even only after every 5th (10th, etc.) MCS/spin, depending on how strongly subsequent states remain correlated. A more formal discussion of these correlations is presented in [1.1].

iii) In a real, very anisotropic magnet, for which (1.5) is a reasonable approximation of the magnetic energy, the (weak)coupling of the spins to the lattice vibrations produces spin flips; hence one MCS/spin on the computer corresponds very roughly to the average time between two such attempts of the thermal motions to flip a given spin. In this sense the Monte Carlo simulation of the Ising model (in this connection also called the Glauber kinetic Ising model [1.20]) can be regarded as the simulation of a truly time-dependent process, where relaxation into equilibrium is observed after sufficiently long times. Note that even then the simulation tells nothing about the scale factor converting 1 MCS/spin to the real physical time scale. The relaxation towards equilibrium is described more realistically if we randomly select the spins to be flipped; but computer time is saved if we go through the lattice in an ordered fashion. Of course, the equilibrium properties are not influenced by this choice.

iv) Since the states of the system are already generated proportional to the desired probability $\exp(-E/k_B T)$, all averages become simple arithmetic averages over the generated states, e.g., $\langle M \rangle \approx \sum_{\nu} M_{\nu}/n$ where n is the number of states. The constant of proportionality involved is not known, however, and hence it is not so straightforward to estimate the partition function $Z = \sum_{\nu} \exp(-E_{\nu}/k_B T)$ and the free energy $F = -k_B T \ln Z$ (Sect.1.2.2).

v) One can save computer time by storing at the beginning of the calculation the small number of different probabilities W for spin flips, instead of evaluating the exponential function again and again. Note also that if a flip is rejected and the old configuration is kept, it is important to count that old configuration again in the averages. Otherwise thermal equilibrium is not described properly. Consider, for example, a single spin in a strong magnetic field: in thermal equilibrium it is nearly always parallel to the field. But should we count in our averages only those configurations which are freshly produced by a spin flip and not also those where the attempt failed, we should get the wrong result of an equal number of up and down orientations in our average. At very low temperatures, where nearly every attempt to flip a spin is bound to fail, this algorithm clearly is rather slow. One can construct a more complicated but much quicker algorithm by keeping track of the number of spins with a given transition probability W_k at each instant of the simulation. Choosing now a spin from the k'th class with a probability proportional to W_k, one can make every attempted spin flip successful (for a justification of this "table method" and more details see [1.1]). Intermediate in effort and time saving is to store for each spin the energy to flip that spin [1.21]. At low temperatures one only looks up this stored energy and does not have to change it if one does not flip the spin.

vi) The disturbance from the boundaries of the system is usually diminished by employing periodic boundary conditions [1.22]. Thus the uppermost and lowermost planes in a three-dimensional lattice are regarded as neighbors in the energy calculation, as are the back and front planes, and also the leftmost and rightmost planes of the lattice. Of course, the periodicity with lattice linear dimensions which is imposed on spin-pair correlation functions, for instance, can still have important disturbing effects ([1.1] and Sect.1.1.5).

vii) For a spontaneous magnetization occurring for temperatures below the Curie temperature T_c, one should start with all spins up (or down), not with their directions chosen randomly. Otherwise one would first get several domains with positive and negative spontaneous magnetization within the simulated system, and the total magnetization behaves rather erratically before a monodomain system is established. (Of course, the kinetics of such a domain-coarsening process may be itself of intrinsic interest and hence one may wish to simulate the kinetics of ordering in exactly this way, Chap.6). For the same reason one should not cool down the system during a simulation from T above T_c to T below T_c; but one may heat it up to use the final state of one temperature as the initial configuration at the slightly higher temperature in order to save computer time. In any case, a number n_0 of states after the start of the simulation will always be affected by the initial state whatever this initial state is, and these states are not yet characteristic of the desired equilibrium state one wants to simulate. Although the influence of these uncharacteristic initial states on the average is negligible when the number n of generated states tends to infinity, for large but finite n it is always advantageous to omit

8

these n_0 initial states from the average, which hence is given by

$$<M> = \sum_{\nu=n_0+1}^{n} M_\nu/(n - n_0) \tag{1.6}$$

in the case of the magnetization, for instance. A more formal discussion of the appropriate choice of n_0 in terms of a "nonlinear relaxation time" is found in [1.1]. Common practice is to follow the relaxation process of M_ν from the initial value to "typical" values in preliminary runs, and thus estimate n_0 by inspection. (Usually one takes n_0 larger than absolutely necessary, in order to be on the safe side. An erroneous estimate of n_0 may result in metastability problems, Sect.1.1.5 and [1.1].)

After these explanations the reader should have no difficulties in understanding the complete FORTRAN program in Table 1.1 for the simple cubic lattice in zero magnetic field at $T = 1.4\ T_c$ (note $J/k_B T_c = 0.221655$ for this model [1.23]). It is obvious how to simplify this program to a two-dimensional square lattice. To incorporate the effects of a magnetic field H one would have to store at the beginning two arrays of transition probabilities, one for up-spins and one for down-spins; they differ from their definition at zero field by a term $\pm 2H/k_B T$ in the exponential. (Further neighbor interactions would make this array still larger.) Figure 1.2 shows for the largest Ising system simulated to our knowledge the relaxation towards equilibrium at $T = 1.4\ T_c$ (using the tricks described in App.1.A [1.24]). This relaxation is exponential for intermediate times, for small times (1 MCS/spin) it is somewhat quicker, and for large times (30 MCS/spin) fluctuations take over, since the average magnetization has now become quite small (at late times we expect the magnetization to fluctuate around zero with an amplitude proportional to $1/\sqrt{N}$). Simulations such as shown in Fig.1.2 are useful for studying the temperature dependence of the relaxation time (characterizing the exponential decay), for instance [1.25].

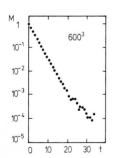

Fig.1.2. Variation of magnetization with "time" (Monte Carlo steps (MCS) per spin) in a simple cubic Ising lattice with $N = 600 \times 600 \times 600$ spins, and periodic boundary conditions [1.24]. The CDC Cyber 76 took about 2 h for this calculation, based on the program in App.1.A

1.1.4 An Example of Continuous Degrees of Freedom: The Heisenberg Model

In the spin 1/2 Ising model each spin points either up or down; in the general spin s Ising model each spin has $2s + 1$ quantum states, but still only its z component enters the interaction energy. In the anisotropic Heisenberg model, however, the interaction Hamiltonian of two spins $\mathbf{S}_i = (S_i^x, S_i^y, S_i^z)$ and $\mathbf{S}_k$ is $-J_x S_i^x S_k^x - J_y S_i^y S_k^y - J_z S_i^z S_k^z$.

9

Table 1.1. Example of a program for a 37*37*37 Ising model of a simple cubic Ising lattice at $T/T_C = 1.4$, calculated for 25 Monte Carlo steps per spin. *Ranset* initializes the random number generator RANF. The energy change is calculated as in (1.5), the flipping probability is taken as exp($-\Delta E/k_B T$). A CDC Cyber 76 took about 3.75 s for this program

```
      DIMENSION IS(37,37,37),EX(13)
      DATA IS/50653*1/
      T=1.40/0.221655
      L=37
      CALL RANSET(1)
      M=L*L*L
      DØ 3 I=1,13,2
    3 EX(I)=EXP(-2*(I-7.)/T)
      DØ 2  ITIME=1,25
      DØ 1 K1=1,L
      K1P1=K1+1
      K1M1=K1-1
      IF(K1.EQ.1)K1M1=L
      IF(K1.EQ.L)K1P1=1
      DØ 1 K2=1,L
      K2P1=K2+1
      K2M1=K2-1
      IF(K2.EQ.1)K2M1=L
      IF(K2.EQ.L)K2P1=1
      DØ 1 K3=1,L
      K3P1=K3+1
      K3M1=K3-1
      IF(K3.EQ.1)K3M1=L
      IF(K3.EQ.L)K3P1=1
      IEN=7+IS(K3,K2,K1)*(IS(K3M1,K2,K1)+IS(K3P1,K2,K1)+IS(K3,K2M1,K1)
    1                 +IS(K3,K2P1,K1)+IS(K3,K2,K1M1)+IS(K3,K2,K1P1))
      IF(EX(IEN).LT.RANF(I)) GØTØ 1
      IS(K3,K2,K1)=-IS(K3,K2,K1)
      M=M+2*IS(K3,K2,K1)
    1 CONTINUE
    2 WRITE (6,4)M,ITIME
    4 FORMAT(2I9)
      STOP
      END
```

Of particular interest is the XY model ($J_x = J_y, J_z = 0$) and the isotropic Heisenberg model ($J_x = J_x = J_z$). The Hamiltonian in the latter case can be expressed in terms of scalar products

$$\mathcal{H} = - \sum_{ik} J_{ik} \mathbf{S}_i \cdot \mathbf{S}_k - H \sum_i S_i^z \quad . \tag{1.7}$$

The development of simulation techniques for this fully quantum-mechanical model is still an area of active research, [1.26] and Chap.4. Here we are concerned only with the classical limit $s \to \infty$, where the quantum nature of the spins is ignored and the Hamiltonian reduces to the internal energy E: after suitable rescaling of J_{ik}, H, the vectors $\mathbf{S}_i$ are regarded as usual three-dimensional unit vectors. (One can also generalize the following remarks to the classical m-vector model where $\mathbf{S}_i$ are unit vectors with m components each).

In principle the simulation follows the above description of the Ising model: one calculates the energy change connected with a change of direction for one spin, and then compares a random number with the transition probability. Since the energy change is now a continuous variable one can no longer calculate all these probabilities at the beginning of the program, unless one approximates the continuum model again by a discrete model, e.g., for m = 2 one may allow only ℓ distinct orientations of the angle φ describing the spin orientation in the plane, $\varphi_\nu = (2\pi/\ell)\nu$, $\nu = 0,1,\ldots,\ell-1$ [1.27]. Due to the gap in the excitation spectrum induced by this discretization of the model one changes its symmetry properties, and hence this approximation is dangerous for both the properties at very low temperatures and near the phase transition. At other temperatures, however, even values of ℓ as small as $\ell = 12$ appear to be a very good approximation, and since again one can use "table methods" significant speeding up of the program seems feasible [1.27]. Since for models other than the two-dimensional classical XY model the accuracy of this approximation has not been studied, we return to the problem of simulating models with truly continuous degrees of freedom. In the XY model, one possibility is to rewrite the Hamiltonian in terms of the angle $\varphi_i : \mathbf{S}_i = (S_i^x = \cos\varphi_i, S_i^y = \sin\varphi_i)$, i.e.,
$\mathcal{H}_{xy} = - \sum_{ik} J_{ik} \cos(\varphi_i - \varphi_k) - H \sum_i \cos\varphi_i$, and for generating a new configuration one selects the angles φ_i uniformly and randomly from the interval $-\pi < \varphi_i \le +\pi$. In the Heisenberg model, a representation in terms of the two polar angles θ_i, φ_i can be used [$S_i^x = \sin\theta_i \cos\varphi_i, S_i^y = \sin\theta_i \sin\varphi_i, S_i^z = \cos\theta_i$], but one must take into account that the angle θ_i is not uniformly distributed in the interval $0 \le \theta_i \le \pi$, but rather with a weight proportional to $\sin\theta_i$. This can be taken into account by writing the spin components $S_i^x = \sqrt{1 - S_i^{z2}} \cos\varphi_i$, $S_i^y = \sqrt{1 - S_i^{z2}} \sin\varphi_i$, S_i^z, and sampling the variables S_i^z, φ_i from the intervals $-1 \le S_i^z \le +1$, $-\pi \le \varphi_i \le +\pi$.

In practice one often prefers to use Cartesian coordinates, but again one must ensure that all possible directions are treated equally. For example, in the Heisenberg case (n = 3) it is not legitimate to determine a new direction simply by first equating each of the three components with one of three random numbers between -1 and +1, and then normalizing the resulting vector to unit length. This method would give preference to those directions where $|S_i^x|, |S_i^y|, |S_i^z|$ are nearly equal, since the random vector before normalization would be distributed homogeneously in a cube, not in a sphere. To avoid that problem, one must omit all vectors having a length larger than unity before normalization.

In all these methods described here so far, each newly tried spin direction would be completely independent of the previous direction. At low temperatures, this is not an efficient way to reach equilibrium. Then it is often better to change the direction of the selected spin only slightly relative to the old direction. In the XY model in polar angle representation, this would mean to determine a trial value φ_i' as $\varphi_i' = \varphi_i + \xi\Delta$, where ξ is uniformly distributed between -1 and +1, and the "distance" Δ is chosen such that on the average one-half of the attempted moves is successful. (For a related scheme for the Heisenberg model, see [1.1].) This method is analogous to the simulation of hard-sphere systems [1.19] where every sphere, when selected for a change, tries to move by a not too large distance of order Δ in an arbitrary direction.

Once a spontaneous magnetization vector **M** has developed in a Heisenberg magnet, its direction changes slowly due to the random addition of random changes of individual spin directions. Due to this "rotational diffusion"-like change of **M** the average of **M** is not a meaningful quantity; in the limit of a large number of configurations n it must tend to zero. This is true even in the Ising model, where in a finite system there is always a nonzero probability that a domain of the phase with opposite orientation of the magnetization is nucleated and would change the magnetization direction by spreading over the whole system. However, in practice, for temperatures sufficiently below T_c these events are so seldom in Ising models that they do not occur for the values of n which are practically accessible and needed for a meaningful average. In the isotropic Heisenberg model in zero field, there is no energy barrier which fixes the orientation of the magnetization in any direction, and thus **M** is not metastable enough to allow any meaningful averaging. This problem then is usually avoided by studying the average length of **M**, i.e., the root mean-square magnetization $M_{rms} = \sqrt{(M^x)^2 + (M^x)^2 + (M^z)^2}$, where $M^\alpha = \sum_i S_i^\alpha/N$.

In the Ising model, the Monte Carlo process could be interpreted as a simulation of the real kinetics of the system. This is no longer true for Heisenberg spins, where isolated spins show precession in a magnetic field, and interacting spins may exhibit spin waves. Such oscillating modes cannot occur in the "time" evolution described by a Monte Carlo process, which takes into account only "friction," i.e., relaxational processes which conserve neither energy nor magnetization. Thus static properties in equilibrium are correctly obtained, but not the correct dynamics of the Heisenberg model. A more general but formal analysis of the dynamics associated with Monte Carlo processes can be found in [1.1,28] and Chap.6.

When one starts a Monte Carlo simulation, one is always faced with the following problems: (i) one must make sure that the algorithm works correctly; (ii) one has to make a judgement concerning the size of the systems N to be used; (iii) one has to make choices concerning the numbers n_0,n of configurations used to equilibrate and to average the system (no equilibrating, of course, is done in percolation-type applications, Sect.1.1.2); (iv) one wishes to estimate statistical errors; (v) one wishes to estimate at least the order of magnitude of systematic errors due to finite size and periodic boundary conditions (other boundary conditions may also sometimes be advantageous, [1.1]); if they exceed the statistical errors, one should try to eliminate them at least in part by using an appropriate extrapolation to the thermo-dynamic limit $N \to \infty$.

It turns out that in all these points there are certain pitfalls to be avoided. Unfortunately there is no general rule for making the judgements involved, but at least some guidelines for them often emerge from a consideration of the physics of the system, as will be outlined below. Thus we comment on all of the above points briefly.

i) *Systematic programming errors* are sometimes easily overlooked, as the Monte Carlo program even then produces results which look reasonable though they are wrong, e.g., suppose in the statement IEN = ... in Table 1.1 one of the coordinate indices would be misprinted. Errors of this type are best identified by comparing results with information from other sources (either other published simulations, or analytical work, e.g., for the three-dimensional nearest-neighbor Ising model exact high- and low-temperature series expansions are available, to which one can compare directly when either $T \gg T_c$ or $T \ll T_c$.) Or suppose in the simulation of an XY model the same random number is used erroneously to perform the change $\varphi \to \varphi'$ and to compare with the transition probability. Such an error would already show up if one runs the program at infinite temperatures $1/T = 0$, where the spins are independent of each other and the problem is exactly soluble. (Note that it is permissible to use the same random numbers for an array of lattices where simulations are executed in parallel, say, for a set of different temperatures and/or different fields, etc. This simple technique may sometimes speed up simulations considerably). In any case it is advisable to check each program carefully by making preliminary runs with small N, n_0, n, where one prints out intermediate steps and configurations to check some steps by hand. And when one changes a correct program to generalize it to calculate more information or to a more general model system, one should make a run with parameters where it reduces to the previous case using exactly the same sequence of random numbers in order to check that the results remain exactly the same.

ii,v) *The choice of the system size* depends on a judgement of *what the character-istic lengths of a problem are*, e.g., if we expect a second-order phase transition to occur, such as in the Ising ferromagnet at T_c, there will be a correlation length

ξ of order parameter fluctuations which diverges at the critical point in the thermo-
dynamic limit. Usually this divergence is of power-law form, $\xi \underset{T \to T_c}{\approx} \xi_0 |1 - T/T_c|^{-\nu}$, while
in certain cases, such as the two-dimensional XY model and related models, an even
stronger exponential divergence $\xi \underset{T \to T_c\uparrow}{\approx} \xi_0 \exp[\text{const}(T_c/T - 1)^{-1/2}]$ is expected. Reliable information on the behavior of the infinite system is obtained only if the
linear dimension L is much larger than the correlation length [1.1,22]. In practice
often the parameters ξ_0, ν (or even T_c) are not known a priori, and then the size dependence of the Monte Carlo data needs to be carefully investigated. Since the size
dependence of physical properties near a critical point is described by finite-size
scaling theory [1.15-17], one can use the size dependence of Monte Carlo results to
study critical properties. While the feasibility and usefulness of this approach has
been demonstrated in several papers [1.22,29,30], only very recently has its full
power seemed to be exploited by using Monte Carlo data in a sort of "phenomenological renormalization" analysis [1.31-36]. Some details about this technique are given
in Sect.1.2.1 and Chap.3 (for related work on the simulation of random walks and percolation problems, see Chaps.5 and 8, respectively).

Apart from studying critical properties, the finite-size scaling approach is also
useful for distinguishing first- and second-order phase transitions. At a first-order
transition, the specific heat in the infinite system has a delta-function singularity
associated with the latent heat of the transition. In a finite system this singularity
is rounded into a smeared broad maximum, just as the power-law divergence of the specific heat at a second-order transition, and hence a distinction of the order of
transitions may be a delicate matter, for which the dependence on system size is a
helpful tool (Sect.1.2.1). Finally we emphasize that finite-size scaling is very useful [1.35,36] for systems with an "ordered" phase without order parameter, where (in
the thermodynamical limit) the order-parameter correlation function decays with distance in a power-law form: the correlation length for all temperatures $T < T_c$ stays
infinite, and hence the boundary conditions disturb the correlation function significantly irrespective of the size of the system. The (particularly difficult) simulation of such systems is discussed in Chap.7.

The correlation length of fluctuations of an order parameter associated with some
phase transition is not the only characteristic length which may become large. For
long period superstructures, as they occur in systems with competing interactions,
the period of the superstructure must be commensurate with the size L of the lattice,
otherwise the order may be disturbed significantly. (This point can be seen already
for the simple antiferromagnetic structure of an Ising spin system, where the lattice is divided into two sublattices only, one sublattice having the majority of
spins-up, the other one having the majority of spins-down. If one would use a linear
dimension with an *odd* number of lattice spacings in one direction, the ordered state
must have at least one interface, since a nonintegral number of unit cells — of size
2a — does not fit together with the periodic boundary conditions. Thus all linear dimensions must be integer multiples of the size of the unit cell). This problem is

14

particularly cumbersome if the ordering (for $N \to \infty$) would be described by a wavelength incommensurate with the lattice spacing a, Chap.3.

Next we emphasize that the *spectrum of long-wavelength excitations* is strongly influenced by the finite size of the system, since only a discrete set of wave vectors $K_{n_x,n_y,n_z} = (2\pi/L)(n_x,n_y,n_z)$ occurs, with n_x, n_y, n_z integers. This discreteness is particularly important for continuous systems, where the energy associated with such excitations vanishes as $k \to 0$, which fact gives rise to so-called Goldstone mode singularities (e.g., the divergence of longitudinal and transverse susceptibilities of a Heisenberg ferromagnet at $T < T_c$ for $H \to 0$ [1.37]). Since the largest wavelength fitting into the system is $2\pi/k_{min} = L$, these singularities are rounded off, similar to the rounding of critical-point singularities [1.38].

The discreteness of k space must also be carefully considered in the Monte Carlo studies of transport phenomena, such as diffusion in lattice gases, etc., since a transport coefficient is usually defined by considering the limiting form of a re-laxation rate as $k \to 0$ [e.g., the "collective diffusion constant" D is defined from the relaxation rate $\Gamma(k)$ by which a concentration fluctuation decays as $D = \lim_{k \to 0} \Gamma(k)/k^2$]. We return to this problem in Sect.1.2.6 (see also Chap.6).

As a last problem where the finite size of the system significantly disturbs the results we mention a system which for $N \to \infty$ should exist in the two-phase coexistence region, e.g., this may happen in the simulation of a fluid at constant density ρ, if the temperature is below critical and ρ is inbetween the densities at the gas branch and the liquid branch of the coexistence curve; or in the simulation of a binary alloy AB at constant relative concentration c_B, if c_B lies inbetween the values of concentration at the respective coexistence curves $c_{coex}^{(1)}, c_{coex}^{(2)}$, etc. Equi-librium then is described by the lever rule: one must have a macroscopic mixture of the two coexisting phases, their respective relative amounts being given by {phase 2} $(c_B - c_{coex}^{(1)})/(c_{coex}^{(2)} - c_{coex}^{(1)})$ and $(c_{coex}^{(2)} - c_B)/(c_{coex}^{(2)} - c_{coex}^{(1)})$, phase 1. In a finite system, there will be a domain of the minority phase surrounded by the majority phase like an island is surrounded by the sea. At most N/2 atoms can participate in the minority domain, and typically of order $N^{1-1/d}$ atoms must be at the surface of that d-dimen-sional domain. Consequently a correction to the internal energy due to this inter-face of relative order $N^{-1/d}$ arises, and similar corrections occur in other physical quantities, too. Often the prefactors in front of the $N^{-1/d}$ factors are quite large, since the interface may be spread out over a larger distance, and it may also affect atoms more distant from the interface if the interaction potential is of longer range. While a simulation of such a two-phase coexistence may be of intrinsic inter-est to study interface free energies of such domains [1.39,40], it is otherwise de-sirable to avoid this situation — also because the size of the domain and its area are both strongly and slowly fluctuating, and good statistical accuracy is hard to obtain. Two-phase coexistence can automatically be avoided if one studies the problem in a statistical ensemble where all independent thermodynamic variables are *intensive* rather than *extensive* ones, i.e., rather than the *canonic ensemble* (given ρ, c_B), one uses the *grand-canonic ensemble* (given chemical potential μ or chemical potential difference $\mu_B - \mu_A$, respectively). In the thermodynamic limit, which one is usually interested in, these ensembles yielded identical answers. In a finite system, the interfacial corrections present in the canonical formulation are absent in the grand-canonical one: the two-phase coexistence region there shows up as a "forbidden region", in which one never observes any stable equilibrium state. If a fluctuation takes the system there, i.e., by nucleating a minority domain, its interface can always shrink until this domain disappears again, or grow until the domain takes over the majority completely. Even if such fluctuations occur frequently, one can identify the two coexisting phases from the peaks of an appropriate probability distribution (Sect.1.2.1). Of course, all results obtained in the grand-canonic ensemble can then be translated back into the canonic ensemble if so desired, using standard thermodynamic relations.

iii) The appropriate *choice of the numbers* n_0,n *of configurations* used to equilibrate and to average depends on a judgement of *what the characteristic times of a problem are*. Again large characteristic times occur at phase transitions: critical slowing down of fluctuations occurs at second-order transitions while metastability may occur at first-order transitions. Often it is not even possible to make the simulation run long enough to see the metastable state actually decaying to the stable one (which is reached by a different choice of initial conditions, for instance). Then a safe judgement which of the two (or more) states is the stable one needs an evaluation of their free energies (Sect.1.2.2): the stable state is that with the lowest free energy. In unfavorable cases, when many metastable states exist and a distribution of energy barriers between them leads to a very broad spectrum of relaxation times, it may be impossible to distinguish safely between slowly relaxing states and true equilibrium. Such a situation is typical for spin glasses, for instance (Chap.8).

iv) To estimate the statistical error of a quantity A, it is possible but delicate [1.1] to use a set of values $\{A_i\}$ obtained for this quantity at different times in the simulation (the time intervals between A_i,A_{i+1} must be large enough such that the $\{A_i\}$ are statistically independent; if they are not independent knowledge of their correlation would be required). Often it is safer to average over a set of variables $\{A_i\}$, where each A_i is the estimate obtained from an individual run. Then the probable statistical error from K independent values $\{A_i\}$ is

$$\Delta A = \left\{ \left[K \sum_i A_i^2 - \left(\sum_i A_i \right)^2 \right] \middle/ K^2 (K - 1) \right\}^{1/2} . \tag{1.8}$$

How do we now proceed in practice, attempting the simulation of a model system about which we do not yet know much? It is then always advisable to make many test runs with small lattice sizes over an amount of time sufficient for the system to reach equilibrium, and apply (1.8). If for the parameters of interest the times necessary to reach a reasonably small error ΔA have been identified, one repeats this procedure for a distinctly larger lattice size to check for finite-size effects. If one is not close to a phase transition, it is most convenient to choose the lattice size so large that the expected systematic error due to the finite size is distinctly smaller than the statistical error. Near a phase transition, of course, this is not possible, and then it is often better to obtain high-precision data for very small systems which are then suitably extrapolated to $N \to \infty$, as described in the next section.

1.2 Special Topics

1.2.1 What can be Learned from Distribution Functions; Finite-Size Scaling

As shown in Sect.1.1.3, one often wishes to use the Monte Carlo method to study phase transitions, such as the transition in the Ising model from a paramagnetic state with zero magnetization M above T_c to a state with nonzero "spontaneous" magnetization ($\pm|M_{Sp}|$ for zero applied field H) below T_c. It is well known, however, that such a spontaneous symmetry breaking can occur in the thermodynamic limit only: $M_{H=0} = (1/N) \sum_{i=1}^{N} \langle S_i \rangle_{T,H=0} = 0$ for all temperatures and any finite, arbitrary large N; a spontaneous magnetization is defined by first taking the thermodynamic limit $N \to \infty$ and then letting H tend to zero, $M_{Sp} \equiv \lim_{H \to 0} \lim_{N \to \infty} M_H$.

Studying M_H for various (large) system sizes and (small) fields one could try to do approximately this double-limiting procedure numerically. Of course, this canonic procedure is hardly ever attempted in practice [1.41]: below T_c a magnetization $\pm M$ is sufficiently metastable for long observation times if N is large, and the sign is basically determined by the initial condition of the Monte Carlo run. Of course, even above T_c some small magnetization is found due to fluctuations in a finite system during a finite observation time; the magnetization fluctuates around zero with some typical amplitude δM. Similarly, below T_c the magnetization fluctuates in a regime $\pm(M \pm \delta M)$, while very close to T_c δM becomes comparable to M.

To extract meaningful estimates of M_{Sp} from such a simulation, it is important to make the above statements more precise. This is done by noting that the distribution function of the magnetization s in a system of linear dimension L is Gaussian if $T > T_c$ and if L is much larger than the correlation length ξ of magnetization fluctuations [1.32]; i.e., for a d-dimensional system we have

$$P_L(s) = L^{d/2}(2\pi k_B T \chi_L)^{-1/2} \exp[-s^2 L^d/(2k_B T \chi_L)] \quad , \quad T > T_c \quad . \tag{1.9}$$

For $T < T_c$ but again $L \gg \xi$ the distribution is peaked near $\pm M_L$ and near those peaks can again be described by Gaussians [1.32] for $H = 0$

$$P_L(s) = 1/2L^{d/2}(2\pi k_B T \chi_L)^{-1/2}\{\exp[-(s - M_L)^2 L^d/(2k_B T \chi_L)]$$

$$+ \exp[-(s + M_L)^2 L^d/(2k_B T \chi_L)]\} \quad , \quad T < T_c \quad . \tag{1.10}$$

Since the static limit of the fluctuation-dissipation theorem relates magnetization fluctuations and susceptibilities

$$k_B T \chi = \lim_{L \to \infty} L^d (\langle s^2 \rangle - \langle s \rangle^2) \quad , \tag{1.11}$$

it is clear that the parameters χ_L determining the widths of the Gaussians in (1.9,10) must tend towards the susceptibility χ as $L \to \infty$. From (1.10) it follows that $P_L(s \approx 0)$ for $T < T_c$ is very small but still nonzero [1.42]: this means that for any $L < \infty$ there is a nonzero probability for a system to move from the region near $+M_L$ to the region near $-M_L$, but the observation time needed to see such transitions will increase with L as $P_L(s = M_L)/P_L(s = 0)$. For large L this ratio of probabilities is so large that for practically realizable observation times one does not yet observe a single transition from $+M_L$ to $-M_L$, and hence one does not sample the full symmetric distribution (1.10), for which $\langle s \rangle_L = \int_{-\infty}^{+\infty} s P_L(s) ds = 0$, but rather only one half of it, e.g., at positive s for which the restricted average $\langle s \rangle_L'$ is nonzero, $\langle s \rangle_L' = \int_0^\infty s P_L(s) ds / \int_0^\infty P_L(s) ds = M_L$. From this consideration it is obvious that M_{Sp} can also be found from the following limits:

$$\lim_{L \to \infty} M_L = \lim_{L \to \infty} \langle s \rangle' = \lim_{L \to \infty} \langle |s| \rangle_L = \lim_{L \to \infty} \langle s^2 \rangle_L^{1/2} = M_{Sp} \quad , \tag{1.12}$$

which are more convenient than the above relation $M_{Sp} = \lim_{H \to 0} \lim_{L \to \infty} \langle s \rangle_L$. Similarly, the susceptibility can also be estimated in several ways,

$$\lim_{L \to \infty} \langle s^2 \rangle_L' L^d/k_B T = \lim_{L \to \infty} P_L^{-2}(0) L^d/(2\pi k_B T) = \lim_{L \to \infty} \frac{(\Delta s)^2 L^d}{8 k_B T \ln 2}$$

$$= \chi, \quad T > T_c \quad , \tag{1.13}$$

Δs being the halfwidth of the distribution, or

$$\lim_{L \to \infty} (\langle s^2 \rangle - \langle |s| \rangle^2) L^d/k_B T = \lim_{L \to \infty} P_L^{-2}(M_L) L^d/(8\pi k_B T)$$

$$= \lim_{L \to \infty} \frac{(\Delta s)^2 L^d}{8 k_B T \ln 2} = \chi \quad , \quad T < T_c \quad . \tag{1.14}$$

Since for finite L the distribution $P_L(s)$ is a Gaussian only approximately, the quantities M_L (peak position), $\langle |s| \rangle_L$, $\langle s^2 \rangle_L^{1/2}$ may deviate distinctly from each other (and from M_{Sp}) for finite L. It is then advisable to perform several of the extrapolations suggested by (1.12-14) simultaneously. A typical example is shown in Fig.1.3, where the various system sizes were obtained simply by dividing one large system (of size 40^3) into subsystems of sizes 2^3, $4^3, 5^3, 8^3, 10^3$ and 20^3. Such division into subsystems is very convenient as one obtains rather complete information on

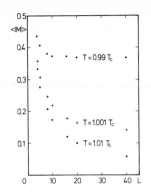

Fig.1.3. Example of spontaneous magnetization and subsystem magnetization plotted vs. linear dimension L of the subsystem in a $40 \times 40 \times 40$ simple cubic Ising ferromagnet with nearest-neighbor interaction. The dots refer to temperatures one percent above and below the critical temperature [1.23], the crosses to the critical point itself, after 10^4 and 10^5 Monte Carlo steps per spin, respectively. An analysis of such and similar data as in [1.32] gave a critical exponent η (describing the correlation function decay at T_c) of $\eta \approx 0.03$ and confirmed T_c within 0.1%

size dependences in one single run! In extrapolating such data one must keep the largest subsystem size distinctly smaller than the size of the total system, of course, since the total system usually has periodic boundary conditions, different from the "subsystem boundary condition" which does not affect any physical correlations directly. Such a consideration of subsystems is also useful when one performs simulations where an extensive thermodynamic variable rather than the conjugate intensive one is treated as an independent variable. In our case, this would be a simulation at fixed total magnetization s = 0 for instance, and then one could no longer estimate χ from fluctuations of the total magnetization; but still s would not be fixed in subsystems, and hence (1.13) would be applicable. Simulations with fixed extensive thermodynamic variables are quite common for fluids where the total number of atoms in a box is held fixed, or mixtures where the relative concentration of two atomic species is held fixed. In such cases the response function analog to χ can be found from the distribution of density (or concentration, respectively) in subsystems of the total system.

Equation (1.9,10) hold when $L \gg \xi$, but in practice ξ often is not yet known. Rather than studying the correlation function $\langle S_i S_j \rangle$ to estimate ξ from its decay with distance, it is much easier to check the Gaussian character of the distribution by estimating the fourth-order cumulant

$$U_L = 1 - \frac{\langle s^4 \rangle_L}{3 \langle s^2 \rangle_L^2} . \tag{1.15}$$

If $U_L \ll 1$ and U_L decreases with increasing L as L^{-d} [1.32] we have $L \gg \xi$ for $T > T_c$ and (1.9,13) are applicable. For $T < T_c$ and $L \gg \xi$, U_L as defined in (1.15) tends to $U = 2/3$. For $L \ll \xi$, on the other hand, U_L will stay more or less constant and close to a (nontrivial) "fixed-point value" U^*.

The behavior in the regime where ξ/L is no longer very small (although L is large) can be understood in terms of the finite-size scaling theory [1.15-17], here also introduced [1.32] from a consideration of the distribution function $P_L(s)$ which near T_c depends on temperatures only via the temperature dependence of ξ. The key assumption then simply is that $P_L(s) = f(L,\xi,s)$ does not depend on the three variables L,ξ, and s separately, but — apart from a power-law prefactor — only the two combinations L/ξ, $s\xi^{\beta/\nu}$ enter, where β,ν are the critical exponents of the spontaneous magnetization and correlation length $\{M_{Sp} \propto (1 - T/T_c)^\beta, \xi \propto |1 - T/T_c|^{-\nu}\}$. Analogously to the free energy $F(T,M)$, which near T_c becomes a homogeneous function of one variable $M\xi^{\beta/\nu}$ only, apart from a regular part $F_{reg}(T,M)$,

$$F(T,M) = F_{reg}(T,M) + \xi^{-(2-\alpha)/\nu}\tilde{f}(M\xi^{\beta/\nu}) \quad, \tag{1.16}$$

the distribution function becomes

$$P_L(s) = f(L,\xi,s) = \xi^{\beta/\nu}P(L/\xi,s\xi^{\beta/\nu}) \quad. \tag{1.17}$$

In this 'ansatz' it is assumed that the scaling function P depends on L only in the combination L/ξ, and $s\xi^{\beta/\nu}$ is just the analogy of $M\xi^{\beta/\nu}$ in (1.16). The extra factor $\xi^{\beta/\nu}$ is necessary for the normalization $\int_{-\infty}^{+\infty}dsP_L(s) = 1$, which must hold independently of L/ξ. Equation (1.17) can as well be rewritten using $sL^{\beta/\nu}$ instead of $s\xi^{\beta/\nu}$ as second variable in addition to L/ξ [1.32],

$$P_L(s) = L^{\beta/\nu}P*(L/\xi,sL^{\beta/\nu}), \quad L \to \infty \quad, \quad \xi \to \infty \quad, \quad L/\xi \quad \text{finite}. \tag{1.18}$$

From (1.18) we immediately derive the finite-size scaling relations for the moments and cumulants of the distribution, again using $\xi = \xi_0|1 - T/T_c|^{-\nu}$ and measuring L in units of the lattice spacing

$$<|s|> = L^{-\beta/\nu}\tilde{M}(L/|1 - T/T_c|^{-\nu}) \tag{1.19}$$

$$\chi(L,T) = L^d(<s^2>_L - <|s|>^2)/k_BT_c = L^{\gamma/\nu}\tilde{\chi}_2(L/|1 - T/T_c|^{-\nu}) \tag{1.20}$$

$$U_L = 1 - \frac{\tilde{\chi}_4(L/|1 - T/T_c|^{-\nu})}{3\tilde{\chi}^2(L/|1 - T/T_c|^{-\nu})} \quad. \tag{1.21}$$

While the susceptibility in the thermodynamic limit diverges at T_c as $\lim_{L\to\infty} \chi(L,T) \propto |1 - T/T_c|^{-\gamma}$, in a finite system this divergence is rounded off to a finite value $L^{\gamma/\nu}\tilde{\chi}(\infty)$, as (1.20) shows. This "finite-size rounding" is another expression of the fact that phase transitions in the strict sense can occur for $L\to\infty$ only. The functions $\tilde{M}$, $\tilde{\chi}_2$, and $\tilde{\chi}_4$ are universal, apart from constant prefactors and a scale factor for their arguments, just as the function $\tilde{f}$ in (1.16) is universal apart from a constant factor and a scale factor for its argument [1.43]. Since the equation of state of a magnet in a field contains two system-dependent scale factors only and is otherwise universal [1.44], the constant prefactors in $\tilde{\chi}_4$ and $\tilde{\chi}_2$ are not independent of each other, and actually cancel out in the ratio $\tilde{\chi}_4/\tilde{\chi}_2^2$. As a result, $U_L(T\to T_c) = 1 - \tilde{\chi}_4(\infty)/3\tilde{\chi}_2^2(\infty) = U*$ is also universal.

Equations (1.18-21) are true not only for the subsystem blocks considered above, but also for finite blocks of size L^d with periodic, free or any other boundary conditions. Note, however, that each boundary condition in finite-size scaling represents a separate "universality class" with respect to the actual scaling function $\tilde{P}$, $\tilde{M}$, $\tilde{\chi}_2$, i.e., on the scale $L/\xi \approx 1$ boundary effects do not yet cancel out. Of course, the exponents β, ν entering here are the bulk exponents of the infinite system and thus the same for all boundary conditions.

From (1.17-21) one recognizes several possibilities to apply finite-size scaling theory to study critical properties.

i) "Data Collapsing" [1.1,3,22,29,30]: treating L as a parameter, one has a family of curves for the susceptibility $\chi(L,T)$ for $T > T_c$, for instance. If we divide χ by a factor $L^{\gamma/\nu}$ and the temperature variable $|1 - T/T_c|$ by a factor $L^{-1/\nu}$, the family of curves "collapses" on a single curve, which represents the scaling function. The disadvantage of this method is that one must try to fit simultaneously the three parameters T_c, γ and ν. In addition, one uses in practice data for which L and ξ are not yet very large, and hence systematic corrections to the finite-size scaling expressions (1.17-20) occur. These corrections prevent complete "collapsing", and may also lead to systematic errors in the estimated quantities. However, the method is clearly useful if one is satisfied with modest accuracy.

ii) "Cumulant Method" [1.31,32,34]: in this method T_c is estimated by considering the ratio U_{bL}/U_L for various scale factors b as a function of temperature; this ratio is unity for $T \to \infty$ where $U_L \to 0$, for $T \to 0$ where $U_L \to 2/3$ and for $T \to T_c$ where $U_L \to U^*$. The "thermal eigenvalue" $1/\nu$ is then estimated from the slope of U_{bL}/U_L at the fixed point for isolated blocks $1/\nu = \ln(\partial U_{bL}/\partial U_L)/\ln b$, and the exponents β/ν and γ/ν are found from (1.19,20),

$$\gamma/\nu = \ln[\chi(bL,T_c)/\chi(L,T_c)]/\ln b \quad . \tag{1.22}$$

One advantage of this method is that T_c and the exponents are estimated in an independent way, and another advantage is that systematic errors due to corrections to finite-size scaling can be analyzed. Such corrections at T_c again have power-law form, e.g., (1.20) then is replaced by

$$\chi(L,T_c) = L^{\gamma/\nu}\tilde{\chi}_2(\infty)(1 + \chi_2^c L^{-x_c} + \ldots) \quad , \tag{1.23}$$

where χ_2^c is another amplitude factor and x_c a correction exponent. Then (1.22) is replaced by

$$\ln \frac{\chi(bL,T_c)}{\chi(L,T_c)} / \ln b = \frac{\gamma}{\nu} - \frac{\chi_2^c L^{-x_c}}{\ln b}(1 - b^{-x_c}) \quad . \tag{1.24}$$

Plotting then the estimates for γ/ν (or $2\beta/\nu$[1.31,32]) versus $(\ln b)^{-1}$, one obtains for each L a different curve, which for $(\ln b)^{-1} \to 0$ should all extrapolate linearly to the same value γ/ν. Unfortunately, the disadvantage of this method is that data of extremely good statistical accuracy are required. This is also true for the third method.

iii) "Phenomenological Renormalization" [1.33]: here one studies two functions $\xi_{L,L'}(T) = \ln\ \chi(L,T)/\chi(L',T)\ /\ln(L/L')$ and $\xi_{L',L''}(T)$. According to (1.22) $\xi_{L,L'}(T)$ and $\xi_{L',L''}(T)$ must intersect at T_c, their value at this intersection is γ/ν. This method does not require knowing cumulants, and is very close in spirit to the phenomenological renormalization method used in conjunction with transfer matrix calculations of two-dimensional $L\times\infty$ strips [1.45].

It is rather straightforward to generalize these methods to problems other than Ising ferromagnets, e.g., in an XY model the variable s in (1.9-21) becomes $\sqrt{s_x^2 + s_y^2}$ $s_x = (1/L^d)\sum_i\ s_i^x$, $s_y = (1/L^d)\sum_i\ s_i^y$, and only the region $0 < s < \infty$ is meaningful [1.34]; in systems where the order is incommensurate with the lattice structure $\chi(L,T)$ in (1.20) is replaced by the wave-vector dependent susceptibility $\chi(q,L,T)$ and the critical wave vector q_c is identified from the intersection which yields the largest value for T_c [1.33], etc.

As a last point of this section, we consider first-order transitions. The simplest example again is provided by an Ising model, treating variation with magnetic field H for $T < T_c$: at $H = 0$ the magnetization jumps for $N\to\infty$ from $+|M_{Sp}|$ to $-|M_{Sp}|$, i.e., the susceptibility has a δ-function singularity at the "coexistence curve". (At first-order transitions induced by temperature variation the internal energy jumps and this "latent heat" involves a δ-function singularity of the specific heat.) In a finite system, these δ-function singularities are rounded off, just as the power-law singularities near second-order transitions. For $L \gg \xi$ one easily generalizes (1.10) to this case [1.46]:

$$P_L(s) = \frac{L^{d/2}(2\ k_B T \chi_L)^{-1/2}}{1 + \exp\left(\frac{-2HM_L L^d}{k_B T}\right)}\left[\exp\left(-\frac{(s - M_L - \chi_L H)^2 L^d}{2k_B T \chi_L}\right)\right.$$

$$\left. + \exp\left(-\frac{2HM_L L^d}{k_B T}\right)\exp\left(-\frac{(s + M_L - \chi_L H)^2 L^d}{2k_B T \chi_L}\right)\right] \tag{1.25}$$

which yields for the magnetization and susceptibility

$$<s>_L = \chi_L H + M_L\ \tanh\left(\frac{HM_L L^d}{k_B T}\right) \tag{1.26}$$

$$\chi(H,L) \equiv \frac{\partial <s>_L}{\partial H} = \chi_L + M_L^2\ \frac{L^d}{k_B T}\ /\ \cosh^2(HM_L L^d/k_B T)\ \ . \tag{1.27}$$

Equation (1.27) shows that the δ-function singularity occurring at $H = 0$ for $L\to\infty$ is rounded into a peak of finite height proportional to L^d; the region of fields ΔH around $H = 0$ where this rounding occurs is of order $\Delta H \propto L^{-d}$. Similarly, at a first-

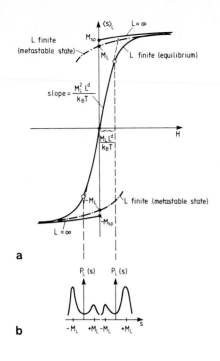

Fig.1.4a,b. Schematic variation of magnetization with magnetic field in an Ising model below T_C (a). The jump at the first-order transition is rounded for finite L. For short observation times one observes a single-peak structure of $P_L(s)$ rather than the correct double-peak structure (b) and the ordered state is even metastable in a weak field of opposite direction (dash-dotted curves) [1.46]

order thermal transition the latent-heat singularity of the specific heat is rounded into a peak of height proportional to L^d over a temperature interval $\Delta T \propto L^{-d}$ [1.46]. These results are in full accord with general arguments on finite-size scaling at first-order transitions [1.17,47].

It must be emphasized, however, that these considerations are relevant for Monte Carlo simulations of first-order transitions only, if the transition is weakly first order: otherwise one observes hysteresis rather than the true equilibrium behavior (Fig.1.4).

1.2.2 Estimation of Free Energy and Entropy

As mentioned in Sect.1.1.3, the importance sampling technique yields estimates for extensive quantities which are averages of observable features of the system, such as energy U, magnetization M, density ρ, etc., but it does not directly yield any information on the partition function Z itself: hence free energy $F = -k_B T \ln Z$ and entropy $S = (U - F)/T$ are also not directly known. This problem has attracted much attention in the literature, and various methods to overcome it have been suggested, some of which are exact and some approximate. Some of these methods are briefly reviewed in [1.1,48]. For more recent work, see Ref. [1.49a]. Here we refrain from giving an exhaustive discussion of this topic, and describe only briefly the thermodynamic integration method, again using Ising magnets as an explicit example. We feel that this method is both simple and versatile, its application is rather straightforward and its usefulness is established by nontrivial examples. The key idea is

the fact that F is a function of the thermodynamic state only, as specified by independent thermodynamic variables (T,H for the Ising system), but does not depend
on the particular integration path in the space of thermodynamic variables, which
is followed reversibly to bring the system from a "reference state" (for which F
and S are known) to the desired state. Consequently, one can choose the most convenient path for the problem under consideration.

The relations which can be explored for a magnetic system are

$$S = -(\partial F/\partial T)_H , \quad M = -(\partial F/\partial H)_T .$$

(1.28)

From $U = -T^2[\partial(F/T)\partial T]_H$, $(\partial S/\partial T)_H = (\partial U/\partial T)_H/T$, one can find S and F by starting the
integration either at $T = 0$ or $T \to \infty$,

$$S(T,H) = S(0,H) + \int_0^T (\partial U/\partial T)_H dT/T ,$$

(1.29a)

$$\frac{F}{k_B T} = \frac{U}{k_B T} - \frac{S(0,H)}{k_B} - \int_0^T (\partial U/\partial T)_H dT/k_B T$$

(1.29b)

or

$$\frac{F}{k_B T} = -\frac{S(\infty,H)}{k_B} + \int_0^{1/k_B T} U d(1/k_B T) ,$$

(1.30a)

$$S(T,H) = S(\infty,H) + U/T - k_B \int_0^{1/k_B T} U d(1/k_B T) .$$

(1.30b)

Note that (1.29) works only for quantum-mechanical systems, such as spin s-Ising systems with finite spin quantum numbers: for a classical system the specific heat
$C_H = (\partial U/\partial T)_H$ is nonzero at $T = 0$, and hence the integral in (1.29) would not exist.
Equation (1.30), on the other hand, is meaningful even then; it is also more convenient to apply, since $S(\infty,H)$ is usually known [e.g., $S(\infty,H) = k_B \ln(2s + 1)$ for a
spin s-Ising model], and first derivatives of F (such as U) are known more accurately
from Monte Carlo sampling than second derivatives of F (such as $\partial U/\partial T$).

Figure 1.5 shows as an example [1.48] internal energy and entropy for fcc Ising
s = 1/2 antiferromagnets. The field chosen is the transition field from the disordered
phase to the ordered phase (where one of the four simple cubic sublattices has all
spins down, the three other sublattices have all spins up) at T = 0. It is seen that
for nonzero next-nearest-neighbor interaction the entropy difference between infinite
and zero temperature has the expected value $\Delta S/k_B = \ln 2$, while for only nearest-neighbor interactions the ground state has a nonzero entropy of $S(0,H_{c2}) \approx (1/3)\ln 2$. This
estimate [1.48] lies inbetween rigorous lower and upper bounds which have more recently been established [1.49b]. This example shows that a relatively small number
of intermediate states (about 12 states in this case) are sufficient to obtain sufficiently accurate entropies from (1.30b), and even to estimate nontrivial ground-state
entropies. Conversely, if S(0,H) is known (1.30b) is very useful for locating first-

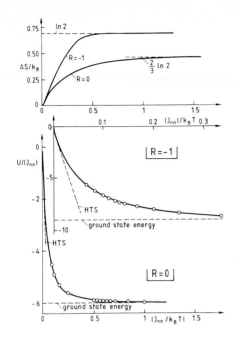

Fig.1.5. Internal energy (lower part) and entropy difference $\Delta S = S(\infty,H) - S(T,H)$ (upper part) plotted versus reciprocal temperature at the upper critical field $H_{c2}/|J_{nn}| = 12$, for both $R \equiv J_{nnn}/J_{nn} = 0$ and $R = -1$. The first term of the high-temperature expansion is also indicated [1.48]

order phase boundaries. One starts [1.48] with a trial assumption $T_0^{(1)}$ for the transition temperature T_0, and then applies (1.30b); then $\Delta S(0,H) = S(\infty,H) - S(0,H)$ will be either somewhat too small or too large by an amount of δS of the order of $\Delta U(1/T_0^{(1)} - 1/T_0)$, where ΔU is the jump in the internal energy at the transition. This consideration then yields a revised trial estimate $T_0^{(2)}$, etc.

The second equation of (1.28) yields [1.48]

$$F(T,H_2) = F(T,H_1) - \int_{H_1}^{H_2} M dH \quad , \tag{1.31a}$$

$$F(T,H) = -\frac{1}{2} \sum_{i(\neq j)} J_{ij} - H - \int_{H}^{\infty} (1 - M) dH \quad . \tag{1.31b}$$

In (1.31b) the "reference state" of the $s = 1/2$ Ising system is the state $H \to \infty$, where all spins are aligned parallel, and hence $S = 0$ in this state. Often it is convenient to reach a state by a combination of integration paths suggested in (1.29-31), particularly in the case of complicated phase diagrams containing several phase boundaries [1.48].

Finally we note that it is straightforward to write equations analogous to (1.31a) for other physical systems: in a fluid, H is replaced by the chemical potential and M by the density; in a binary mixture, H may be replaced by a chemical potential difference, and M by the relative concentration, etc.

1.2.3 Estimation of Intensive Thermodynamic Quantities

If one uses intensive thermodynamic variables as the independent variables of a Monte Carlo simulation, the conjugate extensive variables are obtained from averaging the corresponding observable features in a straightforward way. One sometimes wants to use an extensive variable as an independent variable, e.g., study a fluid at fixed density ρ and then estimate the conjugate variable, the chemical potential μ. (In an Ising magnet, one could work at fixed magnetization M and then sample the magnetic field H.) Such intensive variables cannot be directly expressed as functions of the microscopic degrees of freedom, and thus this problem is not straightforward.

Nevertheless, this problem can be handled. For systems with discrete degrees of freedom a convenient method uses the concept of "local states" [1.50]. Consider, e.g., a lattice gas system, where each lattice site can be either empty or occupied by an atom, and an energy $-\varepsilon$ is won if two neighboring sites are occupied. Hence on a square lattice the interaction energy of an atom can take the five values $E_\alpha = 0$, $-\varepsilon$, -2ε, -3ε, -4ε, which define the local states α of an atom. We define a set of five conjugate states α' by removing the central atom of each state α, with $E_{\alpha'} = 0$. If the frequencies of occurrence (i.e., ensemble average populations) of the local states α and α' are denoted as ν_α and $\nu_{\alpha'}$, the condition of detailed balance requires that [1.50]

$$\nu_\alpha/\nu_{\alpha'} = \exp[(-E_\alpha + \mu)/k_B T] \;, \quad \frac{\mu}{k_B T} = \ln(\nu_\alpha/\nu_{\alpha'}) + E_\alpha/k_B T \;. \tag{1.32}$$

To smooth out fluctuations it is advisable to average μ over all (five) local states. This method can be used to study questions such as what the excess chemical potential is in a system where a droplet coexists with surrounding vapor [1.51], etc.

Obviously, the concept of "local states" is impractical when one deals with a system with continuous degrees of freedom. Then one can use the "particle insertion method" from *Widom* [1.52]: if the introduction of one test particle in a system of N particles involves an energy $U_{t,N}$, the chemical potential μ is

$$(\mu - \mu_0)/k_B T = -\ln\langle\exp(-U_{t,N}/k_B T)\rangle_{T,N,V} \;, \tag{1.33}$$

where μ_0 is the chemical potential of an ideal gas at the same temperature T, particle number N, and volume V. This procedure has been used successfully in several cases [1.53,54]. Of course, related procedures can be formulated for lattice gas systems as well [1.50,55]. Under conditions under which this particle-insertion method works well, it is also possible to perform a successful simulation in the grand-canonical ensemble, even for realistic fluid models, where μ is held fixed and the number of particles fluctuates due to random insertion and removal of particles. For recent examples where this well-known method [1.56] was applied, see Chap.2.

Another intensive variable of interest in simulations of fluids is the pressure p, which in systems with additive pairwise potentials $\phi(r)$ is usually calculated from the formula [1.57]

$$\frac{pV}{Nk_BT} = 1 - \frac{N}{6k_BT} \int_0^\infty g(r)\,\frac{d\phi(r)}{dr}\,4\pi r^2 dr \quad , \tag{1.34}$$

where $g(r)$ is the radial density distribution function. For more details, including the case where $\phi(r)$ has jump discontinuities, see [1.56]. We only mention that using both (1.33,34) allows one to obtain directly the entropy from the relation $TS = pV + U - N\mu$, and hence to avoid performing thermodynamic integrations as described in Sect.1.2.2. For a review of free-energy calculations in fluids, see [1.58].

1.2.4 Interface Free Energy

Usually the interface free energy between coexisting phases is calculated from the profile $\rho(x)$ of the order parameter ρ distinguishing the phases in direction x across the interface, using suitable generalizations of van der Waal's theory [1.59-61]. This procedure is somewhat doubtful for various reasons [1.62]. One reason is that the interface profile is unstable against long-wavelength "capillary-wave" excitations, which over a length L may lead to an interface width $w_d(L) \propto (\ln L)^{1/2}(d=3)$ or $w_d(L) \propto L^{1/2}(d=2)$, Fig.1.6a [1.63]. A recourse to van der Waals-type theories can be avoided by comparing a system with an interface (Fig.1.6b) to another system without an interface but otherwise identical conditions (Fig.1.6c) [1.64]. From the energy difference U_s of these two systems the interface free energy f_s (per unit area) can be found by thermodynamic integration analogous to (1.29,30). This method is clearly useful far below critical [1.61,64], but less useful near T_c where f_s becomes very small, and at the same time the fluctuations of the bulk energy strongly increase.

A recent method [1.63] yields directly the interface free energy near T_c from sampling the (nonzero) minimum of the order-parameter distribution function at ρ_{min} (Fig.1.6d). This minimum is due to configurations where interfaces form spontaneously in a finite system from thermal fluctuations. In principle this would occur at all nonzero temperatures, but in practice is useful near T_c only where f_s is rather small. For an Ising magnet in zero field, $P_L^{(p)}(<\rho_+>) = P_L^{(p)}(<\rho_->)$ in Fig.1.6d just corresponds to the two maxima of the distribution function $P_L(s)$ at M_L considered in Sect.1.2.1. From Fig.1.6d it is plausible that the probability of the minimum will be basically reduced by the Boltzmann factor involving the free-energy cost $\Delta F = 2L^{d-1}f_s$ needed to create two interfaces, and hence [1.63]

$$P_L^{(p)}(\rho_{min}) \propto P_L^{(p)}(<\rho_+>)\exp(-2L^{d-1}f_s/k_BT) = 1/2L^{d/2}(2\pi k_BT\chi_L)^{-1/2}$$

$$\exp(-2L^{d-1}f_s/k_BT) \quad . \tag{1.35}$$

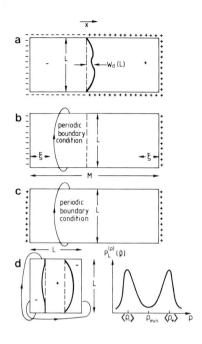

Fig.1.6. (a) Boundary conditions for a two-dimensional Ising system which lead to the formation of an interface below the critical point: spins are fixed at ±1 as indicated at the boundaries. Thick solid line denotes the (coarse-grained) position of the interface between the phases with negative and positive magnetization in a typical spin configuration. (b) Standard boundary conditions for the computer simulation of a system containing an interface. Note the linear dimension $M \gg 2\xi$, where ξ is the bulk correlation length of order parameter fluctuations. (c) Boundary conditions for a reference system without an interface. (d) Finite system with all boundary conditions periodic and its order parameter distribution. The minimum corresponds to a situation with two interfaces, while the maxima correspond to pure phases with order parameters $\langle\rho_-\rangle$ and $\langle\rho_+\rangle$, respectively [1.63]

This result suggests that $f_s/k_B T$ can be found by extrapolating $\ln P_L^{(p)}(\rho_{min})/2L^{d-1}$ or $\ln[P_L^{(p)}(\rho_{min})/P_L^{(p)}(\langle\rho_+\rangle)]/2L^{d-1}$ to $L\to\infty$ as a linear function of $(\ln L)/L^{d-1}$ [1.63]. This method has been checked [1.63] for the Ising model and $d = 2$, where the data in fact converged nicely to the exactly known result [1.65]; also for the $d = 3$ Ising case satisfactory results seem to have been obtained [1.63].

1.2.5 Methods of Locating First-Order Phase Changes

In a Monte Carlo simulation the signature of a first-order transition often is the observation of hysteresis effects. One must be careful, however, since similar behavior is often seen because of critical slowing down: one runs into a regime where the observation times are far too short (which may not be obvious beforehand), and "hysteresis" seen near second-order transitions can be distinguished from the "true" hysteresis associated with a first-order transition only by a careful analysis of the relaxation behavior [1.1,30,66].

If the hysteresis is pronounced and the transition clearly is first order, the problem arises to locate at which parameter value the transition occurs in thermal equilibrium. In the example of the Ising ferromagnet of Fig.1.4, when the magnetic field is varied at constant $T < T_c$, symmetry requires the transition to occur at $H = 0$; but for other problems, such as the Ising antiferromagnet in a uniform field (Sect.1.2.2) there is no longer any symmetry fixing the location of the first-order transitions in the phase diagram. In such a case it may be useful to compare the free energies of the various phases by thermodynamic integration, as described in Sect.1.2.2, and check where they intersect. The intersection point thus locates the

28

first-order transition. This method has been proven easy to apply and practically useful for a variety of models [1.48,67]. An alternative method, used successfully in the context of lattice gauge theories ([1.68] and Chap.9), is to prepare a system composed of both phases, which are separated by flat interfaces in the regime where hysteresis occurs: one then watches which of the two phases tends (on the average) to disappear, and thus checks which of the two phases is more stable as parameters such as temperature, field, etc., are varied.

If for a reasonable observation time hysteresis is not found, it still may be that the transition is first order, albeit weakly. One then sees a change in the distribution function of the energy, for instance; while away from the transition it is a Gaussian, in a regime close to the transition a double-peak structure occurs. The stable phase has a larger weight underneath its peak (Fig.1.4b), and estimating the point where the weights become equal is then a nice method to locate the transition [1.69]. One has to use this method with great care, however. (i) For cases close to a tricritical point, a double-peak structure of the energy already occurs for parameters where the transition still is second order. If one increases the size of the system in this regime, one finds that the distance between the peaks slowly decreases, while on the first-order side their distance is independent of size (but the peaks become sharper with increasing size). Thus, it is necessary to study how the distribution changes with size, and one may use finite-size scaling considerations as described in Sect.1.2.1. (ii) If the transition is strongly first order, it may be impossible to sample the relative weights of the two peaks for large enough systems with reliable accuracy. In fact, there are examples [1.70] where this method based on the equal-weight rule for the peaks failed dramatically, while the method of comparing free energies there gave good results [1.67].

It must be emphasized that the task of distinguishing continuous transitions from those which are weakly first order is among the most delicate goals a simulation may handle, and in any case requires simulating large systems with very good statistics and careful study of the size dependence. In some problems, such as two-dimensional melting (Chap.7), these requirements are clearly very hard to meet, and hence controversial interpretations arise. We emphasize that for transitions which are only very weakly first order, such as the q-state Potts model in two dimensions for q = 5, the correlation length at the transition point is very large though (probably) finite, and "pseudocritical" behavior is seen in many quantities in the simulation [1.71], though in this case it is known exactly that the transition is first order.

Extensive variables (magnetization, or density, etc.) at first-order transitions have a jump. If one uses such a variable in the phase diagram, the "forbidden regime" within the jump is a region of two-phase coexistence. In a simulation of a finite system it is rather hard to observe two-phase coexistence, of course: a phase separation on large scales can be seen only by suitable preparation of the initial state [1.39], and even then interfacial contributions to many observable features will be of importance. Thus a first-order transition with a small two-phase coexistence regime often looks like a second-order transition, when one uses the extensive variable as an independent variable. It is much easier to resolve such a two-phase regime by using the conjugate intensive variable as an independent variable in the simulation, however [1.48]; from the jumps of the extensive variable observed at the transition it is then straightforward to draw the phase diagram in the representation using the extensive variable, of course [1.48]. These considerations will be made more explicit in Chap.3.

1.2.6 Linear Response, Susceptibilities and Transport Coefficients

One often encounters the situation that one wishes to know correlation functions of some local quantity $\rho(\mathbf{r})$ such as $<\rho(0)\rho(\mathbf{r})> - <\rho(0)>_T<\rho(\mathbf{r})>_T$, or their Fourier transform $S(\mathbf{k}) = (1/V)\int dV \exp(i\mathbf{k}\cdot\mathbf{r})[<\rho(0)\rho(\mathbf{r})> - <\rho(0)>_T<\rho(\mathbf{r})>_T]$. Often physical interest is mainly in the behavior of such functions for a few values of $\mathbf{k}$. A related situation occurs when one wishes to study transport coefficients, which are related to the behavior of time-displaced correlation functions $S(\mathbf{k},t) = (1/V)\int dV \exp(i\mathbf{k}\cdot\mathbf{r})[<\rho(0,0)$ $\rho(\mathbf{r},t)> - <\rho(0,0)><\rho(\mathbf{r},t)>]$. For example, in the case of a lattice gas at constant total particle number N, $\rho(\mathbf{r},t)$ is the local density at site $\mathbf{r}$ and time t which may deviate from the average density N/V, and $S(\mathbf{k},t)$ behaves for small $\mathbf{k}$ according to

$$S(\mathbf{k},t) = S(\mathbf{k})\exp(-Dk^2 t) \quad . \tag{1.36}$$

Here D is the (collective) diffusion constant describing how a fluctuation in local concentration spreads out and thereby decays.

In some cases it is more convenient to calculate the associated response functions, which are related to these correlation functions in thermal equilibrium via the well-known fluctuation-dissipation theorem, rather than to sample the correlation of fluctuations. For example, if one formally adds to the Hamiltonian of the system a term describing the coupling of $\rho(\mathbf{r})$ to its conjugate field $\mu(\mathbf{r})$, $\mathcal{H} = \mathcal{H}_0 - \int dV \mu(\mathbf{r})\rho(\mathbf{r})$, one finds for the case of a field $\mu(\mathbf{r}) = \mu(\mathbf{k})\exp(i\mathbf{k}\cdot\mathbf{r})$ that

$$\chi(\mathbf{k}) \equiv \frac{\partial<\rho(\mathbf{r})>}{\partial\tilde{\mu}(k)\exp(i\mathbf{k}\cdot\mathbf{r})} = (1/k_B T)S(\mathbf{k}) \quad ; \tag{1.37a}$$

a similar relation holds if the field was applied in equilibrium and at time t the field was switched off, so that the inhomogeneity $<\rho(\mathbf{r},t)>$ decays to zero for $t > 0$,

$$\chi(\mathbf{k},t) \equiv \frac{\partial<\rho(r,t)>}{\partial\tilde{\mu}(k)\exp(i\mathbf{k}\cdot\mathbf{r})} = (1/k_B T)S(\mathbf{k},t) \quad . \tag{1.37b}$$

In a Monte Carlo simulation, one can apply arbitrary fields, which are not accessible in a real laboratory experiment, and hence one can apply the desired field $\tilde{\mu}(\mathbf{k})$ $\exp(i\mathbf{k}\cdot\mathbf{r})$ for a wave vector of interest in the Hamiltonian. The average amplitude of the wave-like modulation of the density $\rho(\mathbf{r})$ induced by this field is then sampled, and $S(\mathbf{k})$ is obtained from (1.37a), while $S(\mathbf{k},t)$ and hence D is found from (1.37b,36) where the field $\mu(\mathbf{r})$ is switched off after the system was equilibrated. This method is convenient in the simulation of very large systems, where $\mu(\mathbf{k})$ can be chosen small enough such that one is still within the regime of linear response, hence where nonlinear effects may be neglected and (1.37) is valid, and at the same time $<\rho(\mathbf{r})>_{\partial\tilde{\mu}(\mathbf{k})}-N/V$ is far above the "noise level" of the spontaneous thermal fluctuations $\delta\rho(\mathbf{r},t)$, which are of the order of $\sqrt{k_BT\chi(\mathbf{k})/N}$. This method has been applied in the study of diffusion problems in a variety of cases (Chap.6). Due to the periodic boundary condition, only discrete values of k are allowed, of course.

A related method for studying static correlations is to study the effect of a disturbance induced by an inhomogeneity at the boundary of the system, a method which was used for XY models [1.72] and lattice gauge models [1.73], for instance. In the context of simulations using "Brownian dynamics" instead of Monte Carlo techniques, *Parisi* [1.74] has also shown the usefulness of linear response relations.

1.3 Conclusion

In this introductory chapter we have first tried to give an impression to the newcomer to the field what a Monte Carlo simulation is all about, and then we discussed some more special problems of a partially technical nature, where in the last few years a deeper understanding has been achieved. Wherever possible, we have chosen simple lattice problems such as percolation, the Ising model, etc., as an explicit example, but in many cases a generalization to rather different physical systems should be obvious. Again we wish to emphasize that the selection of topics presented here is not at all exhaustive, and other technical aspects will become apparent in other chapters of this book. For instance, a particularly cumbersome problem is the treatment of long-range interactions in conjunction with the finite size of the system. For polar and other fluids this problem is reviewed in [1.75], for dipolar lattice systems in [1.76], and for two-dimensional Coulomb systems in Chap.7. An elegant way of bypassing some of these difficulties is to change the boundary conditions, e.g., rather than treating a two-dimensional system with periodic boundary conditions and Ewald summations, one may put the system on the (closed) surface of a three-dimensional sphere [1.78,77]. This technique has been used for the classical two-dimensional one-component plasma; and compared with other results obtained with the conventional boundary conditions [1.79]. The agreement between both methods seems to be excellent, but the new method clearly requires less computational effort.

It appears that many concepts of the general theory of statistical thermodynamics (subsystems, linear response, etc.) find immediate application in the evaluation of simulation data for certain problems, and it seems likely that additional approaches may find useful applications, and thus the field will rapidly develop further. Particularly fruitful has been the combination of simulations with ideas from renormalization group [1.4] and finite-size scaling, and it is likely that this will be an important direction for future research. The increasing availability of special-purpose machines [1.7] may open the possibility of improving statistical accuracies so significantly that new classes of problems can be tackled. The present chapter should present a useful guideline to the state of the art, and hopefully stimulate research to complement those areas where either the technical availability of performing a Monte Carlo simulation or the analysis of Monte Carlo data are still in an unsatisfactory stage.

Finally we draw attention to the "Brownian dynamics" method, where one numerically solves Langevin equations; this method is intermediate between the "molecular dynamics method" and standard Monte Carlo, since random numbers enter only to represent the random force. Although this method clearly is very useful for a wide variety of problems [1.74,80,81,82], it is beyond the scope of the present treatment.

Appendix.1.A. Multispin Coding

For spin 1/2 Ising models this appendix describes a more complicated program which appreciably saves in computer memory (up to a factor of 20) and computer time (up to a factor of 3). It does not require machine language but one needs the following special FORTRAN elements for bit-by-bit logical operations. In the language used for big CDC computers these are AND, OR, XOR and SHIFT (and preferably also COUNT). On IBM FORTRAN H Extended-Compilers they may be available under option XL in the form LAND (I,K), LXOR(I,K), etc. even if the manual does not mention this possibility.

Let I and K be two computer words, each consisting of m bits; for clarity we assume m = 60. Each bit is either zero or unity. The bit-by-bit exclusive OR, abbreviated as XOR (I,K) gives unity for every bit where the corresponding bit in I is 1 and in K is 0, and also where I has 0 but K has 1. It gives zero for all other bits. I. AND. K is the bit-by-bit logical AND, giving unity for every bit where the corresponding bits of both I and K are unity, and zero elsewhere. Analogously, I.OR.K is the bit-by-bit logical OR, SHIFT (I,3) shifts all bits of the word I circularly to the left by 3 units. That means if the leftmost bit (leading digit) is called the first and the rightmost bit is the 60th bit, then SHIFT(I,3) has as first bit the fourth bit of I, as second bit the fifth bit of I, etc., and in its four last bits are stored bits 60,1,2, and 3 of I. Finally, COUNT(I) counts the number of bits which are unity in the word I (and needs to be declared an integer). These operations are available on many but not all FORTRAN compilers.

Obviously, one bit is sufficient to store one spin in a spin 1/2 Ising model. But computer time is used more efficiently if three or four bits are reserved for each spin. Then the memory requirements are still reduced by about a factor 10 as compared to usual one-spin-one-word storage. But now the program runs much quicker: The trick is to use XOR to find those spin pairs which are antiparallel. Assume that in the 60-bit words I and K we have stored 20 spins, with 3 bits each: three zero bits

(000) correspond to spin down, while (001) corresponds to spin up. Then XOR(I,K) gives 000 wherever two neighbor spins are parallel, and 001 wherever they are anti-parallel. In other words, a single statement XOR(I,K) determines the relative orientations (and thus the interaction energy) of 20 spin pairs at once. Adding XOR(I,K1)+XOR(I,K2)+XOR(I,K3)+XOR(I,K4)+XOR(I,K5)+XOR(I,K6) gives for all 20 spins in I the number of antiparallel neighbors, which fixes the flipping probability (K1 to K6 each contain the 20 nearest neighbors in a simple cubic lattice for the 20 spins in I).

Once the flipping probability is determined, each of the 20 spins has to be investigated separately by comparing the flipping probability with a random number. Since the flip of one spin changes the interaction energy of its neighbors, no spins in the same computer word I must be neighbors. Thus, for 80 spins in one row, we store spins $1,5,9,\ldots,77$ in one 60-bit word, spins $2,6,10,\ldots,78$ in the second word, etc., until we store spins $4,8,12,\ldots,80$ in the fourth word. In general, for L^3 spins on a simple cubic lattice we need an array of dimension LL*L*L where LL = L/20 must be an integer and at least equal to 2, i.e., this program formulation requires L = 40, or L = 60, 80, etc.

A complete program of this multispin coding technique was published by *Zorn* et al. [1.83] (for related work see [1.84,85]). Further improvements particular to lattices with coordination number 6 were published by *Kalle* and *Winkelmann* [1.24] who also explained how they simulate much larger lattices up to $600 \times 600 \times 600$ (Fig.1.2); this technique has been used, for example, to study the kinetics of Ising models near T_c [1.25].

In summary, multispin coding makes a program more complicated and error-prone but may save a lot of memory and some computer time for large systems. The more complicated the interaction is, the less useful is multispin coding, and for continuous degrees of freedom it does not seem to work at all.

References

1.1 K. Binder (ed.): *Monte Carlo Methods in Statistical Physics*, Topics Current Physics, Vol.7 (Springer, Berlin, Heidelberg, New York 1979) p.1
1.2 C. Pangali, M. Rao, B.J. Berne: Chem. Phys. Lett. **55**, 413 (1978);
P.J. Rossky, J.D. Doll, H.L. Friedman: J. Chem. Phys. **69**, 4628 (1978)
1.3 E. Eisenriegler, K. Kremer, K. Binder: J. Chem. Phys. **77**, 6296 (1982)
1.4 R.H. Swendsen: In *Real-Space Renormalization*, ed. by Th.W. Burkhardt and J.M.J. van Leeuwen, Topics Current Physics, Vol.30 (Springer, Berlin, Heidelberg, New York 1982)
1.5 V. Ambegaokar, S. Cochram, J. Kurkijärvi: Phys. Rev. B**8**, 3682 (1973)
1.6 S. Kirkpatrick, E.P. Stoll: J. Comp. Phys. **40**, 517 (1981);
see also C. Kalle, S. Wansleben: Comp. Phys. Comm. **33**, 343 (1984)
1.7 R.B. Pearson, J.L. Richardson, D. Toussaint: J. Comp. Phys. **51**, 243 (1983);
A. Hoogland, J. Spaa, B. Selman, A. Compagner: J. Comp. Phys. **51**, 250 (1983)
1.8 R.B. Pearson: unpublished
1.9 W. Oed: Appl. informatics **24**, 358 (1982);
see also S. Wansleben, J.G. Zabolitzky, C. Kalle: J. Stat. Phys. **37**, 271 (1984);

R.B. Pandey, D. Stauffer, A. Margolina, J.G. Zabolitzky: J. Stat. Phys. **34**, 427 (1984)

1.10 F. James: Repts. Progr. Phys. **43**, 1145 (1980)

1.11 D. Stauffer: Phys. Repts. **54**, 1 (1979); *Introduction to Percolation Theory* (Taylor and Francis, London 1985)

1.12 D. Stauffer: *Lecture Notes in Physics*, Vol.149 (Springer, Berlin, Heidelberg, New York 1981) p.9
D. Stauffer, A. Coniglio, M. Adam: Adv. Polymer Sci. **44**, 103 (1982)

1.13 A. Margolina, H.J. Herrmann, D. Stauffer: Phys. Lett. A**69**, 73 (1983)
A. Margolina, Z. Djordjervic, H.E. Stanley, D. Stauffer: Phys. Rev. B28,1652 (1983)

1.14 B. Nienhuis: J. Phys. A**15**, 199 (1982) and references therein

1.15 M.E. Fisher: In *Critical Phenomena*, ed. by M.S. Green (Academic, New York 1971)

1.16 M.E. Fisher, M.N. Barber: Phys. Rev. Lett. **28**, 1516 (1972)

1.17 M.N. Barber: In *Phase Transitions and Critical Phenomena*, Vol.X, ed. by C. Domb and J.L. Lebowitz (Academic, New York 1983) p.145

1.18 L.D. Landau, E.M. Lifshitz: *Statistical Physics* (Pergamon Press, London 1959)

1.19 N. Metropolis, A.W. Rosenbluth, M.N. Rosenbluth, A.H. Teller, E. Teller: J. Chem. Phys. **21**, 1087 (1953)

1.20 R.J. Glauber: J. Math. Phys. **4**, 294 (1963)

1.21 D.W. Heermann: private communication

1.22 For a systematic study of size and boundary effects in simulations of two- and three-dimensional Ising models with nearest-neighbor ferromagnetic interaction, see D.P. Landau: Phys. Rev. B**13**, 2997 (1976); and B**14**, 255 (1976)

1.23 J.C. Le Guillou, J. Zinn. Justin: Phys. Rev. B**21**, 3976 (1980);
J. Adler: J. Phys. A**16**, 3585 (1983)
S. Pawley, D.J. Wallace, R.H. Swendsen, K.G. Wilson: Phys. Rev. B**29**, 4030 (1984)

1.24 C. Kalle, V. Winkelmann: J. Statist. Phys. **28**, 639 (1982)

1.25 B.K. Chakrabarti, H.G. Baumgaertel, D. Stauffer: Z. Physik B**44**, 333 (1981);
N. Jan, L.M. Moseley, D. Stauffer: J. Statist. Phys. **33**, 1 (1983)

1.26 J.W. Lyklema: Phys. Rev. Lett. **49**, 88 (1982)

1.27 M. Suzuki, S. Miyashita, A. Kuroda: Progr. Theor. Phys. **58**, 701 (1977);
S. Miyashita, H. Nishimori, A. Kuroda, M. Suzuki: Progr. Theor. Phys. **60**, 1669 (1978)

1.28 K. Binder, M.H. Kalos: In *Monte Carlo Methods in Statistical Physics*, ed. by K. Binder, Topics Current Physics, Vol.7 (Springer, Berlin, Heidelberg, New York 1979)

1.29 K. Binder: Thin Solid Films **20**, 367 (1974)

1.30 K. Binder, D.P. Landau: Phys. Rev. B**20**, 1941 (1980)

1.31 K. Binder: Phys. Rev. Lett. **47**, 693 (1981)

1.32 K. Binder: Z. Physik B**43**, 119 (1981);
see also A.D. Bruce: J. Phys. C**14**, 3667 (1981)

1.33 M.N. Barber, W. Selke: J. Phys. A**15**, L617 (1982)

1.34 D.P. Landau: J. Magn. Mag. Mat. **31-34**, 1115 (1983)

1.35 J.W. Lyklema: Phys. Rev. B**27**, 3108 (1983)

1.36 S. Miyashita: Progr. Teor. Phys. **65**, 1595 (1981)

1.37 L. Schäfer, H. Horner: Z. Phys. B**29**, 251 (1978), and references therein

1.38 H. Müller-Krumbhaar: Z. Phys. **267**, 261 (1974)

1.39 H. Furukawa, K. Binder: Phys. Rev. A**26**, 556 (1982)

1.40 K. Binder, M.H. Kalos: J. Statist. Phys. **22**, 363 (1980)

1.41 For an example where this procedure is used, see
A.P. Young, S. Kirkpatrick: Phys. Rev. B**15**, 440 (1982)
In this work the infinite-range model of spin glasses is studied by exact partition function calculations for small N

1.42 In fact, (1.10) is not quantitatively accurate for $s \approx 0$. The behavior of $P_L(s \approx 0)$ is discussed in more detail in Sect.1.2.4

1.43 See, e.g.,
H.E. Stanley: *An Introduction to Phase Transitions and Critical Phenomena* (University Press, Oxford 1971)

1.44 D. Stauffer, M. Ferer, M. Wortis: Phys. Rev. Lett. **29**, 345 (1972)

1.45 M.P. Nightingale: Physica **83**A, 561 (1976);
      R.R. dos Santos, L. Sneddon: Phys. Rev. B**23**, 3541 (1981)
1.46 K. Binder, D.P. Landau: Phys. Rev. B**30**, 1477 (1984)
1.47 N. Berker, M.E. Fisher: Phys. Rev. B**26**, 2507 (1982);
      J.L. Cardy, P. Nightingale: Phys. Rev. B**27**, 4256 (1983)
      Y. Imry: Phys. Rev. B**21**, 2042 (1980)
1.48 K. Binder: Z. Phys. B**45**, 61 (1981)
1.49a S.-K. Ma: J. Stat. Phys. **26**, 221 (1981);
      H. Meirovitch: J. Stat. Phys. **30**, 681 (1983)
1.49b D. Hadjukovic, S. Miloševic: J. Phys. A**15**, 3561 (1982)
1.50 Z. Alexandrowicz: J. Statist. Phys. **13**, 231 (1975); **14**, 1 (1976);
      H. Meirovitch, Z. Alexandrowicz: Molecular Phys. **34**, 1027 (1977)
1.51 H. Furukawa, K. Binder: Ref. [1.38]
1.52 B. Widom: J. Chem. Phys. **39**, 2808 (1963)
1.53 S. Romano, K. Singer: Molecular Phys. **37**, 1765 (1979);
      D.J. Adams: Molecular Phys. **28**, 1241 (1975)
1.54 J.G. Powles: Molecular Phys. **41**, 715 (1980)
1.55 G.E. Murch, R.J. Thorn: Nucl. Metall. **20**, 245 (1976)
1.56 W.W. Wood: In *Physics of Simple Liquids*, ed. by H.N.V. Temperley, G.S. Rushle-
      rooke, and J.S. Rowlinson (North-Holland, Amsterdam 1968)
1.57 T.L. Hill: *Statistical Mechanics* (McGraw Hill, New York 1956)
1.58 J.P. Valleau, G.M. Torrie: In *Statistical Mechanics, Part A*, ed. by B.J. Berne
      (Plenum, New York 1977) p.169
1.59 For a short review, see
      D. Levesque, J.J. Wiess, J.P. Hansen: In *Monte Carlo Methods in Statistical
      Physics*, ed. by K. Binder, Topics Current Phys., Vol.7 (Springer, Berlin,
      Heidelberg, New York 1979) p.112
1.60 F.F. Abraham: Phys. Repts. **53**, 93 (1979)
1.61 C.H. Bennett: J. Comp. Phys. **22**, 245 (1976);
      J. Miyazaki, J.A. Barker, G.M. Pound: J. Chem. Phys. **64**, 3364 (1976)
1.62 B. Widom: In *Phase Transitions and Critical Phenomena*, Vol.II, ed. by C. Domb
      and M.S. Green (Academic, New York 1972) p.79
1.63 K. Binder: Phys. Rev. A**25**, 1699 (1982)
1.64 H.J. Leamy, G.H. Gilmer, K.A. Jackson, P. Bennema: Phys. Rev. Lett. **30**, 601
      (1973);
      see also E. Bürkner, D. Stauffer: Z. Physik B**53**, 241 (1983)
1.65 L. Onsager: Phys. Rev. **65**, 117 (1944)
1.66 D.P. Landau, K. Binder: Phys. Rev. B**17**, 2328 (1978)
1.67 R. Liebmann: Z. Phys. B**45**, 243 (1982)
1.68 M. Creutz, L. Jacobs, C. Rebbi: Phys. Rev. D**20**, 1915 (1979)
1.69 O.G. Mouritsen, S.J. Knak-Jensen, P. Bak: Phys. Rev. Lett. **39**, 631 (1977);
      S.J. Knak-Jensen, O.G. Mouritsen, E. Kjaersgaard Hansen, P. Bak: Phys. Rev.
      B**19**, 5886 (1979)
1.70 O.G. Mouritsen, S.J. Knak-Jensen, B. Frank: Phys. Rev. B**23**, 976 (1981);
      B**24**, 347 (1981)
1.71 K. Binder: J. Stat. Phys. **24**, 51 (1981)
1.72 F. Fucito: Phys. Lett. A**94**, 99 (1983)
1.73 K.M. Mütter, K. Schilling: Nucl. Phys. B**200**, [FS4], 362 (1982)
1.74 G. Parisi: Nucl. Phys. B**180**, [FS2], 378 (1981); B**205** [Fs5], 337 (1982)
1.75 J.P. Valleau, S.G. Whittington: In *Statistical Mechanics, Part A*, ed. by
      B.J. Berne (Plenum, New York 1977) p.137
1.76 R. Kretschmer, K. Binder: Z. Phys. B**34**, 375 (1979)
1.77 J.P. Hansen, D. Levesque, J.J. Weiss: Phys. Rev. Lett. **43**, 979 (1979)
1.78 J.M. Caillol, D. Levesque, J.J. Weiss, J.P. Hansen: J. Statist. Phys. **28**, 325
      (1982)
1.79 S.M. De Leeuw, J.W. Perram: Physica **113**A, 546 (1982)
1.80 R.W. Gerling, A. Hüller: Z. Phys. B**40**, 209 (1980); and J. Chem. Phys. **78**, 446
      (1983);
      R.W. Gerling, B. De Raedt: J. Chem. Phys. **77**, 6263 (1982)

1.81 D.L. Ermak: J. Chem. Phys. **62**, 4189, 4197 (1975);
     T. Schneider, E. Stoll: Phys. Rev. B**27**, 1302 (1978)
1.82 K. Schulten, I.R. Epstein: J. Chem. Phys. **71**, 309 (1979);
     G. Lamm, K. Schulten: J. Chem. Phys. **75**, 365 (1981)
1.83 R. Zorn, H.J. Herrmann, C. Rebbi: Comp. Phys. Comm. **23**, 337 (1981)
     For earlier work see:
     M. Creutz, L. Jacobs, C. Rebbi: Phys. Rev. Lett. **42**, 1390 (1979);
     L. Jacobs, C. Rebbi: J. Comp. Phys. **41**, 203 (1981)
1.84 R. Friedberg, J.E. Cameron: J. Chem. Phys. **52**, 6049 (1970);
     C. Rebbi, R.H. Swendsen: Phys. Rev. B**21**, 4094 (1980)
1.85 M.P. Harding: J. Comp. Phys. **44**, 227 (1981)

2. Recent Developments in the Simulation of Classical Fluids

D. Levesque, J. J. Weis, and J. P. Hansen

The basic aim of the present chapter is to update a similar review in [2.1] by dis-
cussing computer simulation work on classical fluids which has appeared, roughly
speaking, since 1978. This area of research has grown enormously over the last five
years, and the trend has been towards more complex systems, mainly molecular liquids
and solutions, including a vast literature on water. Due to space limitation we restrict the present review to the fluid phases, without explicit reference to work
on phase transitions (melting, nucleation, spinodal decomposition, etc.) and on adsorption phenomena, which are covered elsewhere in this volume. Despite this restriction, the large number of relevant papers leaves us little space to discuss individual contributions in detail. Following [2.1], we examine results from Monte
Carlo (MC) and Molecular Dynamics (MD) simulations on an equal footing, provided
the latter deal with static distribution functions and thermodynamic properties,
i.e., with statistical averages which are accessible through both simulation methods.
This excludes of course any reference to dynamical properties which can be computed
only by the MD method.

This chapter is organized as follows. In Sect.2.1 we discuss some recent developments in methodology. Simple monatomic fluids and Coulombic systems are considered
in Sects.2.2,3, respectively. Section 2.4 is then devoted to the large body of work
on polar and molecular liquids, including water. Solutions of nonelectrolytes, electrolytes and macromolecules are reviewed in Sect.2.5, while fluid surfaces and interfaces are treated in Sect.2.6.

2.1 Some Recent Methodological Developments

Over the years a number of attempts have been made to improve the efficiency of the
original *Metropolis* algorithm [2.2] for sampling the canonical or grand-canonical
distributions. To allow a systematic presentation of these modified algorithms, we
first summarize Metropolis sampling, assuming that the reader is familiar with this
basic algorithm.

Let i,j,..., refer to possible states of a given system, the corresponding Boltzmann factors are denoted by $\pi_i, \pi_j, \ldots$. The transition probabilities p_{ij} between
states i and j define a stochastic matrix which must satisfy the conditions:

$$\sum_j \pi_j p_{ji} = \pi_i \quad \text{and} \quad \sum_j p_{ij} = 1 \quad , \quad \forall i \quad .$$

Starting from an initial state, a set of successive states is generated by a Markov process characterized by this stochastic matrix. To generate states asymptotically according to the Boltzmann distribution π_j, it is sufficient that the p_{ij} satisfy detailed balance

$$\pi_i \, p_{ij} = \pi_j p_{ji} \quad .$$

These conditions are met provided that

$$p_{ij} = q_{ij} \alpha_{ij} \quad ; \quad p_{ii} = 1 - \sum_{j \neq i} p_{ij} \quad ,$$

where q_{ij} is the probability of a trial step from i to j and α_{ij} is the probability of acceptance of this step, equal to $\text{Min}(1, f_{ij})$, with $f_{ij} = \pi_j q_{ji} / (\pi_i q_{ij})$; q_{ij} is positive and $\sum_j q_{ij} = 1 \, \forall i$.

It is clear from this form of p_{ij} that there are many possible choices of q_{ij}. In the original Metropolis algorithm, $q_{ij} = 1/(N \, v_0) \forall i$, where v_0 is the volume of a small domain around one of the N molecules of the system, selected at random. It is essentially this choice which various authors have tried to improve to minimize the asymptotic variance of the averages of the microscopic functions associated with energy, pressure, etc.

This section is divided into three parts which deal successively with modified Metropolis algorithms, new sampling methods in the grand-canonical ensemble and recent attempts to compute the chemical potential directly by MC simulations in the canonical ensemble.

2.1.1 Modified Metropolis Algorithms

Two alternative choices for q_{ij} have been proposed; they are referred to as "preferential sampling" and "force bias sampling." The former method was devised by *Owicki* and *Scheraga* [2.3,4] to overcome the difficulties occurring in the simulation of *dilute* solutions, which are modeled by samples containing $N(\sim 10^2 - 10^3)$ solvent molecules and one solute molecule. In the usual Metropolis sampling the statistics for the crucial solute-solvent interactions are then of order 1/N compared to 1 for solvent-solvent interactions. In the "preferential sampling" scheme, the value of a weighting function w(r) is associated with each molecule in the solvent, r being the distance between that molecule and the solute; w(r) is chosen to be a positive, decreasing function of r. If the system is in a state i, the next molecule (say m) to be moved is selected from the probability distribution

$$W_m(i) = w(r_m) / \sum_{\ell=1}^{N} w(r_\ell) \quad .$$

The q_{ij} are now chosen proportional to $W_m(i)$. Several choices of $w(r)$ have been tried, having the general form $w(r) = 1/r^n$, with $n = 1, 2$ or 4. The method has been tested on a two-dimensional hard disk system [2.4], on n-butane in a Lennard-Jones fluid [2.5] and on aqueous solutions [2.6].

Force bias sampling, suggested by *Kalos* [2.7] was adapted by *Pangali* et al. [2.8] in an attempt to accelerate equilibration in MC simulations of water models. These authors start from the premise that if q_{ij} were proportional to π_j, the efficiency of Metropolis sampling would be optimum, since all steps would then be accepted with probability one. This choice of q_{ij} is not practically feasible, but it can be approximately achieved by biasing the trial moves along the forces and torques acting on the particles. Indeed, if the energy E_j of state j is calculated by a Taylor expansion around the energy of state i limited to first order, $q_{ij} \sim \pi_j \sim \exp\{-\beta E_j\}$ takes the form:

$$q_{ij} = \frac{1}{Q(i)} \exp\left\{\lambda\beta F_m(i)\cdot[R_m(i) - R_m(j)] + \lambda\beta N_m(i)\cdot[\omega_m(i) - \omega_m(j)]\right\} \quad ,$$

where $\beta = 1/k_B T$, $F_m(i)$ and $N_m(i)$ are the total force and torque acting on molecule m in state i, and $R_m(j) - R_m(i)$, $\omega_m(j) - \omega_m(i)$ are the center-of-mass and angular displacements of molecule m from state i to state j. These displacements are limited to fixed domains $D(R)$ and $D(\omega)$ around the initial values $R_m(i)$ and $\omega_m(i)$. Furthermore, λ is a parameter ($0 < \lambda \leq 1$; the choice $\lambda = 0$ leads back to the standard Metropolis algorithm) and $Q(i)$ is a normalization constant.

The efficiency of this algorithm has been examined in [2.6,8-10], which report significant acceleration of the convergence of the random walk as well as a reduction of the variance in the computation of pair distribution functions.

An alternative force bias scheme has been developed by *Rossky* et al. [2.11]; this method, referred to as "smart Monte Carlo," was inspired by the "Brownian dynamics" simulation method [2.12], and also leads to an improved convergence rate for the computed averages. The respective merits of both force bias schemes have been compared by *Rao* and *Berne* [2.10]. Force bias sampling has been extended to the isobaric ensemble by *Mezei* [2.13].

2.1.2 Sampling in the Grand-Canonical Ensemble

The Metropolis algorithm was first extended to the sampling of the grand-canonical distribution by *Norman* and *Filinov* [2.14]. In this scheme the Markov process from state i to j can proceed in three possible steps: a molecule is added, a molecule is removed, or an existing molecule is moved. This basic procedure was generalized by *Valleau* and *Cohen* [2.15], and by *Van Megen* and *Snook* [2.16] to the case of electrolyte solutions, where groups of positive and negative ions must be added or removed at one time to preserve electroneutrality.

Extending an idea of *Rowley* et al. [2.17], *Yao* et al. [2.18] recently proposed a grand-canonical sampling scheme using a fixed maximum number M of particles. For given volume, temperature and chemical potential μ, M is chosen large compared to the expected average number of particles. In the initial configuration the M particles are placed on an fcc lattice, but only N out of these M particles are considered to be "real", i.e., the energy of the system is computed from the positions of these N particles. In an attempt to add a particle, the latter is not inserted at a random position in the volume, but is chosen at random among the M - N "fictitious" particles, while a particle that is removed among the N "real" particles automatically becomes one of the M - N "fictitious" particles. *Yao* et al. [2.18] report excellent agreement for the liquid-gas transition of a Lennard-Jones system with earlier calculations by *Adams* [2.19] which were based on a much larger sample. It should be noted that for a given cell volume, the μVT ensemble permits greater density fluctuations than the NVT ensemble, and is hence particularly suited for the study of phase transitions.

Mezei [2.20] proposed an alternative modification of the Norman-Filinov algorithm, where new particles are not introduced at random into the volume, but rather in favorable domains existing in the fluid, i.e., cavities or "holes." Since the algorithm must be able to detect these cavities and evaluate the probability of finding them, it is more time-consuming than the usual procedure. To be competitive it must consequently accelerate the convergence and reduce the variance of the estimated averages. The results obtained so far by *Mezei* are not very convincing.

For the sake of completeness we quote the work of *Yoon* et al. [2.21], and *Lee* et al. [2.22] based on a formulation of grand-canonical averages in terms of a transfer or Kramers-Wannier matrix; the procedure was applied so far to the two-dimensional lattice gas of dimers, trimers and tetramers [2.21] and to the two-dimensional gas of parallel hard squares [2.22].

2.1.3 Evaluation of the Chemical Potential

Determination of the chemical potential, and consequently of the entropy, is not directly possible in the course of an MC simulation in the canonical ensemble. Recently several schemes have been developed to evaluate these quantities. They are all based on *Widom's* formula [2.23] or its variant by *Shing* and *Gubbins* [2.24]. Let U_{N-1} be the configurational energy of a translationally invariant system of N-1 particles. Into this system we place a test particle, identical to the N-1 other particles; let u_t be the energy of this test particle in the field of the N-1 particles. In an N-particle system let u_r be the contribution of one of the N particles to the total configurational energy U_N. Straightforward manipulation of the configuration integrals Q_{N-1} and Q_N, using translational invariance, then leads directly to the desired results for the excess chemical potential:

$$\mu_{ex} = -k_B T \ln \frac{Q_N}{VQ_{N-1}} = -k_B T \ln \langle \exp(-\beta u_t) \rangle_{N-1} \qquad (2.1)$$

$$\mu_{ex} = k_B T \ln \langle \exp(\beta u_r) \rangle_N \quad . \qquad (2.2)$$

These two formulas can be rewritten by introducing the functions $f_{N-1}(u)$ and $g_N(u)$, which are, respectively, the probability of a randomly placed test particle having energy u, and the probability of one particle in the system of N particles having energy u. Note that $f_{N-1}(u)$ and $g_N(u)$ correspond to systems of N-1 and N particles in the same volume, and consequently not at the same density. With these probability functions, (2.1,2) can be rewritten as

$$\langle \exp(-\beta u_t) \rangle_{N-1} = \int_{-\infty}^{+\infty} f_{N-1}(u) \exp(-\beta u) du \qquad (2.3)$$

$$\langle \exp(\beta u_r) \rangle_N = \int_{-\infty}^{+\infty} g_N(u) \exp(\beta u) du \quad . \qquad (2.4)$$

In (2.1), the main contribution to the average $\langle \ \rangle_{N-1}$ comes from configurations where u_t is negative, while in (2.2) the main contribution is from configurations with positive u_r. These remarks illustrate the difficulties of using (2.1,2) for computing μ_{ex}. In the first case, the energy of a randomly located test particle is almost always positive because it overlaps some of the N-1 particles of the system. In the second case, the particles present in the system rarely have large positive energies, which would correspond to configurations of low probability.

To find a significant number of test particle positions contributing to the average (2.1), *Romano* and *Singer* [2.25] used not one, but 125 test particles placed at fixed positions in the system. The energies of these test particles are calculated every hundred steps in the Markov process; applied to the case of diatomics, the method turned out to be efficient at moderate densities.

To compute μ_{ex} at high density, *Shing* and *Gubbins* [2.24] resort to "umbrella sampling" [2.26]; configurations which make a substantial contribution to (2.1) are sampled preferentially by introducing an appropriate weighting function $W(u_t)$. If the superscript W denotes an average over the weighted distribution $\exp(-U_{N-1})W(u_t)$, then (2.3) is changed to

$$\langle \exp(-\beta u_t) \rangle_{N-1} = \left[\int_{-\infty}^{+\infty} f_{N-1}^W(u) \exp(-\beta u) W(u)^{-1} du \right] / \langle W(u)^{-1} \rangle_{N-1}^W \quad . \qquad (2.5)$$

The shortcoming of (2.5) is that the weight W(u) forbids the evaluation of other useful thermodynamic quantities in the course of a simulation because of the poor sampling of configurations contributing to the corresponding averages. This deficiency can be overcome by calculating the chemical potential from the following, easily proved relation [2.24]:

$$g_N(u) = \exp(\beta\mu_{ex})f_{N-1}(u)\exp(-\beta u) \quad . \tag{2.6}$$

The chemical potential μ_{ex} can then be determined in the energy domain where the probability densities $g_N(u)$ and $f_{N-1}(u)$ overlap; the latter are determined by a standard MC simulation, with the help of the simple weight function ($W(u) = 1, u < u_{max}$; $W(u) = 0, u > u_{max}$), to calculate $f_{N-1}(u)$. Results for pure Lennard-Jones fluids and their mixtures obtained by these methods [2.24,27] agree well with previous determinations of the chemical potential up to liquid densities. *Powles* and collaborators [2.28,29] have tested the efficiency of the preceding formulas for calculating the chemical potential of Lennard-Jones fluids, using configurations generated by MD simulations. These authors do not discuss the limitations of the applicability of the Widom and Shing-Gubbins formulas when configurations are generated at constant energy, as is the case in the microcanonical MD ensemble.

Zollweg [2.30] describes a method for calculating exactly the "free area" of a given configuration of hard disks, i.e., the domain of the total area where a disk of given radius can be inserted without overlapping the other disks of the system. The excess chemical potential of an infinitely dilute hard-disk solute is proportional to the logarithm of this "free" area. This method yields excellent results for solvent densities near the melting transition.

2.1.4 Variations on a Theme

To conclude this section on methodology, we list a number of specific adaptations of the Metropolis algorithm to particular problems. *Coker* and *Watts* [2.31] applied a modification of grand-canonical MC sampling to the study of reacting liquids at equilibrium. *Quirke* and *Jacucci* [2.32] calculated the free energy of liquid nitrogen by "umbrella sampling" [2.26] and by *Bennett's* method [2.33]. *Northrup* and *McCammon* [2.34] examined the relative merits of MC and MD simulations for generating accessible configurations of proteins; *Severin* et al. [2.35] presented an algorithm for sampling the microcanonical distribution. In this algorithm the microcanonical distribution is integrated over the momenta leading to a distribution function for the positions. These are then sampled according to the latter distribution, while the momenta are sampled from the Maxwellian distribution; the momenta are finally scaled to maintain a constant total energy. *Freasier* et al. [2.36] studied a system where part of the degrees of freedom evolve at constant energy, while the others evolve according to a canonical distribution. The MC sampling in this mixed ensemble (microcanonical and canonical) follows [2.35]. *Mezei* [2.37] examined the efficiency of various methods of selecting the particle to be displaced in a single-particle Metropolis trial move. His results indicate that various particle-selection schemes yield essentially equivalent results. He also applied three different free-energy calculation schemes to various water models [2.38].

2.2 Simple Monatomic Fluids

Simple fluids made up of spherical particles interacting through short-range central forces have already been extensively studied by MC and MD simulations in the period prior to 1978 [2.1]. However, a number of interesting studies have been carried out recently, in particular for two-dimensional fluids, or with the purpose of increasing the accuracy of previous computations and testing approximate theories. These are briefly described below.

2.2.1 Hard-Core Systems in Two and Three Dimensions

Uehara et al. [2.39,40] carried out MC calculations of the two- and three-body distribution functions of hard-disk and hard-sphere fluids in the intermediate density range, to test an extension of the Born-Green-Yvon integral equation which involves three-body correlations explicitly (BGY2 equation). The pair-distribution functions of the hard-sphere fluid have been recalculated very accurately in MC simulations of 864-particle samples by *Labik* and *Malijevsky* [2.41] who proposed accurate polynomial fits to their data. The hard-sphere solid was investigated by *Kratky* [2.42] who concluded that the fcc phase is stable near close packing.

Binary mixtures of hard spheres of different diameters have been reconsidered by *Fries* and *Hansen* [2.43] who concentrated on highly dissymmetric mixtures (with diameter ratios $\sigma_2/\sigma_1 = 2$ and 3) in a range of concentrations where both species occupy comparable volume fractions. The paper describes a novel "expulsion" technique for generating an initial high density fluid configuration of large hard spheres in a solvent of smaller spheres.

Frenkel and *Eppenga* [2.44] located the isotropic-nematic phase transition in a three-dimensional fluid of infinitely thin, circular hard platelets of diameter σ. It is found to take place at a reduced density $\rho\sigma^3 \sim 3.8$-4.1.

Young and *Alder* [2.45] determined the phase diagrams of two- and three-dimensional systems of particles interacting via a hard-core plus repulsive step potential. They showed in particular that the melting curve exhibits extrema similar to those observed in cerium or cesium. They also determined the structural phase transitions for the hard core plus attractive square-well potential [2.46]. Further MC simulations of the same system were carried out in the fluid phase for various values of the range of the attractive square well, to test the domain of validity of various perturbation theories [2.47].

2.2.2 Soft-Core and Lennard-Jones Systems in Two Dimensions

A considerable amount of simulation work has been devoted to simple two-dimensional systems, mostly in the hope of detecting a hypothetical "hexatic" phase, intermediate between the isotropic fluid and the solid. Such a phase is characterized by long-range "orientational" order. Much of the work on two-dimensional melting is adequately reviewed in [2.48] and we shall be concerned only with those aspects of the simulations dealing explicitly with the fluid phase. It is, however, fair to state that none of the available simulations has given clear evidence for the existence of the "hexatic" phase.

Henderson [2.49] simulated the Lennard-Jones (LJ) fluid in the density range $0.3 \leq \rho^* = \rho\sigma^2 \leq 0.8$ and in the temperature range $0.3 \leq T^* = k_B T/\varepsilon \leq 2$ (where ε and σ are the usual energy and length parameters of the LJ potential) to provide a systematic test of perturbation theory. He obtained critical parameters which differ somewhat from the earlier results of *Tsien* and *Valleau* [2.50]. *Toxvaerd* [2.51] used MD simulations to obtain the first accurate determination of the fluid-solid coexistence curve and of the triple point of the LJ system. The exact nature of this transition has been investigated by a number of authors [2.51-56]; most of these calculations yield strong evidence for a first-order transition similar to the three-dimensional case. The full phase diagram of the two-dimensional LJ system was determined by *Barker* et al. [2.57] on the basis of a set of accurate MC simulations in the canonical and isobaric ensembles, and using perturbation theory. Their results for the critical parameters are $T_c^* = 0.53$, $\rho_c^* = 0.335$ and $P_c/\rho_c k_B T_c = 0.209$, while the triple-point temperature is $T_t^* = 0.415$.

A number of workers examined the melting transition of "soft core" systems made up of particles interacting via inverse-power potentials $v(r) = \varepsilon(\sigma/r)^n$. *McTague* et al. [2.58] studied the case $n = 6$, *Vashishta* and *Kalia* [2.59] considered the case $n = 3$, while *Broughton* et al. [2.60] concentrated on the power $n = 12$. The latter authors checked the precision of the results obtained by a novel microcanonical MD method at constant pressure and temperature, using the scaling properties of inverse-power potentials.

2.2.3 Rare-Gas Fluids in Three Dimensions

The LJ potential yields a relatively crude description of the interaction between rare-gas atoms, but it is a prototype potential of great theoretical importance. Some efforts has gone into improving the accuracy of earlier simulations [2.1]. *Adams* [2.19] carried out an extensive grand-canonical MC study of the vapor line of the LJ fluid near the critical point. He evaluated the critical parameters to be $T_c^* = 1.30 \pm 0.02$, $\rho_c^* = 0.33 \pm 0.03$ and $P_c^* = 0.13 \pm 0.02$. Several papers on methodological aspects reviewed in Sect.2.1 contain extensive results for the equation of state of LJ fluids [2.13,18,24,25,27]. *Abraham* [2.48] located the supercooled liquid-glass transition of a LJ fluid by MC simulations in the isobaric ensemble.

Much of the recent simulation work deals with more realistic interactions between rare-gas atoms, particularly in the case of krypton. A number of pair potentials for krypton have been tested in the simulations by *Vermesse* and *Levesque* [2.61] and by *Bose* et al. [2.62]. Both studies concluded that all these potentials are inadequate to reproduce the equation of state of krypton in the dense gas and liquid phases.

Egelstaff et al. [2.63] obtained the pair distribution functions of krypton at low densities from two sets of MC simulations; the first set uses *Barker*'s et al. "best" pair potential [2.64]; the second set includes the three-body Axilrod-Teller interaction, in addition to the Barker pair potential. The effect of the three-body interaction is found to be small at low densities. However, their influence on internal energy and the pair-distribution function becomes significant at moderate and high densities, both in two [2.65] and in three [2.66] dimensions. *Vermesse* and *Levesque* [2.67] have shown that including the three-body Axilrod-Teller interactions leads to a realistic description of the equation of state and the transport coefficients of krypton and xenon over a wide range of temperatures and densities.

2.2.4 Binary Mixtures of Simple Fluids

Simulation work on binary mixtures of inverse-power or LJ atoms is generally motivated by a possible phase separation of the two components. *Shiotani* et al. [2.68] examined binary mixtures of "soft-core" particles interacting via the inverse-power potentials $\epsilon(\sigma_{\alpha\beta}/r)^n$, with nonadditive diameters: $\sigma_{12} \neq (\sigma_{11} + \sigma_{22})/2$. They located the phase separation for the case $2\sigma_{12}/(\sigma_{11} + \sigma_{22}) = 1.2$ by MC simulations. *Toukubo* et al. [2.69-71] presented extensive results of simulations for a LJ fluid composed of 107 identical solvent atoms and one solute atom. The purpose of this work was to examine the variation of the structure around the solute atom and its diffusion when the solute mass and the potential parameters of the solute-solvent interaction are changed.

They extended their investigations to the case of equimolar LJ mixtures [2.71-73], calculating the local fraction of the two components around an atom of given species when the strength of the potential between atoms of different species is varied [$\epsilon_{12} = \epsilon_{11}$, $(\epsilon_{11}\epsilon_{22})^{1/2}$ or ϵ_{22}]. They also obtained results for the thermodynamic properties and transport coefficients of these mixtures. *Nakanishi* et al. [2.74] used "umbrella sampling" [2.26] to extract the free energy of such mixtures from MC simulations. They discussed the stability of the mixtures, but located the phase separation only qualitatively as a function of the strength parameter of the LJ potential at given temperature and density.

Helium-xenon mixtures, modeled by a mixture of LJ fluids, were simulated by *Fiorese* and *Pittion-Rossillon* [2.75], who determined the critical locus curve and the phase transition for a xenon concentration of 75.56%. *Hoheisel* and *Deiters* [2.76] examined a binary LJ mixture as a model for neon-krypton mixtures at high pressure and over the whole concentration range. Their MD results are in good agreement with experimental data which indicate a phase separation of the mixture; these results exhibit a structural rearrangement of the mixture near the critical point.

2.3 Coulombic Systems

2.3.1 Boundary Conditions

The simulation of systems with electrostatic interactions among particles raises the traditional problem of the truncation of long-range forces [2.1]. The present overview is restricted to purely ionic systems, while the case of multipolar inter-actions is examined in Sect.2.4.

If the usual periodic boundary conditions are used, there are essentially three different ways of handling the long-range Coulomb potential:

a) spherical truncation: the potential is simply set equal to zero for inter-particle distances larger than a given radius r_c;

b) minimum-image (MI) convention, according to which each particle interacts with the images of all other particles contained within a cell centered on the given particle;

c) Ewald summation of the interactions of a given particle not only with the nearest image of each of the system's N-1 other particles, but with all their images in the infinite periodic array.

Spherical truncation must be rejected outright since it violates local charge neutrality: ions of the opposite sign to the central ion remain in the truncation sphere while ions of the same sign tend to move outside.

The nearest-image convention automatically satisfies charge neutrality and leads to reasonable results as long as the Coulomb coupling parameter [$\Gamma \sim Z^2 e^2 / (a k_B T)$, where a is some characteristic interionic spacing and Ze the ionic charge] is not too large, i.e., typical for electrolytes. For $\Gamma \gg 1$ (e.g., in the molten salt re-gime), the MI convention leads to a strong N dependence of the computed averages, and to very unphysical structures. Even at low and moderate couplings ($\Gamma \lesssim 1$), the MI results exhibit some N dependence and lead to a slight distortion of the fluid structure when the usual cubic shape is chosen for the basic cell ("cube-corner effect"). To overcome this difficulty, *Adams* [2.77] suggested using a periodically repeated truncated octahedron (the Wigner-Seitz cell of a bcc lattice) which is more spherical than the cube.

The most widely used treatment of long-range forces is the Ewald summation over the (generally cubic) infinite superlattice of periodically repeated cells. The to-tal Coulomb energy of a given configuration of N ions is

$$V_N = \frac{1}{2} \sum_{\mathbf{n}} \sum_i \sum_j{}' \frac{Z_i Z_j e^2}{|\mathbf{r}_i - \mathbf{r}_j - \mathbf{n}L|} \quad , \tag{2.7}$$

where $\sum_{\mathbf{n}}$ is a sum over all vectors of integer components, L is the edge length of the basic cubic cell, and the primed summation implies $i \neq j$ if $\mathbf{n} = (0,0,0)$. The N de-pendence of the simulation results obtained with the Ewald procedure is generally very weak; in the strong coupling regime ($\Gamma \gg 1$) it does not lead to the unphysical

46

features which spoil the MI convention. The objections raised by *Valleau* and *Whittington* [2.78] against the Ewald method, in particular the possible enhancement of expanded lattice-like configurations, have been shown to be largely without foundation, through experience and by *Adam*'s careful comparative study [2.77].

It is well known that the slowly (in fact conditionally) convergent sum (2.7) can be transformed into two rapidly convergent sums, one over direct space and one over reciprocal space, by an Ewald transformation. This procedure has recently been examined in more detail by *De Leeuw* et al. [2.79], who demonstrated the importance of specifying the order of summation of the conditionally convergent series (2.7) through an appropriate convergence factor. If the series is summed over a sequence of spherical shells inside an infinite sphere surrounded by a continuum of dielectric constant ε', the result is:

$$V_N = \frac{1}{L} \sum_{i<j} Z_i Z_j e^2 \psi_E(\mathbf{r}_{ij}/L) + \frac{2\pi}{2\varepsilon'+1} \frac{M^2}{V} \qquad (2.8)$$

with

$$\psi_E(\mathbf{r}) = \sum_{\mathbf{n}} \frac{\mathrm{erfc}(\alpha|\mathbf{r}+\mathbf{n}|)}{|\mathbf{r}+\mathbf{n}|} + \frac{1}{\pi} \sum_{\mathbf{n}\neq 0} \frac{\exp(-\pi^2|\mathbf{n}|^2/\alpha^2)}{|\mathbf{n}|^2} \exp(2\pi i \mathbf{r}\cdot\mathbf{n}) \quad , \qquad (2.9)$$

where α is a disposable convergence factor, and $\mathbf{M} = \sum_i Z_i e \mathbf{r}_i$ is the total dipole moment of the sample. Most simulations have been carried out without the last term of (2.8); this corresponds to a situation where the infinite sphere is surrounded by a conducting medium ($\varepsilon' = \infty$), a necessary condition to obtain a nonzero electrical conductivity (closed circuit). The effect of varying ε' has been examined by *De Leeuw* and *Perram* [2.80] and by *Smith* et al. [2.81] on systems of 64 and 216 oppositely charged hard or soft spheres in the molten salt regime. The thermodynamic properties and pair distribution functions are practically identical for $\varepsilon' = 1$ and $\varepsilon' = \infty$; however the fluctuation of the total dipole moment ($<|\mathbf{M}|^2> - |<\mathbf{M}>|^2$) increases considerably between $\varepsilon' = 1$ and $\varepsilon' = \infty$.

The Ewald method has been extended to semi-infinite lattices [2.82] and to two-dimensional electrostatics [2.83,84]. The Ewald summations of Coulomb interactions in a periodic system can be avoided for two-dimensional systems by confining the particles to a spherical surface, thus eliminating the boundaries. This procedure was first tested successfully for two-dimensional electron layers [2.85]. The obvious extension to d-dimensional systems confined to (d + 1)-dimensional spheres was put forward by *Kratky* [2.86].

2.3.2 The One-Component Plasma (OCP)

The OCP is the simplest model for continuous Coulomb systems, comprising N identical point charges Ze, charge neutrality being ensured by a uniform background of opposite charge ("jellium"). Simulation results prior to 1978 are reviewed in [2.1,87]. Very extensive and highly accurate MC results for the equation of state and the pair structure of the three-dimensional OCP have recently been reported by *Slattery* et al. [2.88,89]. They have considerably improved the cubic harmonics representation of the Ewald potential (2.8) and used very long Markov chains ($\sim 10^7$ configurations) to obtain results in the fluid and solid phases in the range $1 \leq \Gamma = Z^2 e^2/(ak_B T) \leq 300$ (where $a = (3V/4\pi N)^{1/3}$). They carefully studied the very weak N dependence ($128 \leq N \leq 1024$) and extrapolated their results to $N = \infty$, thus obtaining an accurate estimate of the small free-energy difference between the two phases, allowing a precise location of the phase transition which occurs for somewhat stronger coupling ($\Gamma_m \simeq 178$) than previously estimated. This first-order phase transition is characterized by a latent heat, but occurs with zero volume change [2.90].

The thermodynamic pressure of the OCP is known to become negative for $\Gamma > 3$. *Navet* et al. [2.91] have shown that one can define a virial kinetic partial pressure of the point ions which is independent of the properties assumed for the background, and which is always positive since it is related to the density of the OCP near the wall confining the particles. They carried out some MC simulations of 256 point ions confined inside a sphere in the range $\Gamma \leq 10$; the estimated values of the kinetic pressure have large statistical uncertainties, but appear to decrease rapidly towards zero as Γ increases.

In two dimensions, Poisson's equation leads to a logarithmic interaction between charges:

$$v_{\alpha\beta}(r) = -Z_\alpha Z_\beta e^2 \ln(r/L) \quad ,$$

where L is an arbitrary scale factor. The properties of the corresponding OCP depend on the dimensionless coupling constant $\Gamma = Z^2 e^2/k_B T$. For this model MC results have been obtained for $104 \leq N \leq 256$ particles confined to a spherical surface, and interacting along the chords joining the particles on the sphere, over the range $0.5 \leq \Gamma \leq 200$ [2.92]. The calculated energies and pair distribution functions agree within statistical errors with MD results obtained for the corresponding periodic planar system using Ewald summations [2.93]. The fluid-solid phase transition occurs at $\Gamma \simeq 140$.

2.3.3. Two-Dimensional Electron Layers

Layers of electrons trapped at the surface of liquid helium represent a nearly perfect realization of the classical two-dimensional electron gas, a model of point particles interacting through the three-dimensional Coulomb potential $v(r) = e^2/r$. *Totsuji* [2.94] carried out MC simulations of this model, using periodic samples of 36 and 81 electrons and Ewald sums; his calculations cover the fluid phase in the range $0.15 \leq \Gamma = e^2/(ak_B T) \leq 50$ and show the emergence of short-range order for $\Gamma \geq 3$. His computations were extended by *Gann* et al. [2.95] who examined the fluid and solid phases over the range $1 \leq \Gamma \leq 300$. Melting occurs at $\Gamma = 125 \pm 15$, in good agreement with experiment; the transition appears to be first-order, in contradiction with the λ-type transition observed in the earlier MD simulation of *Hockney* and *Brown* [2.96]. This conclusion is supported by an MD study by *Kalia* et al. [2.97].

An MD simulation of samples of $N = 104$ and 400 electrons confined to a spherical surface [2.85] yielded pair distribution functions at $\Gamma = 36$ and 90 which are indistinguishable from the results for the corresponding planar system [2.95].

2.3.4 Liquid Metals

Metals can be viewed as two-component ion-electron plasmas, in which the degenerate gas of conduction electrons is weakly coupled to the ionic plasma. This coupling can be treated by second-order perturbation theory, resulting in a screened effective ion-ion potential. If ion size and electron screening effects were ignored, a metal would essentially reduce to the OCP. In fact, the OCP turns out to be a better starting point than the traditional hard-sphere model for the study of the pair structure and thermodynamics of alkalis. *Ross* et al. [2.98] carried out an extensive MC study of liquid lithium over a wide range of temperatures and densities to demonstrate this point convincingly. An MD simulation of a large ($N = 1458$) sample of liquid lithium using a more refined ion-ion potential yielded excellent agreement with neutron scattering data of the static structure factor [2.99]. Further MC simulations based on similar first principles potentials led to overall good agreement with X-ray or neutron scattering data for Li, Na, K and Rb [2.100,101], but such comparisons can hardly be regarded as stringent tests of the validity of the effective ion-ion potentials, since the OCP model leads to excellent results for alkalis near the triple point, except at small wave numbers [2.102].

The temperature and density dependence of these effective potentials and of the resulting pair structure is a more delicate question which has been investigated in the case of liquid Rb by the MC calculations of *Mountain* [2.103] and the MD simulations by *Tanaka* [2.104] covering several high-temperature states of the expanded liquid. *Mountain* [2.105] has also examined the metastable supercooled liquid phase of Rb.

Among the nonalkali simple metals, only Al and Pb have been examined in some detail. Reasonable agreement with the experimental structure factor data was found by *Michler* et al. [2.106] and by *Jacucci* et al. [2.107] in the case of liquid Al, provided the best available first-principles potential is used. The Al potential has the unusual characteristic of being positive in the first-neighbor region and exhibiting large amplitude Friedel oscillations. These long-range ($\sim \cos(2k_F r)/r^3$) oscillations call for a proper Ewald summation which allows the construction of a damped effective potential [2.108].

Giró et al. [2.109] and *Mitra* [2.110] carried out some MD simulations of the pair structure of liquid Pb to test an effective pair potential obtained from the experimental structure factor data by an inversion procedure based on thermodynamic perturbation theory.

2.3.5 Primitive Model Electrolytes

The "primitive model" of electrolyte solutions ignores the discrete nature of the solvent which is replaced by a continuum of macroscopic dielectric constant ε. The ions are generally modeled by oppositely charged hard spheres of diameter σ_α, valence Z_α and stochiometric coefficient ν_α ($\alpha = +$ or $-$), such that $\nu_+ Z_+ + \nu_- Z_- = 0$ (charge neutrality). The "restricted" version of the primitive model (RPM), characterized by $Z_+ = -Z_-$ and $\sigma_+ = \sigma_-$, has been extensively simulated both in the electrolyte and molten salt regimes, prior to 1978 [2.1]. The more recent extensions of this work have branched in several directions: grand-canonical MC calculations, investigations of ion clustering and the simulation of dissymmetric versions of the primitive model (different absolute ionic valences and/or diameters).

Larsen and *Rogde* [2.111] examined ergodic problems in the MC simulation of the RPM, linked to metastability problems near the "liquid-gas" first-order phase transition. Subsequently [2.112] they studied the relation between the RPM and a fluid of ions interacting via the screened Coulomb potential:

$$v_{\alpha\beta}(r) = \begin{cases} \infty & r < \sigma \\ \dfrac{Z_\alpha Z_\beta e^2}{\varepsilon r} e^{-\alpha(r-\sigma)} \, , & r > \sigma \end{cases}$$
$$(\alpha,\beta = +,-) \, . \tag{2.10}$$

Following the *Ceperley* and *Chester* work [2.113] on the OCP, they developed a perturbation theory for the pair structure of the RPM, choosing the screened Coulomb fluid as a reference system.

Van Megen and *Snook* [2.16] and *Valleau* and collaborators [2.15,114] applied the grand-canonical MC method to the primitive model for 1:1, 2:2 and 3:1 aqueous electrolytes with equal ion sizes. The grand-canonical scheme briefly described in our introduction is generalized by attempting to add or delete an electrically neutral combination of $\nu = \nu_+ + \nu_-$ ions. The chemical potential of the electrolyte $\mu = \nu_+\mu_+ + \nu_-\mu_-$ is fixed, and from the calculated mean number of ions, one deduces the mean ionic activity coefficient $\gamma = \exp(\mu - \mu_{ideal})/\nu k_B T$, a quantity of central importance in electrochemistry. *Valleau* and collaborators adopted the minimum image (MI) convention, and extrapolated the N dependence of their results linearly in 1/N. Their calculations for 2:2 electrolytes [2.114] show good overall agreement between canonical and grand-canonical MC results and hypernetted chain (HNC) theory. They also found a short-range enhancement of the pair distribution functions $g_{++}(r) = g_{--}(r)$ near $r = 2\sigma$, indicating the formation of linear triple ion clusters ($+ - +$ or $- + -$ trimers). Subsequent canonical MC calculations by *Rogde* and *Hafskjold* [2.115] on the same electrolyte using the Ewald method show that the MI convention leads to small errors above concentrations of about 2 M (at room temperature) and that HNC theory overestimates the amount of triple-ion clustering. A similar conclusion had been reached earlier by *Rossky* et al. [2.116] who studied ionic association for a model 2:2 electrolyte with a soft-sphere repulsion:

50

$$v_{\alpha\beta}(r) = A_{\alpha\beta}\left(\frac{\sigma_\alpha + \sigma_\beta}{2r}\right)^n + \frac{Z_\alpha Z_\beta e^2}{\varepsilon r} \quad . \tag{2.11}$$

These authors chose $n = 9$, $\sigma_+ = \sigma_-$, and used a biased MC scheme to enhance the sampling of close encounters of like ions.

Hafskjold and *Weis* [2.117] carried out some canonical MC calculations for charged hard spheres of different diameters in the molten salt regime. Their data for the partial pair distribution functions have been used by *Abramo* et al. [2.118] to test the mean spherical approximation (MSA) and its generalization (GMSA). They found evidence for the formation of linear trimers.

Vorontsov-Vel'yaminov and collaborators [2.119] used MC simulations of the restricted primitive model in the NPT ensemble to determine the boundary between the Debye regime and the "ion pair" region where each ion is essentially shielded by a single ion of opposite charge. They later obtained similar results, replacing the hard-sphere repulsion by the soft-sphere repulsion in (2.11) (with $n = 8$) [2.120].

2.3.6 Simple Models of Polyelectrolytes

A realistic description of solutions of large globular or chainlike macro-ions and small counterions (polyelectrolytes in a broad sense) is still outside the possibilities of present-day simulation techniques. However a number of highly simplified models have been studied. *Van Megen* and *Snook* [2.121] examined the structure and ordering of dilute dispersions of charged spherical particles. They modeled such colloidal dispersions using equally charged hard spheres interacting via screened Coulomb potential of the form (2.10), which results from the interaction of two charged double layers. Their MC simulations disclose a considerable amount of structure, in agreement with light-scattering experiments, and the formation of a colloidal crystal at relatively low packing fractions ($\eta \gtrsim 0.001$).

Wennerström et al. [2.122] performed some simulations of cell models corresponding to various geometries, in particular for spherical micelles in a spherical cell; the Poisson-Boltzmann equation is found to overestimate systematically the osmotic pressure. *Vlachy* and *Dolar* [2.123] examined a cylindrical cell model for chains of 10-100 monomers carrying discrete charges to determine the influence of polymerization on the concentration dependence of the osmotic coefficient.

2.3.7 Molten Salts and Superionic Conductors

Computer simulation of simple molten salts was a lively research area in the seventies. Work prior to 1978 is summarized in [2.1] and in the excellent review by *Sangster* and *Dixon* [2.124]. Interionic potential models fall into two classes according to whether ion polarization effects are neglected (rigid ion models) or explicitly included via a shell model (polarizable ion models). Since ion polarization is essentially dynamic in nature and depends on the instantaneous local electric field acting on a given ion, it cannot be treated by the MC method, but only by MD simulations. Note that the simplest "rigid ion" model is the primitive model which has already been discussed in Sect.2.3.5.

a) KCl

Since the pioneering work of *Woodcock* and *Singer* [2.125] KCl is the most widely studied molten salt because it can be reasonably well described by rigid ion potentials of the Born-Huggins-Mayer form:

$$v_{\alpha\beta}(r) = A_{\alpha\beta} \exp(-a_{\alpha\beta}r) + \frac{Z_\alpha Z_\beta e^2}{r} - \frac{C_{\alpha\beta}}{r^6} + \frac{D_{\alpha\beta}}{r^8} \quad . \tag{2.12}$$

The potential parameters are usually chosen to be those determined from solid-state data. *Adams* [2.126] made a systematic MC investigation of the volume dependence of the pair structure and thermodynamics of KCl along the isotherm T = 1700 K (13 cm^3/mole $\leq$ V $\leq$ 80 cm^3/mole). His most interesting finding is that under increasing pressure, the structure of the liquid gradually changes from an open, charge-ordered arrangement, to one more nearly resembling that of a simple nonionic mixture. *Schäfer* and *Klemm* [2.127] used MD simulations to examine the triplet distribution functions of molten KCl and found large deviations from the standard superposition approximation. *Woodcock* et al. [2.128] made an MD investigation of isobaric and isothermal vitrification of KCl, as well as of BeF$_2$, ZnCl$_2$ and SiO$_2$.

b) KCN

The effect of replacing the Cl$^-$ ion by the anisotropic ion (CN)$^-$ has been analyzed in the molecular dynamics calculations of *Miller* and *Clarke* [2.129] who used a three-center potential of the form (2.12), and an ionic charge distribution of the form K^{+1}C$^{-0.45}$N$^{-0.55}$. This substitution leads to an increase in molar volume but to no significant change in liquid structure.

c) Alkali Chlorides

Dixon and *Gillan* [2.130] made a systematic comparison of rigid ion and polarizable ion potential models for all alkali chlorides, showing that the effect of ion polarizability is small whenever the two ions have similar sizes (e.g., KCl), but is substantial when the cation is relatively small and unpolarizable (e.g., NaCl). For NaCl the introduction of polarizable-ion potentials yields significantly better agreement with experiment, especially as regards the substantial lowering and broadening of the first peak in the Na-Na pair distribution compared with the Cl-Cl distribution function [2.131]. On the other hand, for KCl, RbCl and CsCl the polarizability effects are much weaker. The MD results for the rigid ion models have been compared to the predictions of HNC theory by *Abernethy* et al. [2.132], who concluded that this theory is considerably superior to the "mean spherical approximation".

d) Rb Halides

The polarizability of the Rb$^+$ ion is comparable to that of the negative halid ion, and hence a realistic calculation must allow anions as well as cations to be polarized; the corresponding MD simulations were carried out by *Dixon* and *Sangster* [2.133] for all Rb halides. They subsequently reexamined the case of RbCl with a larger sample (216 instead of 64 ions) [2.134] and concluded that rigid and polarizable ion models lead only to small differences of the pair structure of this molten salt.

e) Cs Halides

Dixon and *Sangster* [2.135] extended their systematic MD study of polarizable ion models to Cs halides to investigate the effect of the larger Cs$^+$ ion on the structural properties of the melt. Their calculations indicate the emergence of a subsidiary peak in the like-particle functions $g_{++}(r)$ and $g_{--}(r)$ for CsCl, CsBr and CsI.

f) Alkaline Earth Halides

De Leeuw performed NPT ensemble MC calculations [2.136] on simple rigid ion models for SrCl$_2$ and CaF$_2$ at several temperatures below and above melting. He observed a striking increase of the anion-anion coordination number n_{--} (from 6 to about 10) upon melting. In CaF$_2$ he observed a premelting effect of the anion sublattice just below the melting temperature; this effect was already reported by *Rahman* [2.137] and is linked to superionic conduction. In subsequent MD work, *De Leeuw* computed the partial static structure factors of liquid SrCl$_2$ [2.138].

g) Molten Salt Mixtures

Saboungi and *Rahman* [2.139] tested the conformal ionic solution theory on the basis of MD simulations of binary chloride mixtures, with KCl chosen as reference salt. *Adams* and *McDonald* [2.140] performed systematic NPT ensemble MC calculations for the alkali chlorides and equimolar mixtures (Li-K)Cl, (Li-Rb)Cl and (Na-K)Cl. They found fair agreement between the calculated and measured enthalpies and volumes of mixing and only qualitative agreement with conformal solution theory. *Caccamo* and *Dixon* [2.141] investigated the pair structure of a rigid-ion model for (Li-K)Cl mixtures for several concentrations. These MD simulations suggest that Li$^+$ ions remain fourfold coordinated by Cl$^-$ ions throughout the concentration range while the K$^+$ ions are about sixfold coordinated. A similar size effect had been disclosed by the earlier MD simulations of the (Li-K)Br mixture by *Lantelme* and *Turq* [2.142]; it is also apparent in the MD results for (Li-Rb)Cl obtained by *Okada* et al. [2.143].

h) Superionic Conductors

Certain crystalline salts exhibit an electrical conductivity comparable to that of the molten phase, due to a delocalization of one of the two ionic species [2.144]. Such superionic conductors have been extensively simulated, mostly by MD.

In his pioneering paper, *Rahman* [2.137] dealt with a rigid-ion model of CaF_2 and showed that the Ca^{2+} ions form a stable lattice, whilst the F^- ions are delo-calized, about one-half occupying the octahedral sites of the fcc Ca^{2+} sublattice. Subsequently, *Jacucci* and *Rahman* [2.145] analysed diffusion of the F^- ions which does not exhibit fluid-like Gaussian behavior. Such behavior was confirmed by a simi-lar study by *Dixon* and *Gillan* [2.146a] who, however, came to the conclusion that in spite of their very high mobility, the anions do not occupy the octahedral intersti-tial sites to any appreciable extent. This conclusion was corroborated by a study of superionic CaF_2 by *Hiwatari* and *Ueda* [2.147], based on the very simple potential model of (2.11) (with $n = 12$, $\varepsilon = 1$).

Gillan and *Dixon* [2.146b] made an extensive study of another superionic conduc-tor, $SrCl_2$. They were able to show from ion density maps that the defect concentra-tion is low and that the anions spend most of their time vibrating about their regu-lar lattice sites. Results which are qualitatively similar to those of $SrCl_2$ and CaF_2 have been obtained by *Walker* et al. [2.148] for PbF_2 and by *Walker* and *Catlow* [2.149] for UO_2. All of these materials are "Type II" superionic conductors exhibit-ing a behavior reminiscent of a second-order phase transition (specific heat anomaly).

Class I superionic conductors, on the other hand, undergo a first-order struc-tural phase transition accompanied by a change from an insulating to a conducting state. The best-known example of this class is AgI, whose superionic α phase has been simulated by *Vashishta* and *Rahman* [2.150] and by *Fukumoto* et al. [2.151]. Both sets of MD calculations were based on different potential models and the computed Ag^+ density maps turn out to be rather sensitive to details of the interionic po-tentials. A comparative MC study of both types of superionic conductors has been re-ported by *Hiwatari* and *Ueda* [2.152,153], done on the basis of the simple potential model (2.11) which yields a qualitatively correct description of such systems.

2.4 Molecular Liquids

This long section is entirely devoted to fluids of particles interacting through anisotropic forces. These fall essentially into two classes; anisotropic short-range forces accounting for the nonspherical shape of molecules (steric effects), and ani-sotropic long-range forces due to the electric multipoles which characterize the nonspherical charge distribution of the molecular orbitals. Since [2.1], there has been an exponential growth of the literature devoted to the simulation of molecular fluids, with a tendency to examine systems of increasingly complex molecules. As a result, the simulation data obtained for molecular liquids are often of a more quali-tative nature, and do not yet lead to a complete, quantitative picture of these sys-tems comparable to the present situation for the simpler fluids considered so far. We shall review molecular liquids roughly in the order of increasing complexity, the last section being devoted to the very extensive literature on water.

2.4.1 Hard Nonspherical Particles

a) Hard Spherocylinders

Monson and *Rigby* [2.154], and *Nezbeda* et al. [2.155,156] reported MC calculations
for prolate hard spherocylinders (of packing fraction y and length - to - breadth
ratio γ) supplementing those quoted in [2.1]. These simulations cover the ranges
$1.4 \leq \gamma \leq 3$ and $0.2 \leq y \leq 0.5$. In [2.156], all available results for the compressibility
factor corresponding to $\gamma = 2$ and 3 are summarized, and compared to an accurate semiempirical equation of state. The compressibility factors obtained by *Monson* and *Rigby*
replace erroneous earlier results [2.157] due to the use of an incorrect pressure
formula. Their new results are in good agreement with MD calculations by *Rebertus*
and *Sando* [2.158] for $\gamma = 2$, $y = 0.2$ and 0.4. The influence of elongation and density
on the structural properties of spherocylinder fluids has been investigated by means
of a spherical harmonics expansion [2.154,159]; comparison with a direct calculation
of the pair distribution function for specific orientation shows the poor convergence
of such an expansion for reduced separations shorter than γ [2.159].

b) Mixtures of Hard Spheres and Hard Spherocylinders

Equimolar mixtures of hard spheres and hard spherocylinders with $\gamma = 2$ have been
examined in the cases where the hard-sphere diameter equals the breadth of the
spherocylinders [2.160,161], and where the volumes of the spheres and of the spherocylinders are equal [2.160]. The compressibility factors are well reproduced by an
equation of state proposed by *Pavlíček* et al. [2.162]. The average pair structure of
the spherocylinders is not greatly affected by the presence of the hard spheres,
but the spatial arrangement of the latter differs significantly from that of an
equivalent mixture of hard spheres having volumes equal to those of the two components
in the mixture under consideration [2.160,161]. Similar conclusions remain valid for
mixtures with low or high concentrations of spherocylinders [2.163].

c) Hard Diatomics

Fused hard spheres of diameters σ_A and σ_B, and elongation ℓ (i.e., hard "dumbbells")
are a popular model system for diatomic molecules. Inconsistencies between the compressibility factors obtained from MC simulations of homonuclear hard diatomics by
Aviram et al. [2.164] and *Freasier* et al. [2.165,166], as well as the failure of their
results to satisfy some conjectured inequalities for the pressures [2.167], prompted
both groups to improve the accuracy of their calculations [2.168,169]. The Table
[Ref.2.1, Table 2.3] should hence be discarded. Extensive new MC simulations by
Tildesley and *Streett* [2.168] cover the range of elongations $0.2 \leq \ell/\sigma \leq 1$, and densities $0.2 \leq \rho d^3 \leq 0.9$, where d is the diameter of a sphere having the same volume as
the dumbbell. These authors propose an equation of state which reproduces their MC
results within the uncertainties of the latter. The pair distribution function for

three fixed molecular orientations has been obtained by *Cummings* et al. [2.170] for one state ($\ell/\sigma = 0.6$, $\rho\sigma^3 = 0.5$), with particular emphasis on the short-range part; the results are compared to the spherical harmonics expansion [2.171]. *Streett* and *Tildesley* [2.172] also report MC results for the pressure and atom-atom distribution functions of heteronuclear hard diatomics, for diameter ratios $0.5 \leq \sigma_B/\sigma_A \leq 1$, and elongations $0.05 \leq \ell/\sigma_A \leq 0.75$. Equimolar mixtures of homonuclear hard diatomics with equal diameters and elongations in the ratios 0.2/0.4 and 0.2/0.6 have been studied by *Aviram* and *Tildesley* [2.173].

2.4.2 Two-Center Molecular Liquids

The hard dumbbell model considered in the preceding section can be made more realistic if the fused hard spheres are replaced by Lennard-Jones (LJ) atoms. *Wojcik* et al. [2.174] report results of extensive MC simulations of this model; they obtained thermodynamic properties including the excess chemical potential for low density, high-temperature states and over the range of elongations $0.329 \leq \ell/\sigma \leq 0.793$, where σ now denotes the LJ range parameter. This paper contains a convenient summary of previous calculations.

At high pressures, the LJ repulsion is known to be too stiff for a realistic representation of the pair interactions in most diatomic fluids. *Fiorese* [2.175-177] carried out extensive MC calculations of the equation of state of highly compressed hydrogen (H_2) and nitrogen (N_2), using the much softer exponential -6 potential for the atom-atom interactions; the calculated pressures agree reasonably well with experimental Hugoniot plots.

2.4.3 Simple Dipolar and Multipolar Liquids

The models considered so far in this section contain anisotropic short-range forces only, to account for the shape of the molecules. We now turn our attention to models including long-range multipolar interactions. In the simpler versions of polar liquid models, the molecules are assumed to be spherical, with embedded point dipoles or higher-order multipoles. A fair amount of simulation work has been devoted to these models, with special emphasis on dielectric properties. The recent theoretical and simulation work on the statistical mechanics of polar systems has been reviewed by *Stell* et al. [2.178] and by *Alder* and *Pollock* [2.179].

Due to the long-range character of the dipolar interaction ($\sim 1/r^3$), the structure and dielectric properties of polar systems are always affected by the boundary conditions, as demonstrated in the very pedagogical paper by *Neumann* and *Steinhauser* [2.180]. The purpose of computer simulations of simple polar models, like dipolar hard spheres or Stockmayer particles, being an accurate calculation of the thermodynamics and dielectric properties of such systems, the fundamental question is that of the best choice of boundary conditions for a system of a few hundred particles to simulate a macroscopic polar fluid. The systems under consideration are generally

56

(but not always) chosen to be periodic, but, just as in the case of ionic fluids (Sect.2.3), there are different ways of handling long-range dipolar interaction:

a) The potential is arbitrarily truncated beyond a given distance (spherical cut-off SC).

b) The exact tail of the dipolar interaction is replaced by a reaction field (RF) calculated on the assumption that particles situated beyond a certain distance of a given particle act on the latter like a dielectric continuum.

c) An Ewald summation is carried out over all periodic images of the particles in the basic simulation cell.

The SC method has the obvious disadvantage of replacing the true dipolar inter-action by a finite-range interaction; hence the results for the correlation func-tions can, at best, be trusted for short distances only.

The RF method introduces surface effects at a distance where discrete particles are replaced by a dielectric continuum; the exact nature of these effects is very difficult to estimate.

The Ewald summation is a consistent manner of handling a periodic system, but it introduces long-range correlations not present in a truly macroscopic dipolar sys-tem. The situation is probably worse than in the case of ionic fluids, because of the absence of screening in dipolar systems.

To avoid the difficulties associated with the SC, RF and Ewald methods, some authors simulated finite samples surrounded by vacuum. Large samples must be used to minimize surface effects in this case; the method has so far been applied to two-dimensional systems only.

a) Dipolar Hard-Sphere Systems

This very simple model of polar fluids had already received much attention prior to 1978 [2.1]. More recently extensive calculations have been reported by *Patey* et al. [2.181], *Adams* [2.182] and *De Leeuw* et al. [2.79]. The essential purpose of [2.181] was a systematic comparison between the computer-generated pair distribution functions and approximate ones, obtained from a numerical solution of integral equations (HNC equation and its variants); SC conditions were used throughout to al-low a meaningful comparison. Good agreement was found for the short-range part of the correlation functions and for internal energies, with dipole moments μ such that $1 \lesssim \mu^2/\sigma^3 k_B T \lesssim 2.75$ and for densities $0.15 \lesssim \rho\sigma^3 \lesssim 0.80$. *Adams* [2.182] used the RF and Ewald methods in extensive simulations at $\rho\sigma^3 = 0.8$ and $\mu^2/\sigma^3 k_B T = 2.75$; the results for the correlation functions are compared to those obtained with the SC method. The case of the Ewald summation method is also considered in the work of *De Leeuw* et al. [2.79], who reported computations for the dipolar hard-sphere fluid at $\rho\sigma^3 = 0.8$ and $\mu^2/\sigma^3 k_B T = 2.0$. The effect of surrounding an infinite sphere of replicae of the simulation cell by a medium of dielectric constant ε is examined in detail. The main conclusion drawn from the detailed comparisons of the three types of boun-dary conditions [2.79,2.181,182] for the calculation of the dielectric constant is that the values of ε derived from MC simulations have substantial uncertainties when ε is large (typically 10% or 15% with 10^6 configurations generated, when $\varepsilon \simeq 20$).

The liquid-vapor coexistence line of the same model has been investigated by *Ng* et al. [2.183]. Combining MC data and the results of approximate theories, they located the critical point at $\mu^2/\sigma^3 k_B T_c = 4.0 \pm 0.1$ and $(\rho_c \sigma^3)^{-1} = 3.65 \pm 0.1$.

Patey et al. [2.184] studied a fluid of hard spheres with dipoles and linear quadrupoles. As in [2.181], the MC results with SC boundary conditions are compared to the predictions of integral equations. The quadrupolar interaction has a strong effect on the correlations between dipoles, leading to a decrease of the estimated dielectric constant compared to the value of a purely dipolar hard-sphere fluid with the same dipole moment.

b) Two- and Three-Dimensional Stockmayer Fluids

In the Stockmayer model, particles interact via a central Lennard-Jones potential and carry point dipoles. The two-dimensional version has been extensively studied by *Bossis* et al. [2.185-188] and *Caillol* et al. [2.189]. In the work of Bossis et al., the sample is placed in vacuum and surrounded by a steep circular barrier to avoid evaporation of the particles. The advantage of this boundary condition is that macroscopic electrostatics can be used to derive a relation between the fluctuations of the total dipole moment of small domains of the system and the dielectric constant, and also to calculate the asymptotic behavior of the correlation functions. The simulation results obtained for the finite sample in vacuum are compared to those obtained for a periodic system with SC boundary conditions. The former calculations lead to correlation functions which, at large distances, agree with the asymptotic behavior derived from macroscopic electrostatics, so justifying the relation between the dielectric constant and dipole moment fluctuations of small domains. *Bossis* and *Brot* [2.188] studied the effects of adding a quadrupole moment to the two-dimensional Stockmayer fluid, or decentering the point dipole: both modifications lead to a decrease of the calculated ε. *Caillol* et al. [2.189] compared the correlation functions derived from numerical solutions of integral equations to the MC data of *Bossis* et al. [2.186] and to their own MC results for a system of Stockmayer particles placed on the surface of a three-dimensional sphere. This study confirms that the integral equations overestimate the dielectric constant for large dipole moments, and explains the limitations of the RF method.

Three-dimensional Stockmayer fluid has also been the object of extensive investigations [2.190-192]. The work of *Adams* et al. [2.190] is essentially a methodological study concerning the use of various boundary conditions in MD simulations. This work is extended in [2.192] similarly to the simulations of the dipolar hard-sphere fluid [2.182]. Extensive comparisons of results with different boundary conditions are presented; the behavior of the local field and the dielectric saturation of the Stockmayer fluid in applied fields are examined in detail. *Pollock* and *Alder* [2.191] used Ewald summations to calculate the dielectric constant of the Stockmayer fluid in the range $\mu^2/\varepsilon\sigma^3 \leq 3$. The accuracy of the calculated dielectric constant is well established through a comparison of the results for different numbers of particles in the sample.

Shing and *Gubbins* [2.193] used "umbrella sampling" to determine the free energy and the vapor-liquid equilibrium curve of a system of LJ particles with a quadrupolar interaction ($Q^2/\varepsilon\sigma^5 = 2$).

c) Systems of Polarizable Particles

In a more realistic description of polar systems the particles must be considered as polarizable, i.e., in addition to the permanent dipole moment μ, they carry an induced dipole $\boldsymbol{\mu}_i = \alpha\boldsymbol{E}_i$, where α is the scalar polarizability of spherical particles and $\boldsymbol{E}_i$ is the total electric field acting on the given particle. *Vesely* [2.194] and *Pollock* et al. [2.195] considered the case of polarizable Stockmayer particles, whereas *Patey* et al. [2.196,197] simulated fluids of polarizable dipolar hard spheres. *Vesely* [2.194] used the RF method in MD simulations to calculate the pressure and internal energy of polarizable Stockmayer fluids; perturbation theory underestimates the effects of polarizability. *Pollock* et al. [2.195] used the Ewald method to simulate the same system in the range $0.25 \leq \mu^2/\varepsilon\sigma^3 \leq 2.25$. The dielectric constant is found to increase with increasing polarizability, while the oscillations in the dipole-dipole correlation function are slightly shifted. In their MC simulations of polarizable dipolar hard spheres, *Patey* et al. [2.196,197] showed that the induced forces make a sizeable contribution to the thermodynamic properties; the latter are well reproduced by an approximation of the free energy of this system, proposed by *Wertheim* [2.198].

d) Steric Effects in Polar Fluids

While the models considered so far in this subsection relate to spherical particles, it is clear that the shape of the molecules must have a strong influence on the dielectric properties of polar systems. *Steinhauser* [2.199] examined this problem in the case of a three-center LJ model of CS_2. The long-range part of the dipolar interaction is treated by the RF method. The qualitative conclusion of this work is that the steric effects overwhelm the electrostatic force in this model. *Morris* and *Cummings* [2.200,201] considered the polar hard dumbbell fluid, calculating the dielectric constant as a function of the elongation ℓ of the dumbbell and found that ε decreases as ℓ increases.

2.4.4 Realistic Models of Molecular Liquids

Due to the rapid evolution of computer technology and development of reliable first-principles intermolecular potentials, computer simulation techniques are entering an era where "realistic" models of chemical interest can be studied. Table 2.1 lists

Table 2.1. MC or MD simulations of molecular fluids (not including water)

	Temperature	Volume V [cm³/mol] mass density ρ_m [g/cm³], or pressure p [atm]	Simulation method	Number of particles	Potential model	Ref.
Hydrogen H_2	300 K 1000 K	V=7.622-22.653	MC(NVT)	108	Two-site model fitted to ab initio SCF calculations+dispersion term	2.175, 176
Nitrogen N_2	67.6-97.1 K	V=33.22-39.86	MC(NVT)	108	Ab initio MO calculations fitted by two- and three-site models + quadrupolar term	2.202
	102-3732 K	V=8.82-61.8	MC(NVT)		Two-site exp-6 potential	2.177
Carbon dioxide CO_2	218 K 273 K	V=37.55 V=47.35	MD	256	Two- and three-center LJ (m,6) + point quadrupole interaction	2.203
Carbon disulphide CS_2	281 K	ρ_m=1.26	MD	512	Three-center LJ potential	2.204
	193-395 K	ρ_m=1.09-1.42	MD	256	Three-center LJ potential	2.205
Carbon tetrachloride CCl_4	300 K	V=96.51	MD	64	Four-center LJ potential	2.206
	285 K 553 K	V=96.52 V=201.7	MD	108	Five-center LJ potential	2.207
Vanadium tetrachloride VCl_4	303 K	ρ_m=1.816	MD	108	Five-center LJ potential	2.208
Methane CH_4	124-525 K	V=39.57-100	MD	108-500	Five-center exp-6 potential	2.209
	130 K	V=40.15	MC(NVT)	125	EPEN/2	2.210
Ethane CH_3CH_3	105 K 181 K	V=47.42 V=54.25	MC(NVT)	125	1) Two-site exp-6 potential 2) Model described in [2.211]	2.211

Table 2.1 (cont.)

	Temperature	Volume V [cm³/mol] mass density ρ_m [g/cm³], or pressure p [atm]	Simulation method	Number of particles	Potential model	Ref.
n-butane $CH_3CH_2CH_3$	274 K 291.5 K 200 K	V=99.5 V=99.7 V=86.7	MD	64	Semi-rigid skeleton model LJ(12,6) interaction between carbon atoms. Torsional potential associated with bond rotation from *Scott* and *Scheraga*	2.212, 213
	291.5 K		Brownian dynamics	1	Same model as above [2.212, 213]	2.214
	76.4-745.35 K	ρ_m=0.2888-0.72199	MD	100	LJ (9,6) interactions between carbon atoms. Intramolecular potential allowing for bond rotation, angle bending and stretch vibrations	2.215
	-0.5°C	ρ_m=0.602	MC(NVT)	128	TIPS revised Scott-Scheraga potential for internal rotation	2.216-218
	-0.5°C	p =1,500,15000	MC(NpT)	128		
	-0.5°C	p =1	MC(NpT)	216	TIPS rotational potential function from molecular mechanics (MM2) calculations	2.219
n-octane	305-589 K	ρ_m=0.7036	MD	50	Potential functions similar to those for n-butane [2.215]	2.220
n-decane	481 K	V=236.5	MD	27	Potential functions similar to those for n-butane [2.213]	2.213
Methanol CH_3OH	25°C	ρ_m=0.78664	MC(NVT)	128	Ab initio STO-3G fitted by 12-6-1 potential+dispersion correction	2.221
	25°C	ρ_m=0.78664	MC(NVT)	128	1) Three-site TIPS model 2) Six-site model treating the methyl hydrogens explicitly	2.222

61

Table 2.1 (cont.)

	Temperature	Volume V [cm³/mol] mass density ρ_m [g/cm³], or pressure p [atm]	Simulation method	Number of particles	Potential model	Ref.
Methanol CH₃OH	25°C	p=1, 5000, 15000	MC(NpT)	128	Three-site TIPS model	2.223
(cont.)						
Ethanol CH₃CH₂OH	25°C	ρ_m=0.78509	MC(NVT)	128	Four-site TIPS potential Potential function for internal rotation around CO bond from ab initio MO calculations	2.224
Neopentane C(CH₃)₄	25°C	ρ_m=0.592	MC(NVT)	256	Three empirical five-site interaction models	2.225
1,2 Dichloroethane Cl CH₂CH₂Cl	25°C	ρ_m=1.246	MC(NVT)	128	Four-site TIPS model	2.216, 217
1,2 Dichloropropane CH₃CHClCH₂Cl	25°C	p=1, 5000	MC(NpT)	216	Five-site TIPS model Empirical rotational potential for internal rotation about C_1-C_2 bond	2.226
Dimethyl ether CH₃OCH₃	-24.8°C -24.8°C	ρ_m=0.737 p =1	MC(NVT) MC(NpT)	128 128	TIPS potential with interaction sites located on the oxygen and carbon atoms. Intramolecular potential based on ab initio MO calculations	2.227
Methyl-ethyl ether CH₃OC₂H₅	7.35°C	ρ_m=0.720	MC(NVT)	128		
Diethyl ether C₂H₅OC₂H₅	25°C	ρ_m=0.708	MC(NVT)	128		
Tetrahydrofuran CH₂CH₂OCH₂CH₂	25°C	p=1	MC(NpT)	128	Five-site TIPS potential Intramolecular potential function for pseudorotation of the ring from molecular mechanics calculations	2.228

Table 2.1 (cont.)

	Temperature	Volume V [cm³/mol] mass density ρ_m [g/cm³], or pressure p [atm]	Simulation method	Number of particles	Potential model	Ref.
Ammonia NH₃	196 K 277 K	$\rho_m = 0.731$	MD	108	Central force model	2.229
	273 K	V=26.5	MD	108	1) Modified central force model 2) Gas-phase model 3) Solid-state model	2.230
	270 K	V=26.5	MD	108	SCF-MO calculations fitted by a five-site model + dispersion correction	2.231, 232
	−33.35°C	$\rho_m = 0.682$	MC(NVT)	128	Ab initio STO-3G calculations fitted to a five-site model with 12-6-3-1 interactions	2.233
	277 K	V=26.92	MC(NVT)	125	EPEN/2	2.234
Hydrogen fluoride HF	0°C	$\rho_m = 1.015$	MC(NVT)	64 108	Ab initio 6-31G calculations fitted to a two-site model with 12-6-3-1 interactions	2.235
	0°C	$\rho_m = 1.015$	MC(NVT)	108	Ab initio STO-3G fitted by three-site model with 12-6-3-1 interactions	2.236
	0°C	$\rho_m = 1$	MD	216	1) Modified rigid central force model 2) Ab initio near Hartree-Fock calculations fitted by a three-site model + dispersion correction	2.237

Table 2.1 (cont.)

Temperature	Volume V [cm^3/mol] mass density ρ_m [g/cm^3], or pressure p [atm]	Simulation method	Number of particles	Potential model	Ref.
Hydrogen chloride HCl					
	coexistence line and supercritical isochore	MD MD	165,180	Two-center LJ potential Two-center truncated LJ potential + point dipole and quadrupole interactions	2.238, 239
171-201 K 345 K	V=30.74 V=86.08	MD	216	Three-site interaction models fitted to ab initio 4-31G calculations, (model B), second virial coefficients (model C), solid-state data (model D)	2.240
	V=43.64	MD	216	Modified model C [2.240]	2.241
	296 K				
Acetylene C$_2$D$_2$					
191.9 K 196.5 K	V=42.06	MD	108	Two-site model with exp-6 + Coulomb interactions	2.242
Phosphoric acid H$_3$PO$_4$					
200°C	ρ_m=1.73378	MC	27	Polarization model	2.243

representative systems which have been simulated, including alkanes [2.209-220], alcohols [2.221-224], ethers [2.227,228], alkyl chlorides [2.216,217,226] and associated liquids, like ammonia [2.229-234], hydrogen fluoride [2.235-237] and hydrogen chloride [2.238-241]. Not surprisingly, water is the most studied molecular liquid, so a special section will be devoted to it.

In all these simulations, the primordial problem is the selection of a sufficiently realistic intermolecular potential. There are two basic classes of such potentials: purely empirical and ab initio potentials, the former being essentially fitted to selected experimental data, while the latter are derived from extensive quantum-mechanical computations; in many cases potentials are determined by combining both methods.

Empirical potentials are specified by a distribution of interaction sites on each monomer, not necessarily coinciding with the atomic nuclei, and by a set of Coulombic or non-Coulombic potential functions between pairs of interaction sites belonging to different molecules. For many purposes the monomers can be considered as rigid or semirigid, in particular when torsional rotation around certain bonds is the only internal motion relevant to conformational aspects. More generally, the intramolecular potential may include vibrational bond stretching and bending [2.215,244].

An alternative route to intermolecular potential amounts to performing ab initio quantum-mechanical calculations of the dimerization energies for a large number of monomer configurations; the results are fitted to some convenient analytical form. Such calculations can be carried out at various levels of sophistication, depending on the choice of basis set and on the way correlation energy corrections are included (minimal basis set, Hartree-Fock (HF) or configuration interaction (CI) level).

A conceptually simple potential model for deformable molecules has been proposed by *Lemberg* and *Stillinger* [2.245] in the form of a central force model: each monomer is resolved into its atomic constituents (effective point charges); any pair of constituents, whether belonging to the same or different monomers, interact through a pair-wise additive central potential which is consistent with the formation of a stable monomer. Such a model permits explicit consideration of intramolecular vibrations, distortion or dissociation. Central force models have been worked out for water [2.245-247], hydrogen fluoride [2.229,248], and ammonia [2.230,248], though not yet in optimized form. *Stillinger*'s and *David*'s polarization model [2.249] extends the central force model by including nonadditive interactions due to the polarizability of the constituents.

If interactions between complex molecules, e.g., macromolecules, are considered, it is necessary to develop "transferable" potentials between atoms or molecular fragments to describe the interactions between many different molecules. The EPEN (empirical potential based on the interaction of electrons and nuclei) [2.250,251] and TIPS (transferable intermolecular potential functions) [2.252] models belong to this category.

Table 2.2. Thermodynamic properties of water as a function of density (or pressure for the isobaric MC calculations) and temperature for various potential models (Table 2.3). SC and RF refer to spherical cutoff and reaction field methods, respectively

Mass density [g/cm³] or pressure	Temperature [°C]	Potential model	Simulation method	Boundary conditions	Number of particles	Potential energy [kcal/mol]	$\frac{p}{\rho k_B T}$ or mass density ρ_m [g/cm³]	References
0.374	600	MCY-CI	MC(NVT)	SC	108	− 2.87		2.253
0.865	600	MCY-CI	MC(NVT)	SC	108	− 4.61		2.253
0.865	200	MCY-CI	MC(NVT)	SC	108	− 6.93		2.253
1.000	− 31	MCY-CI	MD	Ewald	64	− 8.93		2.254
1.000	− 31	MCY-CI	MD	Ewald	125	− 9.22	9.1	2.255
1 atm	− 30	TIPS2	MC(NpT)	SC 7.5 Å	125	−10.95	ρ_m=0.994	2.256
1.000	− 8	BNS	MD	SC 9.165 Å	216	−10.09		2.257
1.000	− 3	ST2	MD	SC 8.46 Å	216	−10.70		2.258
exp. density	− 2	MCY-CI	MD	SC 9.0 Å	343	− 8.96	0.76	2.259
1.00	0	Watts	MC	SC 8.1 Å	216			2.260
exp. density	4	HF	MC(NVT)	SC	64	− 8.55		2.261
exp. density	4	HF + D	MC(NVT)	SC 7–7.5 Å	64	− 9.25		2.262
exp. density	4	PE	MC(NVT)	SC	216			2.263
1.000	9	MCY-CI	MD	Ewald	64	− 8.46		2.264
1.000	9	MCY-CI	MD	Ewald	125	− 8.72	8.1	2.255
1.000	10	ST2	MD	SC 8.46 Å	216	−10.41	0.53	2.258
1.000	10	ST2	MD	SC 8.46 Å	216	−10.6		2.265
1.000	10	ST2	MC(NVT)	SC 7.75 Å	216	−10.62		2.9,
0.997		ST2	MC(NVT)	SC 4.65 Å	125	−10.59	0.977	2.260
			MC(NVT)	Ewald	27	−10.62		
1.000	21	MCY-CI	MD	Ewald	125	− 8.58	8.5	2.255
1.000	22	RSL	MD	Ewald	216	− 9.20		2.246
0.997	25	RWL	MC(NVT)	SC 6.2 Å	64	−10.14		2.266,267
0.997	25	BNS	MC(NVT)	SC	216	− 9.78		2.267
				SC	64	− 9.58		
					216	− 9.28		
					500	− 8.38		
0.997	25	ST2	MC(NVT)	SC	216	−10.47		2.268
1.000	25	HF	MC(NVT)	SC	125	− 6.9		2.264
0.997	25	HF + D	MC(NVT)	SC	64	− 8.98		2.262
0.997	25	MCY-CI	MC(NVT)	SC	343	− 8.38		2.269
					343	− 8.58		

Table 2.2 (cont.)

Mass density [g/cm³] or pressure	Temperature [°C]	Potential model	Simulation method	Boundary conditions	Number of particles	Potential energy [kcal/mol]	$\frac{p}{\rho k_B T}$ or mass density ρ_m [g/cm³]	References
1.000	25	MCY-CI	MC(NVT)	SC 7.75 Å	125	− 8.58		2.264
0.997	25	MCY-CI	MC(NVT)	SC 6.35 Å	125	− 8.64		2.265
1 atm	25	MCY-CI	MC(NpT)	SC	64	− 9.38		2.270
1.000	25	STO-3G	MC(NVT)	SC	100	− 9.07		2.271
	25	STO-3G	MC(NVT)		125	− 8.9	$\rho_m = 0.756$	2.272
1.000	25	TIPS	MC(NVT)	SC 7.5 Å	125	− 9.04		2.252
1 atm	25	TIPS2	MC(NpT)	SC	125	− 8.31		2.256
1 atm	25	EPEN/2	MC(NpT)	SC	100	− 10.04	$\rho_m = 0.997$	2.273
1.000	25	EPEN/2 (revised)	MC(NVT)	SC	100	− 10.26	$\rho_m = 1.095$	2.274
0.997	25	Watts	MC(NVT)	SC	216	− 8.75		2.268
0.997	25	RWK1	MC(NVT)	SC	216	− 11.28	− 0.91	2.268
0.997	25	RWK2	MC(NVT)	SC	216	− 10.02	− 0.94	2.268
1.000	27	Watts	MD		250	− 8.82		2.275
1.000	27	SPC	MD	SC 8.5 Å	216	− 10.08		2.276
exp. density	29.5	RSL2	MD	Ewald	343	− 9.478	− 0.36	2.247
1.000	31	MCY-CI	MD	SC 9.0 Å	125	− 8.54	0.1	2.259
1.000	31	SPC	MD	Ewald	216	− 9.184		2.277
0.993	34.3	BNS	MD	SC 9.165 Å	216	− 8.57		2.278
1.000	37	MCY-CI	MC(NVT)	SC	216	− 9.83		2.279
0.988	41	ST2	MD	SC 8.46 Å	64	− 8.03	0.37	2.258
1.000	49	MCY-CI	MD	Ewald	216	− 8.40		2.254
	50	MCY-CI	MC(NVT)	SC	216			2.279
exp. density	50	PE	MC(NVT)	SC	64	− 6.3		2.263
exp. density	75	HF + D	MC(NVT)	SC	125	− 9.3		2.262
1 atm	75	TIPS2	MC(NpT)	SC 7.5 Å	125	− 7.81	$\rho_m = 0.960$	2.256
1.000	87	MCY-CI	MD	Ewald	343	− 7.78		2.255
exp. density	104	MCY-CI	MD	SC 9.0 Å	64	− 7.43		2.259
1.000	104	MCY-CI	MD	Ewald	216	− 8.78		2.254
1.000	118	ST2	MD	SC 8.46 Å	216	− 8.79	0.66	2.280
1.000	120	ST2	MD	RF	216	− 8.59		2.280

Table 2.2 (cont.)

Mass density [g/cm^3] or pressure	Temperature [°C]	Potential model	Simulation method	Boundary conditions	Number of particles	Potential energy [kcal/mol]	$\dfrac{p}{\rho k_B T}$ or mass density ρ_m [g/cm^3]	References
1.000	137	Watts	MD	Ewald	216	− 7.88		2.281
1.000	300	MCY-CI	MC(NVT)	SC	108	− 6.35		2.253
1.000	314.8	BNS	MD	SC 8.46 Å	216	− 5.91		2.257
1.000	600	MCY-CI	MC(NVT)	SC	108	− 4.63		2.253
1.000	900	MCY-CI	MC(NVT)	SC	108	− 3.34		2.253
1.346	57	ST2	MD	SC 7.7 Å	216	− 9.92	4.7	2.282
1.346	68	MCY-CI	MD	Ewald	64	− 6.69		2.254
1.346	97	ST2	MD	SC 7.7 Å	216	− 9.40	4.9	2.282
1.346	148	ST2	MD	SC 7.7 Å	216	− 8.95	5.0	2.282

Thermodynamic and structural properties for all fluids listed in Tables 2.1,2 have been obtained by MC or MD simulations. The published results include atom-atom pair distribution functions, probability distribution functions [2.283] for pair energies, bonding energies, coordination numbers, hydrogen bonding numbers, hydrogen bond angles, as well as stereoscopic pictures of instantaneous configurations, etc. These results are mostly qualitative and allow some insight into two important mechanisms: hydrogen bonding and conformational equilibrium.

In hydrogen fluoride [2.235-237], methanol [2.221-223] and ethanol [2.224], where each molecule engages on the average in only two hydrogen bonds, the predominant structure of the liquid at $0°C$ is characterized by long winding hydrogen-bonded chains. The precise structural arrangement of the chain may depend crucially on the potential, as exemplified by HF for which several potential models have been investigated [2.235,237].

Hydrogen chloride [2.240,241] and ammonia [2.229-231,233] appear to be weakly associated so that the structural properties are likely to result from the balance of hard-core packing effects and the tendency to form hydrogen bonds [2.231]. This explains the great sensitivity of the structural properties to potential models.

As regards conformational equilibrium, a specific question addressed by several authors is the shift of conformer populations upon going from the gas phase to the pure liquid [2.213,215-220,222,224,226-228]. A study of this kind poses severe convergence problems in MC simulations, due to the infrequent crossings of intramolecular potential barriers when these are much higher than k_BT (as in the case of dichloropropane [2.226]). The problem has been satisfactorily solved by the use of importance sampling over "chopped" barriers [2.284,5,219]. Furthermore, MC calculations for liquid n-butane [2.219], dimethyl ether, and methyl ethyl ether [2.227] do not show any evidence of change in the trans and *gauche* populations from their gas-phase values. Earlier simulations for n-butane remained somewhat inconclusive, due to insufficient statistics [2.212,213,215-217]. Upon compression from 1 to 15000 atm, the *gauche* population of n-butane increases only slightly [2.218]. Contrasting with this situation, a large increase in the *gauche* population relative to the gas has been observed in simulations of liquid dichloro-ethane [2.216,217] and dichloropropane [2.226]; this can be traced back to electrostatic effects. Recent, more complex topics investigated by simulations include the pseudorotation of rings in tetrahydrofuran and its effects on intermolecular structure and thermodynamics [2.228], and the mixing of enantiomeric liquids, e.g., 1,2 dichloropropane, during a search for chiral discrimination [2.226].

In concluding this section it must be stressed that any MC work pretending to quality should give careful consideration to questions like convergence, ergodicity, error estimates or dependence on sample size and boundary conditions. This is particularly crucial for systems with strongly angle-dependent and long-range dipolar forces, as in the case of water, which will be considered next.

2.4.5 Liquid Water

A summary of computer simulations dealing essentially with structural and thermodynamic properties of water, published before December 1982, is given in Table 2.2.
Potential models of water are listed in Table 2.3, with reference to the MC or MD
simulations where they were used. Potential models are summarized in [2.268,297],
while reviews on water simulations have been published in [2.298-302]. Small clusters
of water molecules [2.303,304] and simplified models which do not aim at a realistic
description of water [2.305,306] will not be discussed here.

Table 2.3. Potential models of water

Acronym	Type	Number of interaction sites	Ref.	Reference to MC or MD calculations using the potential
RWL	empirical	5	2.285	2.266,267,286
BNS	empirical	5	2.287	2.267,268,257
ST2	empirical	5	2.258	2.9,13,258,260,265,279,280, 282,288-290
HF	ab initio	4	2.291	2.291,264
HF + dispersion correction	ab initio	4	2.262	2.262
MCY-CI	ab initio	4	2.292	2.13,253-255,259,264,269, 270,277,293,294
STO-3G	ab initio	4	2.295	2.271,272
TIPS	empirical	3	2.252	2.252
TIPS2	empirical	4	2.256	2.256,293
SPC	empirical	3	2.276	2.13,276,277
EPEN/2	empirical	7	2.265	2.265
EPEN/2 revised	empirical	7	2.274	2.274
RSL	central force		2.246	2.246
RSL2	central force		2.247	2.247
Watts	central force	3	2.244	2.244,275,296
RWK1	semiempirical	4	2.268	2.268
RWK2	semiempirical	4	2.268	2.268
SD	polarization		2.249	
PE	polarization		2.263	2.263

Most of our qualitative understanding of the structural behavior of computer-
simulated water stems from the extensive calculations of *Rahman* and *Stillinger* [2.258,
282,288,289] on the ST2 model of water. Similar calculations based on different potential models (predominantly MCY-CI) essentially confirmed or supplemented the
Rahman-Stillinger work, and a fairly coherent picture has emerged for the structure
of water [2.307].

A helpful guide to the identification of water structure is provided by the known
ice structures. In hexagonal ice (I_h), each H_2O molecule is surrounded by four
nearest neighbors to which it is hydrogen bonded with tetrahedral symmetry. The hydrogen bonds connecting neighboring molecules form a space-filling three-dimensional
network. Upon melting, the network persists, still showing a preference for local
tetrahedral geometry, but it exhibits substantial bending and large-scale randomness.
The first evidence for such a residual structure in the liquid is provided by a

second-nearest-neighbor peak observed in the oxygen-oxygen (O-O) pair distribution function at a distance where it is expected to occur in a tetrahedrally ordered hydrogen-bond arrangement [2.258,269,308]; the width of the peak, however, indicates considerable disorder.

A more detailed characterization of hydrogen-bond patterns requires a working definition of the hydrogen bond. Both energetic [2.278] and geometric [2.272,309] definitions have been put forward. In the former case, a pair of molecules is considered to be hydrogen bonded whenever their interaction energy is less than some negative value V_{HB} [2.288]. With any reasonable definition of V_{HB}, most molecules are engaged in 3 to 4 hydrogen bonds at room temperature and atmospheric pressure, and the distribution of molecules versus the number of hydrogen bonds in which they participate is unimodal; the latter observation rules out "interstitial" theories of water [2.310-313]. The fraction of nonbonded molecules is negligibly small whenever the mean number of hydrogen bonds exceeds 2 [2.309]. Simulations show the hydrogen bonds to be bent by 20° on the average [2.256,272,279], in agreement with a recent value inferred from vibrational spectroscopic data [2.314]. Enumeration of "nonshort-circuited" polygons in the course of simulations eliminates a number of simple lattice models [2.288,315-318].

The extent of hydrogen bonding has been studied by *Geiger* et al. [2.289] for ST2 water and by *Mezei* and *Beveridge* [2.309] for MCY-CI water. At room temperature, *Geiger* et al. [2.289] observed a percolation threshold where clusters of bonded molecules suddenly produce a large, space-filling random network as soon as the average number of hydrogen bonds per molecule reaches a critical value of about 1.3. With increasing temperature, the second peak in the O-O pair distribution function decreases, indicating weakening of tetrahedral order without appreciable hydrogen-bond rupture [2.254,256,257]. Under compression the hydrogen-bond network tends to become more compact, with a corresponding distortion of hydrogen bonds and interpenetration of parts of the network [2.254,282].

Discrimination between various potential models of water has been based primarily on the confrontation between computer-generated pair structures and the results of radiation scattering experiments. Such a comparison was until very recently tainted with a substantial degree of uncertainty for the following reason: while computer simulations readily provide the three partial pair distribution functions g_{OO}, g_{OH}, g_{HH}, and their Fourier transforms, the partial structure factors, experimental techniques, whether X-ray [2.319-321], neutron [2.322-335] or electron [2.336] diffraction, yield a single scattering function, which is a weighted sum of the 3 partial structure factors. A comparison of this scattering function with the corresponding combination of computer-generated partial structure factors is not very instructive, because the complicated mixing of all characteristic intermolecular O-O, O-H, H-H distances does not lend itself to unambiguous interpretation. The O-O pair distribution function was obtained by *Narten* and *Levy* [2.320] under the assumption that X-ray scattering is dominated by spherical scattering from oxygen atoms.

Although this is a fairly good approximation, contributions from O-H and H-H correlations are not entirely negligible [2.277,333]. Subsequently O-H and H-H distribution functions were extracted by *Narten* [2.324] from a combination of X-ray and neutron-scattering experiments on D_2O. But additional assumptions on the short-range structure had to be made to disentangle the 3 partial pair distribution functions, so that the results for g_{O-H} and g_{H-H} are subject to much larger uncertainties than g_{O-O}. Very recently three independent neutron diffraction measurements based on the isotopic substitution technique allowed the extraction of the 3 partial structure factors without explicit reference to a structural model [2.334,337], thus offering prospects for more precise comparisons with simulation data. Measurements of the isochoric temperature derivative of the neutron scattering function will test potential models further [2.333].

Prior to these recent measurements, comparison between computer simulations and experiment focused largely on g_{OO} (HF potential [2.261,264,291], HF + D [2.262], MCY-CY [2.254,259,265,269], STO-3G [2.271], RWL [2.264], ST2 [2.265], TIPS [2.252], TIPS2 [2.256], SPC [2.276]). Such comparisons, made at room temperature, were already sufficient to reveal the weaknesses of several potential models, and to improve upon them. Thus the ST2 potential superseded the earlier potential models by Rowlinson and Ben-Naim and Stillinger, the MCY-CI potential was recognized as the most accurate among ab initio potentials, a revised central-force model (RSL2) gave better pressures than RSL, TIPS2 replaced TIPS and SPC, etc.

Even among the more successful models, none achieves complete agreement with Narten's g_{OO} at room temperature: the ST2, MCY-CI, RSL2, RWK2, TIPS2 models predict correctly the position of the first peak, but its height is overestimated by MCY-CI and RWK2 by roughly 10%, and more dramatically by TIPS by 20%, ST2 and RSL2 by 30%. Although MCY-CI, ST2, TIPS2, RSL2 predict second and third peaks in agreement with experiment, the second maximum is shifted to shorter distances for MCY-CI and TIPS, while ST2 overestimates the amplitude of the second peak at its correct position. The structural changes brought about by varying temperature have been studied for the BNS [2.257], ST2 [2.258], MCY-CI [2.254,259,277,279] and TIPS [2.256] potentials and compared to Narten's X-ray data. Similarly, the pressure variation of the structure has been investigated for the ST2 [2.282] and the MCY-CI [2.254,277] models, but these data still await detailed comparison with experiment [2.331,332,335].

Further tests of the potential models are provided by the predicted ice structures [2.268,297] and by the predicted molar volume of the liquid at atmospheric pressure. Although the MCY-CI potential underestimates the latter quantity by 24% [2.270], this model turns out to give the overall most accurate description of the hydrogen bond network. Provided the MCY-CI potential can be regarded as a sufficiently realistic pair potential, the differences observed between the simulations based on this model and experiment can give valuable information on many-body effects, which are known to be important in water [2.338-345]. The implications of this point of view have recently been examined by *Egelstaff* and *Root* in an analysis of their scattering data [2.346].

Methodological aspects concerning convergence characteristics of thermodynamic and structural properties of water and their dependence on boundary conditions and system size have been considered by *Pangali* et al. [2.9,260], *Mezei* et al. [2.6,265] and by *Jorgensen* [2.293]. Their main conclusions can be summarized as follows:

a) Standard Metropolis sampling reveals slowly converging thermodynamic properties. Variation of potential energy with the number of generated configurations consists of a rapidly fluctuating part superimposed on a slowly varying part of a small amplitude [2.9,265]. Most MC calculations reported in Table 2.2 are insufficient in length to sample these slow fluctuations, and consequently underestimate heat capacities as well as error estimates of the statistical averages.

b) Thermodynamic quantities calculated from fluctuation formulas (heat capacities, isothermal compressibility, thermal expansion coefficient, dielectric constant, etc.) converge several times more slowly than the internal energy.

c) Importance sampling, e.g., the force bias method discussed in Sect.2.1.1, improves the rate of convergence of the internal energy by roughly a factor of two [2.6,9].

The properties most sensitive to boundary conditions and system size are those related to dielectric properties (mean square polarization and dipole-dipole correlation functions) [2.260,267,278,280,286,347]. The problems raised by the long-range nature of the dipolar interaction have been extensively studied for simple prototype systems (Sect.2.4.3) and the conclusions drawn from such investigations are directly applicable to the water models. Atomic pair distribution functions and thermodynamic properties, on the other hand, are considerably less sensitive to sample size and boundaries, and simple spherical truncation of the intermolecular potential is generally sufficient for their evaluation. Calculation of the free energy of water is partly still an open problem [2.38,270,294,348].

2.5 Solutions

Solvation of nonpolar, polar or ionic solutes in both aqueous and nonaqueous solvents has been studied for small prototype systems, with the ultimate goal of simulating systems of biological interest. A common problem in the simulation of dilute solutions is the slow convergence of energetic and structural solute-solvent properties, which may impede accurate evaluation of many quantities of interest, e.g., partial molar internal energies, in a run of reasonable length. The problem can be overcome by a combination of the preferential sampling and force bias methods described in Sect.2.1.1. Solutions of nonelectrolytes, ionic solutions and the solvation of large dipoles will be examined in succession.

The two aspects of solvation which have been more closely examined in computer simulations are the rearrangement of water molecules in the vicinity of a single nonpolar solute molecule (hydrophobic hydration), and the solvent-induced interaction between two or more nonpolar solute molecules (hydrophobic interaction). Prototype solutes include Lennard-Jones particles [2.349-354], argon [2.355] or methane [2.356-360], but small molecules incorporating both nonpolar and polar functional groups have been considered as well [2.361,362]. A list of representative systems is given in Table 2.4.

The structure of water in the immediate vicinity of a nonpolar solute molecule is entirely different from the structure of pure water because the nonpolar solute does not participate in hydrogen bonding. Since it is energetically advantageous for the water molecules in the first hydration shell to have all four bond directions engaged in interactions with neighboring solvent molecules, they adopt preferentially a "straddling" configuration, where three of the hydrogen bonds are roughly tangent to a sphere surrounding the solute molecule, with the fourth bond pointing away form that molecule [2.260,350,354,355]. In this arrangement the solute molecule has of the order of 20 nearest neighbors [2.350,353-358,360], in marked contrast with the 4-5 neighbors of the tetrahedral-like structure of pure water. The facts that the average number of hydrogen bonds per water molecule in the first hydration shell does not differ appreciably from that in pure water [2.349,351,353,362], the lowering of the binding energy [2.349,351,353,355,357,358] and the observed increase of orientational correlations in the first hydration shell support the current view of solvent "structuring" by a nonpolar solute. Another indication of this structuring is the negative entropy difference between an aqueous methane solution and pure water, estimated by the "umbrella sampling" technique [2.350]; due to large errors bars this interesting result should, however, be accepted with caution.

Polar solutes, on the other hand, can participate in hydrogen bonding, so consequently the structural arrangement of water molecules around polar groups resembles much more closely that of bulk water [2.361-363].

The potential of mean force between two solute molecules at infinite dilution provides a measure of solvent-induced interactions. *Pangali* et al. [2.352] used the force bias MC method in conjunction with importance sampling to determine the potential of mean force between two LJ particles dissolved in ST2 water, as a function of their separation. The calculated potential of mean force exhibits an oscillatory structure; a first minimum corresponds to the solute particles roughly at contact, while the second minimum describes configurations where the two solute atoms are separated by one layer of water molecules. The latter conformation appears to be the more probable one, in contradiction with the conventional view of hydrophobic interaction (i.e., a tendency of solute particles to aggregate). The latter view is also contradicted by the MD simulations by *Rapaport* and *Scheraga* [2.354] and by *Geiger* et al. [2.349]. The potential of mean force calculated by *Pangali* et al. [2.352] agrees

Table 2.4. List of MC or MD simulations representative of nonelectrolyte dilute aqueous solutions

Solute	Temperature	Volume V [cm³/mol] or mass density ρm [g/cm3] or pressure p [atm]	Simulation method	Number of solute molecules	Number of water molecules	Solute-solute interaction	Solute-water interaction	Water-water interaction	Ref.
Argon	25°C	V = 18.19	MC(NVT)	1	124		Ab initio SCF calculations	MCY-CI	2.355
Methane	300 K	ρm = 1.	MC(NVT)	1	63		cf. [2.356]	cf. [2.356]	2.356
	298 K	p = 1.	MC(NpT)	1	100		Ab initio SCF calculations	MCY-CI	2.357
	25°C	ρm = 1.	MC(NVT)	1	124		Ab initio 6-31G MO calculations	MCY-CI	2.358
	300 K	V = 17.98	MC(NVT)	1	202		Ab initio SCF calculations	MCY-CI	2.360
Methane Isobutane Ethane Pentane	298.15 K	V = 18.07	MC(NVT)	1 1 1 1	63 63 63 63		LJ	ST2	2.350, 351
n-butane	25°C	p = 1.	MC(NpT)	1	216		TIPS	TIPS2	2.219
LJ particles	305,5 K	ρm = 1.	MD	2	214	LJ	LJ	ST2	2.349
LJ particles	298 K	ρm = 1.	MC	2	214	LJ	LJ	ST2	2.352, 353
LJ particles	30°C	density of pure water	MD	4	339	Truncated LJ	Truncated LJ	MCY-CI	2.354
Formaldehyde	25°C	ρm = 1.	MC(NVT)	1	124		Ab initio 6-31G MO calculations	MCY-CI	2.363
Methanol	300 K	V = 18.34	MC(NVT)	1	198		Ab initio quantum mechanical	MCY-CI	2.361
Alanine dipeptide	300 K	ρm = 1.004	MD	1	195		Empirical cf. [2.364]	ST2 (flexible)	2.362, 364

75

well with a semiempirical theory by *Pratt* and *Chandler* [2.365], but disagrees
strongly with earlier MC calculations by *Dashevsky* and *Sarkisov* [2.356] who used a
method subject to large statistical errors. A similar remark holds for the solvent-
averaged force obtained by *Swaminathan* and *Beveridge* [2.359]. The potential of mean
force between two LJ atoms in water has been obtained on an absolute scale by *Goldman* [2.366].

While the mean force calculations involving atomic solutes do not yield any clear
evidence of hydrophobic interaction, the latter is more clearly apparent in conformational changes of molecules upon solvation in water. A significant increase in the
gauche population has been observed upon transferring n-butane from the ideal gas to
aqueous solution [2.219], whereas no population change occurs when n-butane is dissolved in the pure liquid [2.219] or in a nonaqueous solvent [2.5,284].

2.5.2 Solvation of Ions

Computer simulations have been reported for a single ion (Li^+, Na^+, K^+, F^-, Cl^-) in
water [2.367-370], for 2.2 molal solutions of LiCl [2.371-373], CsCl [2.372-374],
NaCl [2.372,373,375-377], LiI [2.372,373,378], CsF [2.372,373], NH_4Cl [2.379],
$MgCl_2$ [2.380,381], for large clusters of 50-200 water molecules containing either a
single ion [2.382-386], or an ion pair [2.387,388], and for water-ion microclusters
[2.389-394]. The most frequently used water models are ST2 [2.367,369,371,372,374-
376,379], MCY-CI [2.368,383,385], RSL2 [2.377,380,395], TIPS2 [2.370], HF [2.382,
388], HF+D [2.387,391]. The ion-water interactions are either based on ab initio
quantum-mechanical calculations [2.368,382,383,385,387,388,391-393] or chosen to be
simple LJ+Coulomb potentials [2.367,369,375-377,379]. Quantitative predictions for
the ion-water structure depend on the potential model, boundary conditions, ion concentration, presence and nature of counterions, etc. General qualitative aspects can
be summarized as follows. In the first hydration shell, the water structure is dominated by the strong ion-water interactions. The preferential ordering of the water
molecules near a cation is intermediate between the "dipole" configuration (with the
water dipole moment antiparallel to the oxygen-ion vector) and the "lone pair" configuration (where the lone pair is oriented towards the ion) [2.367,368,372,377,379,
380,390,395]. The ST2 potential favor the "lone pair" configuration [2.372,379],
while the central force model favors the "dipole" configuration [2.377,380,395].
Pronounced preferential orientations subsist in the second hydration shell of Li^+
and Mg^{2+} [2.380,395].

The local structure around anions is dominated by ion-water hydrogen bonding
[2.367,368,377,379,380,383,391,395]. A comparative study of the water structure
around ions in "bulk" water and in clusters shows that the local structure is not
significantly affected by the boundary conditions [2.369]. Coordination numbers
[2.378,384,396] for aqueous solutions and ion-water clusters are conveniently summarized in [2.397] and compared with experiment.

The simulation data for the structure have been compared to X-ray diffraction measurements for aqueous solutions of NaCl [2.376,377], NH_4Cl [2.398], LiI [2.378, 399] and $MgCl_2$ [2.400]. The total scattering function is, however, a complex quantity, involving ten partial structure factors, and hence is difficult to analyze [2.397]. More direct information on the ion-water structure is obtained from neutron first-order difference scattering functions [2.368,397,401].

2.5.3 Solvation of Large Dipoles

Neumann et al. [2.402,403] reported MD simulation of a system of 500 Stockmayer particles, one of which carries a very large dipole moment, with the purpose of examining the structural arrangement of the solvent particles around the solute. In a similar spirit *Costa Cabral* et al. [2.404] simulated solutions composed of a single solute molecule fixed at the origin in a solvent of dipolar hard spheres. The aim of the calculation was to compare the solute-solvent interaction energy to the predictions of the cavity model.

2.5.4 Solvation of Biological Molecules

Solvation studies of biomolecules obviously combine all aspects of hydrophobic, hydrophilic and ionic hydration. Representative computer simulations deal with amino acids [2.405-410], proteins [2.411-414], phosphate groups [2.393,415], bases of nucleic acids and base pairs [2.416], and DNA [2.414,417-423]. This area of research is expanding rapidly, due to the steady increase in computer power and to the crucial importance of such studies in biophysics.

2.6 Surfaces and Interfaces

This last section is devoted to computer studies of inhomogeneous fluids: liquid-vapor and liquid-solid interfaces, fluids against a wall or between parallel plates, etc. The features common to all these systems is the breaking of translational invariance in at least one direction.

2.6.1 Liquid-Vapor Interface of Simple Fluids

The aim of the earlier computer simulations of coexisting gas and liquid phases was the determination of the density profile, i.e., the variation of the microscopic density from the bulk liquid to the bulk gas. For simple fluids this variation has been shown to be monotonic and very steep near the triple point where the profile changes abruptly from liquid to gas over one or two atomic diameters. Most of this work is reviewed in [2.1,424,425]. Much of the more recent work described here is discussed in [2.426].

Recent articles on the liquid-vapor coexistence contain detailed information on the properties of the fluids near the interface, like the two-body distribution functions and the location of the surface of tension. These calculations provide explicit tests of approximate theories. *Ebner* et al. [2.427] examined the structure of the interface in a one-dimensional LJ fluid immersed in an external potential (to induce an inhomogeneity) and found good agreement between simulations and theories derived from approximate free-energy functionals of the local density. *Kalos* et al. [2.428] carried out more ambitious three-dimensional computations to test the hypothesis that the density profile is an ensemble average of quasi-planar sharply separated gas-liquid regions; this hypothesis leads to an approximation for the two-body distribution functions near the interface, which is strongly supported by the simulation results. This work confirms the monotonic behavior of the density profile and contains interesting results on the transverse correlation function associated with the probability of finding two particles at a given distance in a plane parallel to the interface. The low -k values of the Fourier transform of this correlation function are enhanced in the interface region. In general, the results of this investigation agree with the picture of the liquid-gas region as a sharp interface modulated by capillary waves. *Rao* et al. [2.429] offered further justifications of the analysis presented in [2.428]. These authors simulated the liquid-vapor interface in the presence of a hard wall; if the wall is placed in the transition region, the density profile becomes oscillatory.

Rao and *Berne* [2.430] examined the variation of the normal (P_N) and transverse (P_T) components of the pressure tensor. As expected, P_N turns out to be constant, but P_T becomes negative near the interface where the density profile changes rapidly. Because the surface tension is the integral of the difference P_N-P_T, the computation by *Rao* and *Berne* establishes that the surface tension is localized in the region of rapid variation of the density profile.

The liquid-gas coexistence in the two-dimensional LJ system was simulated by *Abraham* [2.431]. The density profile turns out to be monotonic, as in three dimensions, but the interfacial width extends over 6-7 atomic diameters.

2.6.2 Liquid-Vapor Interface of Molecular Fluids

Thompson and *Gubbins* [2.432,433] studied the orientational order near the liquid-vapor interface of fluids of diatomic homonuclear molecules modeling chlorine and nitrogen. In these molecular fluids the density profile depends on two variables: the distance to the interface and the average angle between the molecular axis and the normal to the interface. When averaged over this angular dependence, the density profile turns out to be monotonic; the molecules are preferentially oriented perpendicularly to the interface in the liquid phase and parallel in the gas phase. These angular correlations are weakened by strong quadrupolar interactions.

The surface properties of fluids of more complex molecules have been examined by *Weber* and *Helfand* [2.434] and by *Lee* and *Scott* [2.435,436]. The former simulated a model of flexible n-octane molecules; their results are essentially qualitative and show that the molecules are preferentially aligned perpendicular to the interface. *Lee* and *Scott* attempted to calculate the surface tension of pure water [2.435] and of water in the presence of a layer of hydrated phospholipid head groups at the interface [2.436]; this was achieved via a method of *Miyazaki* et al. [2.437] based on an estimation of the change in free energy between a fluid with periodic boundary conditions and a free surface. The calculated surface tension of ST2 water turns out to be larger than the experimental value.

2.6.3 Density Profiles of the One-Component Plasma and Liquid Metals

The surface density profile of a few hundred ions embedded in a rigid background uniformly distributed over a sphere (the OCP model) was obtained in the MC simulations of *Badiali* et al. [2.438]. For low temperatures, the ionic density profile at the surface of the sphere is strongly oscillatory, but it becomes monotonic when the temperature is increased. These authors also examined the influence of a hard wall at the boundary of the spherical background on the density profile; the latter does not change at low temperature.

D'Evelyn and *Rice* [2.439] simulated a cluster of 256 atoms interacting by the effective potential between pseudoatoms in liquid sodium, reporting an oscillatory density profile in such a microscopic droplet.

2.6.4 Liquid-Wall Interfaces

Computer simulations of liquid-wall interfaces essentially have a twofold objective: for simple models to obtain qualitative results on the changes of the fluid structure in the vicinity of a wall or a rigid body, and to calculate the solvation forces between two planar surfaces separated by a fluid. The wall can be modeled either by an infinite potential barrier ("hard wall") or by a soft potential acting on the particles in the fluid and which depends on their distance from the wall ("soft wall"). Such an external potential results from the interaction between atoms on fixed regular lattice positions in the wall and the atoms in the fluid. The actual potential due to the "frozen" atoms in the wall acting on the fluid particles is replaced, for computational efficiency, by approximate expressions of the form:

$$\psi = \sum_\alpha a_\alpha z^{-\alpha} \quad ,$$

where $z(>0)$ is the distance measured from the wall. The external potentials ψ are characterized by the values of the powers α. Thus an LJ interaction between wall and fluid atom leads to external potentials which can be approximated by (9-3), (10-4) or (10-4-3) potentials.

The simulated systems are placed in a slab, limited by two hard or soft walls along the z direction and with periodic boundaries along the x and y directions. Many simulations concern relatively small samples (N~200), but recently systems as large as 7000 particles have been considered. The reason for simulating larger systems is to ensure that the calculated density profiles are close to their thermodynamic limit. A good criterion for this condition is to check that the density of the fluid is uniform in the central part of the slab and that the correlation functions are identical to those in the bulk fluid at the same density. Good statistics (several millions of configurations) are needed to obtain the density profile with a precision of the order of 1%.

Abraham and *Singh* [2.440-442] computed the density profiles of the hard-sphere fluid near a hard wall and of the LJ fluid near a soft wall. They examined the changes in density profile when the characteristics of the soft-wall potential Ψ are varied, and concluded that the form of the external potential strongly influences the shape of the density profile.

Snook and *Henderson* [2.443] made an extensive MC study of the structure of hard-sphere and hard-disk fluids near a hard wall, for several values of the mean density (defined as the total number of particles divided by the volume of the slab). The density profiles are maximum at the wall, in accordance with the well-known relation between the pressure and the density of the fluid in contact with the wall. The simulations show that the motion of the hard disks and hard spheres close to the wall is constrained to be essentially one- (respectively, two-) dimensional. This conclusion is supported by the MD computations of *Grigera* [2.444] for hard spheres. *Snook* and *Van Megen* [2.445] found a layered structure of the hard sphere or LJ fluids between two hard or soft walls at sufficiently high bulk densities ($\rho\sigma^3 \sim 0.8$). *Toxvaerd* [2.446] examined the effect of introducing additional terms in the wall potential Ψ to obtain a more realistic description of the discrete lattice structure of the wall; he found that such terms lead to very small modifications of the density profile but sizeable variations of the surface tension. *Cape* [2.447] considered the case of a fluid of soft spheres (inverse -12 potential) in contact with a hard wall and found that the layer of particles near the wall has the structure of a two-dimensional fluid. The calculations of *Sullivan* et al. [2.448] on systems of hard spheres with a truncated attractive LJ tail reveal the sensitivity of the density profile to pressure variations in the bulk fluid.

In a series of articles, *Van Megen* and *Snook* present a detailed analysis of the solvation force between two hard walls separated by an LJ fluid, both in the canonical [2.449] and in the grand-canonical [2.450-452] ensembles. The solvation force per unit area A of the wall is defined by:

$$f_s(h) = \frac{1}{A}\left[\left\langle \sum_{j=1}^{N} \frac{\partial\Psi(z_j)}{\partial z_j}\right\rangle_h - \left\langle \sum_{j=1}^{N} \frac{\partial\Psi(z_j)}{\partial z_j}\right\rangle_\infty\right] \quad ,$$

where $< >_h$ denotes the ensemble average of a system enclosed in a slab of width h. The solvation force provides a potential barrier which opposes close contact of the surfaces [2.452]. It is an oscillatory function of h which vanishes only for h larger than about 6 atomic diameters at the temperatures considered in the simulation work ($k_B T/\varepsilon \sim 1$).

Molecular liquids against a hard wall have been examined by *Sullivan* et al. [2.453] and by *Jönsson* [2.454]. *Sullivan* et al. considered a fluid of hard diatomics while *Jönsson* simulated a water model from *Matsuoka* et al. [2.292]. *Sullivan* et al. found that the diatomics preferentially align parallel to the wall. *Jönsson* also found a preferential orientation of water molecules near the wall, with a hydrogen density profile which is maximum at the wall, whereas the oxygen density profile goes through a maximum at about 1 Å from the wall.

2.6.5 Liquid-Solid Coexistence

Toxvaerd and *Praestgaard* [2.455] made an MD investigation of the interface of the co-existing liquid and solid phases of the LJ system. They initialized their computation by dividing the sample into two subsystems of 672 and 1008 atoms; these subsystems are taken at different densities and are then relaxed towards the same pressure and temperature. These authors report excellent agreement with a previous determination of the pressure and densities of the coexisting solid and liquid phases [2.456], and show that the transition region extends over 3-4 atomic diameters. *Cape* and *Woodcock* [2.457] used a similar procedure to simulate the coexistence of the two phases for a system of 7680 atoms interacting by an inverse -12 potential. They present a detailed analysis of the variation of the pair distribution function across the interface. *Ladd* and *Woodcock* [2.458] determined the density profile of the LJ system at its triple point. The profile clearly exhibits the coexistence of the three phases (solid, liquid and gas) under pressure and density conditions which are very close to their previous estimate [2.456]. A similar calculation has been reported by *Borštnik* and *Ažman* [2.459].

2.6.6 The Electrical Double Layer

The nonuniform distribution of ions in electrolyte solutions near the surface of an electrode or near the particles of an electrically stabilized colloidal dispersion is referred to as an electrical double layer. The charge separation gives rise to a potential difference across the interface between two polarizable media. Such electrical double layers have been simulated in the framework of the primitive model of electrolytes defined in Sect.2.3.5. The simulated samples usually have the shape of a rectangular prism with periodic boundaries along the x and y directions, terminated by two impenetrable walls at z = 0 and z = L. One or both these walls carries a superficial charge density and the medium extending beyond the walls may have the same or a different dielectric constant than the solvent in the slab. Since the Ewald summation over the periodic replica of the central cell may introduce unphysical correlations, the charge distribution in the surrounding cells is taken to be the nonuniform average charge distribution of the central cell. This external charge distribution acts on the ions in the sample via an external potential which is easily computed from macroscopic electrostatics.

Torrie et al. [2.460-463] used the grand-canonical MC method to simulate electric double layers near a single-charged wall under the boundary conditions described above. The surface charge is compensated by an excess of negative ions. Ions of opposite sign carry equal [2.460-461] or different [2.462] absolute charges. The effect of surface polarization, embodied by image forces, is examined in [2.463]. The results of the MC simulations are compared to the predictions of Gouy-Chapman, modified Poisson-Boltzmann MFB [2.464] and HNC theories. The MBP theory appears to work surprisingly well, particularly where the ions carry equal absolute charges. The simulations show that at high concentration and charge densities the counterions are packed closely at the surface and begin to show a layered structure.

Similar MC computations were carried out by *Snook* and *Van Megen* [2.465,466], who examined the case of a primitive model electrolyte contained between two uniformly or discretely charged planes. Their calculations clearly establish that the minimum image convention is insufficient to yield a realistic description of electrical double layers, and that long-range corrections of the type described above are essential if small samples are being simulated.

Jönsson et al. [2.467] reported simulations of systems of equally charged mono- or divalent counterions constrained to move between parallel plates carrying a compensating surface charge. The MC calculations were carried out in the canonical ensemble and confirm the validity of the Poisson-Boltzmann equation for lamellar geometry.

2.7 Conclusion

The art of simulating classical fluids has made enourmous progress since the pioneering work of the late 1950s and 1960s. Current studies treat the microscopic structure and dynamics of increasingly complex systems, including molecular liquids, solutions and inhomogeneous fluids. The present state of the art can be summarized as follows: MC and MD simulations have led to a quantitative understanding of simple atomic, ionic and polar fluids and their mixtures, and to a qualitative picture of the structure of complex molecular systems, particularly of water and aqueous solutions. Further progress in that direction appears to hinge at present on the availability of sufficiently realistic and yet tractable inter- and intramolecular potential models and on the presence of potential barriers and very different time scales. Some of the recent methodological developments have partially solved the latter problem, but more effort will have to go into a satisfactory answer to the former.

The authors are indebted to E. Clementi and M. Mezei for sending preprints of their work.

References

2.1   D. Levesque, J.J. Weis, J.P. Hansen: In *Monte Carlo Methods in Statistical Physics*, ed. by K. Binder, Topics Current Phys., Vol.7 (Springer, Berlin, Heidelberg, New York 1979)
2.2   N. Metropolis, A.W. Rosenbluth, M.N. Rosenbluth, A.N. Teller, E. Teller: J. Chem. Phys. **21**, 1087 (1953)
2.3   J.C. Owicki, H.A. Scheraga: Chem. Phys. Lett. **47**, 600 (1977)
2.4   J.C. Owicki: In *Computer Modelling of Matter*, ACS Symposium Series **86**, 159 (1978)
2.5   B. Bigot, W.L. Jorgensen: J. Chem. Phys. **75**, 1944 (1981)
2.6   P.K. Mehrotra, M. Mezei, D.L. Beveridge: J. Chem. Phys. **78**, 3156 (1983)
2.7   D. Ceperley, G.V. Chester, M.H. Kalos: Phys. Rev. B**16**, 3081 (1977)
2.8   C. Pangali, M. Rao, B.J. Berne: Chem. Phys. Lett. **55**, 413 (1978)
2.9   M. Rao, C. Pangali, B.J. Berne: Mol. Phys. **37**, 1773 (1979)
2.10 M. Rao, B.J. Berne: J. Chem. Phys. **71**, 129 (1979)
2.11 P.J. Rossky, J.D. Doll, H.L. Friedman: J. Chem. Phys. **69**, 4628 (1978)
2.12 D.L. Ermak: J. Chem. Phys. **62**, 4189, 4197 (1975)
2.13 M. Mezei: Mol. Phys. **48**, 1075 (1983)
2.14 G.E. Norman, V.S. Filinov: High. Temp. **7**, 216 (1969)
2.15 J.P. Valleau, L.K. Cohen: J. Chem. Phys. **72**, 5935 (1980)
2.16 W.J. Van Megen, I.K. Snook: J. Chem. Phys. **73**, 4656 (1980)
2.17 L.A. Rowley, D. Nicholson, N.G. Parsonage: J. Comp. Phys. **17**, 401 (1975)
2.18 J. Yao, R.A. Greenkorn, K.C. Chao: Mol. Phys. **46**, 587 (1982)
2.19 D.J. Adams: Mol. Phys. **37**, 211 (1979)
2.20 M. Mezei: Mol. Phys. **40**, 901 (1980)
2.21 K. Yoon, D.G. Chae, T. Ree, F.H. Ree: J. Chem. Phys. **74**, 1412 (1981)
2.22 Y.S. Lee, D.G. Chae, T. Ree, F.H. Ree: J. Chem. Phys. **74**, 6881 (1981)
2.23 B. Widom: J. Chem. Phys. **39**, 2808 (1963)
2.24 K.S. Shing, K.E. Gubbins: Mol. Phys. **46**, 2109 (1982)
2.25 S. Romano, K. Singer: Mol. Phys. **37**, 1765 (1979)
2.26 G.M. Torrie, J.P. Valleau: In *Modern Theoretical Chemistry*, Vol.5, ed. by B.J. Berne (Plenum, New York 1977)
2.27 K.S. Shing, K.E. Gubbins: Mol. Phys. **43**, 717 (1981)
2.28 J.G. Powles: Chem. Phys. Lett. **86**, 335 (1982)
2.29 J.G. Powles, W.A.B. Evans, N. Quirke: Mol. Phys. **46**, 1347 (1982)
2.30 J.A. Zollweg: J. Chem. Phys. **72**, 6712 (1980)
2.31 D.F. Coker, R.O. Watts: Mol. Phys. **44**, 1303 (1981)
2.32 N. Quirke, G. Jacucci: Mol. Phys. **45**, 823 (1982)
2.33 C.H. Bennett: J. Comp. Phys. **22**, 245 (1978)
2.34 S.H. Northrup, J.A. McCammon: Biopolymers **19**, 1001 (1980)
2.35 E.S. Severin, B.C. Freasier, N.D. Hamer, D.L. Jolly, S. Nordholm: Chem. Phys. Lett. **57**, 117 (1978)
2.36 B.C. Freasier, D.L. Jolly, N.D. Hamer, S. Nordholm: Chem. Phys. **38**, 293 (1979)
2.37 M. Mezei: J. Comp. Phys. **39**, 128 (1981)
2.38 M. Mezei: Mol. Phys. **47**, 1307 (1982)
2.39 Y. Uehara, T. Ree, F.H. Ree: J. Chem. Phys. **70**, 1876 (1979)
2.40 Y. Uehara, Y.T. Lee, T. Ree, F.H. Ree: J. Chem. Phys. **70**, 1884 (1979)
2.41 S. Labik, A. Malijevsky: Mol. Phys. **42**, 739 (1981)
2.42 K.W. Kratky: Chem. Phys. **57**, 167 (1981)
2.43 P.H. Fries, J.P. Hansen: Mol. Phys. **48**, 891 (1983)
2.44 D. Frenkel, R. Eppenga: Phys. Rev. Lett. **49**, 1089 (1982)
2.45 D.A. Young, B.J. Alder: J. Chem. Phys. **70**, 473 (1979)
2.46 D.A. Young, B.J. Alder: J. Chem. Phys. **73**, 2430 (1980)
2.47 D. Henderson, O.H. Scalise, W.R. Smith: J. Chem. Phys. **72**, 2431 (1980)
2.48 F.F. Abraham: Phys. Repts. **80**, 341 (1981)

2.49 D. Henderson: Mol. Phys. **34**, 301 (1977)
2.50 F. Tsien, J.P. Valleau: Mol. Phys. **27**, 177 (1974)
2.51 S. Toxvaerd: J. Chem. Phys. **69**, 4750 (1978)
2.52 D. Frenkel, J.P. McTague: Phys. Rev. Lett. **42**, 1632 (1979)
2.53 F.F. Abraham: Phys. Rev. Lett. **44**, 463 (1980)
2.54 S. Toxvaerd: Phys. Rev. Lett. **44**, 1003 (1980)
2.55 F. Van Swol, L. Woodcock, J.N. Cape: J. Chem. Phys. **73**, 913 (1980)
2.56 J. Tobochnik, G.V. Chester: Phys. Rev. B**25**, 6778 (1982)
2.57 J.A. Barker, D. Henderson, F.F. Abraham: Physica **106**A, 226 (1981)
2.58 J.P. McTague, D. Frenkel, M.P. Allen: In *Ordering in Two Dimensions*, ed. by
     S.K. Sinha (North Holland, Amsterdam 1981)
2.59 R.K. Kalia, P. Vashishta: J. Phys. C**14**, L643 (1981)
2.60 J.Q. Broughton, G.H. Gilmer, J.D. Weeks: J. Chem. Phys. **75**, 5128 (1981)
2.61 J. Vermesse, D. Levesque: Phys. Rev. A**19**, 1801 (1979)
2.62 T.K. Bose, W. Brostow, J.S. Sochanski: Phys. Chem. Liq. **11**, 65 (1981)
2.63 P.A. Egelstaff, A. Teitsma, S.S. Wang: Phys. Rev. A**22**, 1702 (1980)
2.64 J.A. Barker, R.O. Watts, J.K. Lee, T.P. Schaefer, Y.T. Lee: J. Chem. Phys.
     61, 3081 (1974)
2.65 W. Schommers: Phys. Rev. A**16**, 327 (1977)
2.66 C. Hoheisel: Phys. Rev. A**23**, 1998 (1981)
2.67 J. Vermesse, D. Levesque: High Pressure Sci. Technol. **2**, 642 (1979)
2.68 T. Shiotani, T. Ishimura, A. Ueda: J. Phys. C**8**, 313 (1980)
2.69 K. Toukubo, K. Nakanishi: J. Chem. Phys. **65**, 1937 (1976)
2.70 K. Toukubo, K. Nakanishi, N. Watanabe: J. Chem. Phys. **67**, 4162 (1977)
2.71 K. Nakanishi, K. Toukubo, N. Watanabe: J. Chem. Phys. **72**, 3089 (1980)
2.72 K. Nakanishi, K. Toukubo: J. Chem. Phys. **70**, 5848 (1979)
2.73 K. Nakanishi, M. Narusawa, K. Toukubo: J. Chem. Phys. **72**, 3089 (1980)
2.74 K. Nakanishi, S. Okazaki, K. Ikari, T. Higuchi, H. Tanaka: J. Chem. Phys. **76**,
     629 (1982)
2.75 G. Fiorese, G. Pittion-Rossillon: Chem. Phys. Lett. **70**, 597 (1980)
2.76 C. Hoheisel, U. Deiters: Mol. Phys. **37**, 95 (1979)
2.77 D.J. Adams: Chem. Phys. Lett. **62**, 329 (1979)
2.78 J.P. Valleau, S.G. Whittington: In *Statistical Mechanics*, Pt. A, ed. by B.J.
     Berne (Plenum, New York 1977)
2.79 S.W. De Leeuw, J.W. Perram, E.R. Smith: Proc. Roy. Soc. (London) A**373**, 27,
     57 (1980)
2.80 S.W. De Leeuw, J.W. Perram: Physica **107**A, 179 (1981)
2.81 E.R. Smith, C.C. Wright, C.S. Hoskins: Mol. Phys. **43**, 275 (1981)
2.82 S.W. De Leeuw, J.W. Perram: Mol. Phys. **37**, 1313 (1979)
2.83 J.W. Perram, S.W. De Leeuw: Physica A**109**, 237 (1981)
2.84 E.R. Smith: Mol. Phys. **45**, 915 (1982)
2.85 J.P. Hansen, D. Levesque, J.J. Weis: Phys. Rev. Lett. **43**, 979 (1979)
2.86 K.W. Kratky: J. Comp. Phys. **37**, 205 (1980)
2.87 M. Baus, J.P. Hansen: Phys. Repts. **59**, 1 (1980)
2.88 W.L. Slattery, G.D. Doolen, H.E. De Witt: Phys. Rev. A**21**, 2087 (1980)
2.89 W.L. Slattery, G.D. Doolen, H.E. De Witt: Phys. Rev. A**26**, 2255 (1982)
2.90 J.D. Weeks: Phys. Rev. B**24**, 1530 (1981)
2.91 M. Navet, E. Jamin, M.R. Feix: J. Physique Lett. **41**, L-69 (1980)
2.92 J.M. Caillol, D. Levesque, J.J. Weis, J.P. Hansen: J. Stat. Phys. **28**, 325
     (1982)
2.93 S.W. De Leeuw, J.W. Perram: Physica A**113**, 546 (1982)
2.94 H. Totsuji: Phys. Rev. A**17**, 399 (1978)
2.95 R.C. Gann, S. Chakravarty, G.V. Chester: Phys. Rev. B**20**, 326 (1979)
2.96 R.W. Hockney, T.R. Brown: J. Phys. C**8**, 1813 (1975)
2.97 R.K. Kalia, P. Vashishta, S.W. De Leeuw: Phys. Rev. B**23**, 4794 (1981)
2.98 M. Ross, H.E. De Witt, W.B. Hubbard: Phys. Rev. A**24**, 1016 (1981)
2.99 G. Jacucci, M.L. Klein, R. Taylor: Solid State Commun. **19**, 657 (1976)
2.100 R.S. Day, F. Sun, P.H. Cutler: Phys. Rev. A**19**, 328 (1979)
2.101 F. Sun, R.S. Day, P.H. Cutler: Phys. Lett. **68**A, 236 (1978)
2.102 H. Minoo, C. Deutsch, J.P. Hansen: J. Physique Lett. **38**, L-191 (1977)
2.103 R.D. Mountain: J. Phys. F**8**, 1637 (1978)
2.104 M. Tanaka: J. Phys. F**10**, 2581 (1980)

2.105 R.D. Mountain: Phys. Rev. A**26**, 2859 (1982)
2.106 E. Michler, H. Hahn, P. Schofield: J. Phys. F**6**, L319 (1976)
2.107 G. Jacucci, R. Taylor, A. Tenenbaum, N. van Doan: J. Phys. F**11**, 793 (1981)
2.108 M.S. Duesbery, G. Jacucci, R. Taylor: J. Phys. F**9**, 413 (1979)
2.109 A. Girô, M. Gonzâles, J.A. Padrô, V. Torra: J. Chem. Phys. **73**, 2970 (1980)
2.110 S.K. Mitra: J. Phys. C**11**, 3551 (1978)
2.111 B. Larsen, S.A. Rogde: J. Chem. Phys. **68**, 1309 (1978)
2.112 B. Larsen, S.A. Rogde: J. Chem. Phys. **72**, 2578 (1980)
2.113 D.M. Ceperley, G.V. Chester: Phys. Rev. A**15**, 755 (1977)
2.114 J.P. Valleau, L.K. Cohen, D.N. Card: J. Chem. Phys. **72**, 5942 (1980)
2.115 S.A. Rogde, B. Hafskjold: Mol. Phys. **48**, 1241 (1983)
2.116 P.J. Rossky, J.B. Dudowicz, B.L. Tembe, H.L. Friedman: J. Chem. Phys. **73**, 3372 (1980)
2.117 B. Hafskjold, J.J. Weis: Unpublished
2.118 M.C. Abramo, C. Caccamo, G. Pizzimenti: J. Chem. Phys. **78**, 357 (1983)
2.119 V.P. Chasovskikh, P.N. Vorontsov-Vel'yaminov: Teplefizika Vysokikh Temperatur **14**, 379 (1976)
2.120 P.N. Vorontsov-Vel'yaminov: Teplofiz Vys. Temp. **15**, 1137 (1977)
2.121 W. Van Megen, I.K. Snook: J. Chem. Phys. **66**, 813 (1977)
2.122 H. Wennerström, B. Jönsson, P. Linse: J. Chem. Phys. **76**, 4665 (1982)
2.123 V. Vlachy, D. Dolar: J. Chem. Phys. **76**, 2010 (1982)
2.124 M.J.L. Sangster, M. Dixon: Adv. Phys. **25**, 247 (1976)
2.125 L.V. Woodcock, K. Singer: Trans. Faraday Soc. **67**, 12 (1971)
2.126 D.J. Adams: J. Chem. Soc. Faraday Trans II **72**, 1372 (1976)
2.127 L. Schäfer, A. Klemm: Z. Naturforsch. **34**a, 993 (1979) and **36**a, 584 (1981)
2.128 L. Woodcock, C.A. Angell, P. Cheeseman: J. Chem. Phys. **65**, 1565 (1976)
2.129 S. Miller, J.H.R. Clarke: J. Chem. Soc. Faraday Trans II **74**, 160 (1978)
2.130 M. Dixon, M.J. Gillan: Phil. Mag. B**43**, 1099 (1981)
2.131 M. Dixon, M.J.L. Sangster: J. Phys. C**9**, L-5 (1976)
2.132 G.M. Abernethy, M. Dixon, M.J. Gillan: Phil. Mag. B**43**, 1113 (1981)
2.133 M. Dixon, M.J.L. Sangster: J. Phys. C**9**, 3381 (1976)
2.134 M. Dixon, M.J.L. Sangster: Phil. Mag. **35**, 1049 (1977)
2.135 M. Dixon, M.J.L. Sangster: J. Phys. C**10**, 3015 (1977)
2.136 S.W. De Leeuw: Mol. Phys. **36**, 103 (1978)
2.137 A. Rahman: J. Chem. Phys. **65**, 4845 (1976)
2.138 S.W. De Leeuw: Mol. Phys. **36**, 765 (1978)
2.139 M.L. Saboungi, A. Rahman: J. Chem. Phys. **65**, 2392 (1976) and **68**, 2773 (1977)
2.140 D.J. Adams, I.R. McDonald: Mol. Phys. **34**, 287 (1977)
2.141 C. Caccomo, M. Dixon: J. Phys. C**13**, 1887 (1980)
2.142 F. Lantelme, P. Turq: Mol. Phys. **38**, 1003 (1979)
2.143 I. Okada, R. Takagi, K. Kawamura: Z. Naturforsch. **35**a, 493 (1980)
2.144 S. Geller (ed.): *Solid Electrolytes*, Topics Appl. Phys., Vol.21 (Springer, Berlin, Heidelberg, New York 1977);
     M.B. Salamon: *Physics of Superionic Conductors*, Topics Current Phys., Vol.15 (Springer, Berlin, Heidelberg, New York 1979)
2.145 G. Jacucci, A. Rahman: J. Chem. Phys. **69**, 4117 (1978)
2.146 M. Dixon, M.J. Gillan: J. Phys. C**11**, L-165 (1978) and C**13**, 1901 (1980)
2.147 Y. Hiwatari, A. Ueda: J. Phys. Soc. Japan **49**, 2129 (1980)
2.148 A.B. Walker, M. Dixon, M.J. Gillan: Sol. State Ionics **5**, 601 (1981)
2.149 J.R. Walker, C.R.A. Catlow: J. Phys. C**14**, L979 (1981)
2.150 P. Vashishta, A. Rahman: Phys. Rev. Lett. **40**, 1337 (1978)
2.151 A. Fukumoto, A. Ueda, Y. Hiwatari: Sol. State Ionics 3/4, 115 (1980)
2.152 Y. Hiwatari, A. Ueda: J. Phys. Soc. Japan **48**, 766 (1980)
2.153 Y. Hiwatari, A. Ueda: Sol. State Ionics 3/4, 111 (1981)
2.154 P.A. Monson, M. Rigby: Chem. Phys. Lett. **58**, 122 (1978)
2.155 I. Nezbeda, T. Boublík: Czech. J. Phys. B**28**, 353 (1978)
2.156 I. Nezbeda, J. Pavlíček, S. Labík: Collection Czech. Chem. Comm. **44**, 3555 (1979)
2.157 G.A. Few, M. Rigby: Chem. Phys. Lett. **20**, 433 (1973)
2.158 D.W. Rebertus, K.M. Sando: J. Chem. Phys. **67**, 2585 (1977)
2.159 I. Nezbeda: Czech. J. Phys. B**30**, 601 (1980)
2.160 P.A. Monson, M. Rigby: Mol. Phys. **39**, 977 (1980)

2.161 T. Boublík, I. Nezbeda: Czech. J. Phys. B**30**, 121 (1980)
2.162 J. Pavlíček, I. Nezbeda, T. Boublík: Czech. J. Phys. B**29**, 1061 (1979)
2.163 I. Nezbeda, T. Boublík: Czech. J. Phys. B**30**, 953 (1980)
2.164 I. Aviram, D.J. Tildesley, W.B. Streett: Mol. Phys. **34**, 885 (1977)
2.165 B.C. Freasier, D. Jolly, R.J. Bearman: Mol. Phys. **31**, 255 (1976)
2.166 B.C. Freasier: Chem. Phys. Lett. **35**, 280 (1975)
2.167 I. Nezbeda, W.R. Smith, T. Boublík: Mol. Phys. **37**, 985 (1979)
2.168 D.J. Tildesley, W.B. Streett: Mol. Phys. **41**, 85 (1980)
2.169 B.C. Freasier: Mol. Phys. **39**, 1273 (1980)
2.170 P. Cummings, I. Nezbeda, W.R. Smith, G. Morris: Mol. Phys. **43**, 1471 (1981)
2.171 W.B. Streett, D.J. Tildesley: Proc. Roy. Soc. A**348**, 485 (1976)
2.172 W.B. Streett, D.J. Tildesley: J. Chem. Phys. **68**, 1275 (1978)
2.173 I. Aviram, D.J. Tildesley: Mol. Phys. **35**, 365 (1978)
2.174 M. Wojcik, K.E. Gubbins, J.G. Powles: Mol. Phys. **45**, 1209 (1982)
2.175 G. Fiorese: J. Chem. Phys. **73**, 6308 (1980)
2.176 G. Fiorese: J. Chem. Phys. **75**, 1427 (1981)
2.177 G. Fiorese: J. Chem. Phys. **75**, 4747 (1981)
2.178 G. Stell, G.N. Patey, J.S. Høye: *Advances in Chemical Physics*, Vol.48, (Wiley, New York 1981) pp.183-328
2.179 B.J. Alder, E.L. Pollock: Ann. Rev. Phys. Chem. **32**, 311 (1981)
2.180 M. Neumann, O. Steinhauser: Mol. Phys. **39**, 437 (1980)
2.181 G.N. Patey, D. Levesque, J.J. Weis: Mol. Phys. **38**, 219 (1979)
2.182 D.J. Adams: Mol. Phys. **40**, 1261 (1980)
2.183 K.C. Ng, J.P. Valleau, G.M. Torrie, G.N. Patey: Mol. Phys. **38**, 781 (1979)
2.184 G.N. Patey, D. Levesque, J.J. Weis: Mol. Phys. **38**, 1635 (1979)
2.185 G. Bossis: Mol. Phys. **38**, 2023 (1979)
2.186 G. Bossis, B. Quentrec, C. Brot: Mol. Phys. **39**, 1233 (1980)
2.187 C. Brot, G. Bossis, C. Hesse-Bezot: Mol. Phys. **40**, 1053 (1980)
2.188 G. Bossis, C. Brot: Mol. Phys. **43**, 1095 (1981)
2.189 J.M. Caillol, D. Levesque, J.J. Weis: Mol. Phys. **44**, 733 (1981)
2.190 D.J. Adams, E.M. Adams, G.J. Hills: Mol. Phys. **38**, 387 (1979)
2.191 E.L. Pollock, B.J. Alder: Physica **102**A, 1 (1980)
2.192 D.J. Adams, E.M. Adams: Mol. Phys. **42**, 907 (1981)
2.193 K.S. Shing, K.E. Gubbins: Mol. Phys. **45**, 129 (1982)
2.194 F.J. Vesely: Chem. Phys. Lett. **56**, 390 (1978)
2.195 E.L. Pollock, B.J. Alder, G.N. Patey: Physica **108**A, 14 (1981)
2.196 G.N. Patey, J.P. Valleau: Chem. Phys. Lett. **58**, 157 (1978)
2.197 G.N. Patey, G.M. Torrie, J.P. Valleau: J. Chem. Phys. **71**, 96 (1979)
2.198 M.S. Wertheim: Mol. Phys. **37**, 83 (1979)
2.199 O. Steinhauser: Mol. Phys. **46**, 827 (1982)
2.200 G.P. Morris, P.T. Cummings: Mol. Phys. **45**, 1099 (1982)
2.201 G.P. Morris: Mol. Phys. **47**, 833 (1982)
2.202 B. Jönsson, G. Karlström, S. Romano: J. Chem. Phys. **74**, 2896 (1981)
2.203 C.S. Murthy, K. Singer, I.R. McDonald: Mol. Phys. **44**, 135 (1981)
2.204 O. Steinhauser, M. Neumann: Mol. Phys. **37**, 1921 (1979)
2.205 D.J. Tildesley, P.A. Madden: Mol. Phys. **42**, 1137 (1981)
2.206 O. Steinhauser, M. Neumann: Mol. Phys. **40**, 115 (1980)
2.207 I.R. McDonald, D.G. Bounds, M.L. Klein: Mol. Phys. **45**, 521 (1982)
2.208 S. Murad, K.E. Gubbins: Mol. Phys. **39**, 271 (1980)
2.209 S. Murad, D.J. Evans, K.E. Gubbins, W.B. Streett, D.J. Tildesley: Mol. Phys. **37**, 725 (1979)
2.210 R.H. Kincaid, H.A. Scheraga: J. Phys. Chem. **86**, 838 (1982)
2.211 S.I. Sandler, M.G. Lombardo, D.S.H. Wong, A. Habenschuss, A.H. Narten: J. Chem. Phys. **77**, 2144 (1982)
2.212 J.P. Ryckaert, A. Bellemans: Chem. Phys. Lett. **30**, 123 (1975)
2.213 J.P. Ryckaert, A. Bellemans: Discuss. Faraday Soc. **66**, 95 (1978)
2.214 W.F. Van Gunsteren, H.J.C. Berendsen, J.A.C. Rullmann: Mol. Phys. **44**, 69 (1981)
2.215 T.A. Weber: J. Chem. Phys. **69**, 2347 (1978)
2.216 W.L. Jorgensen: J. Am. Chem. Soc. **103**, 677 (1981)
2.217 W.L. Jorgensen, R.C. Binning, Jr., B. Bigot: J. Am. Chem. Soc. **103**, 4393 (1981)
2.218 W.L. Jorgensen: J. Am. Chem. Soc. **103**, 4721 (1981)
2.219 W.L. Jorgensen: J. Chem. Phys. **77**, 5757 (1982)

2.220 T.A. Weber: J. Chem. Phys. **70**, 4277 (1979)
2.221 W.L. Jorgensen: J. Am. Chem. Soc. **102**, 543 (1980)
2.222 W.L. Jorgensen: J. Am. Chem. Soc. **103**, 341 (1981)
2.223 W.L. Jorgensen, M. Ibrahim: J. Am. Chem. Soc. **104**, 373 (1982)
2.224 W.L. Jorgensen: J. Am. Chem. Soc. **103**, 345 (1981)
2.225 D.S.-H. Wong, S.I. Sandler: Mol. Phys. **45**, 1193 (1982)
2.226 W.L. Jorgensen, B. Bigot: J. Phys. Chem. **86**, 2867 (1982)
2.227 W.L. Jorgensen, M. Ibrahim: J. Am. Chem. Soc. **103**, 3976 (1981)
2.228 J. Chandrasekhar, W.L. Jorgensen: J. Chem. Phys. **77**, 5073 (1982)
2.229 M.L. Klein, I.R. McDonald: J. Chem. Phys. **71**, 298 (1979)
2.230 M.L. Klein, I.R. McDonald, R. Righini: J. Chem. Phys. **71**, 3673 (1979)
2.231 A. Hinchliffe, D.G. Bounds, M.L. Klein, R. Righini: J. Chem. Phys. **74**, 1211
       (1981)
2.232 M.L. Klein, I.R. McDonald: J. Chem. Phys. **74**, 4214 (1981)
2.333 W.L. Jorgensen, M. Ibrahim: J. Am. Chem. Soc. **102**, 3309 (1980)
2.234 R.H. Kincaid, H.A. Scheraga: J. Phys. Chem. **86**, 833 (1982)
2.235 W.L. Jorgensen: J. Am. Chem. Soc. **100**, 7824 (1978)
2.236 W.L. Jorgensen: J. Chem. Phys. **70**, 5888 (1979)
2.237 M.L. Klein, I.R. McDonald: J. Chem. Phys. **71**, 298 (1979)
2.238 J.G. Powles, W.A.B. Evans, E. Grath, K.E. Gubbins, S. Murad: Mol. Phys. **38**,
       893 (1979)
2.239 S. Murad, K.E. Gubbins, J.G. Powles: Mol. Phys. **40**, 253 (1980)
2.240 I.R. McDonald, S.F. O'Shea; D.G. Bounds, M.L. Klein: J. Chem. Phys. **72**, 5710
       (1980)
2.241 M.L. Klein, I.R. McDonald: Mol. Phys. **42**, 243 (1981)
2.242 M.L. Klein, I.R. McDonald: Chem. Phys. Lett. **80**, 76 (1981)
2.243 F.H. Stillinger, T.A. Weber: J. Chem. Phys. **76**, 3131 (1982)
2.244 R.O. Watts: Chem. Phys. **26**, 367 (1977)
2.245 H.L. Lemberg, F.H. Stillinger: J. Chem. Phys. **62**, 1677 (1975)
2.246 A. Rahman, F.H. Stillinger, H.L. Lemberg: J. Chem. Phys. **63**, 5223 (1975)
2.247 F.H. Stillinger, A. Rahman: J. Chem. Phys. **68**, 666 (1978)
2.248 F.H. Stillinger: Israel J. Chem. **14**, 130 (1975)
2.249 F.H. Stillinger, C.W. David: J. Chem. Phys. **69**, 1473 (1978)
2.250 L.L. Shipman, A.W. Burgess, H.A. Scheraga: Proc. Natl. Acad. Sci. **72**, 543
       (1975)
2.251 J. Snir, R.A. Nemenoff, H.A. Scheraga: J. Phys. Chem. **82**, 2497 (1978)
2.252 W.L. Jorgensen: J. Am. Chem. Soc. **103**, 335 (1981)
2.253 S.F. O'Shea, P.R. Tremaine: J. Phys. Chem. **84**, 3304 (1980)
2.254 R.W. Impey, M.L. Klein, I.R. McDonald: J. Chem. Phys. **74**, 647 (1981)
2.255 R.W. Impey, P.A. Madden, I.R. McDonald: Mol. Phys. **46**, 513 (1982)
2.256 W.L. Jorgensen: J. Chem. Phys. **77**, 4156 (1982)
2.257 F.H. Stillinger, A. Rahman: J. Chem. Phys. **57**, 1281 (1972)
2.258 F.H. Stillinger, A. Rahman: J. Chem. Phys. **60**, 1545 (1974)
2.259 D.C. Rapaport, H.A. Scheraga: Chem. Phys. Lett. **78**, 491 (1981)
2.260 C. Pangali, M. Rao, B.J. Berne: Mol. Phys. **40**, 661 (1980)
2.261 H. Kistenmacher, H. Popkie, E. Clementi, R.O. Watts: J. Chem. Phys. **60**, 4455
       (1974)
2.262 G.C. Lie, E. Clementi: J. Chem. Phys. **62**, 2195 (1975
2.263 P. Barnes, J.L. Finney, J.D. Nicholas, J.E. Quinn: Nature **282**, 459 (1979)
2.264 S. Swaminathan, D.L. Beveridge: J. Am. Chem. Soc. **99**, 8392 (1977)
2.265 M. Mezei, S. Swaminathan, D.L. Beveridge: J. Chem. Phys. **71**, 3366 (1979)
2.266 J.A. Barker, R.O. Watts: Chem. Phys. Lett. **3**, 144 (1969)
2.267 R.O. Watts: Mol. Phys. **28**, 1069 (1974)
2.268 J.R. Reimers, R.O. Watts, M.L. Klein: Chem. Phys. **64**, 95 (1982)
2.269 G.C. Lie, E. Clementi, M. Yoshimine: J. Chem. Phys. **64**, 2314 (1976)
2.270 J.C. Owicki, H.A. Scheraga: J. Am. Chem. Soc. **99**, 7403 (1977)
2.271 W.L. Jorgensen: J. Am. Chem. Soc. **101**, 2016 (1979)
2.272 W.L. Jorgensen: Chem. Phys. Lett. **70**, 326 (1980)
2.273 R.A. Nemenoff, J. Snir, H.A. Scheraga: J. Phys. Chem. **82**, 2504 (1978)
2.274 F.T. Marchese, P.K. Mehrotra, D.L. Beveridge: J. Phys. Chem. **85**, 1 (1981)
2.275 P.H. Berens, D.H.J. Mackey, G.M. White, K.R. Wilson: J. Chem. Phys. **79**, 2375
       (1983)

2.276 H.J.C. Berendsen, J.P.M. Postma, W.F. van Gusteren, J. Hermans: "Interaction models for water in relation to protein hydration", in *Intermolecular Forces*, ed. by B. Pullmann (Reidel, 1981) pp.331-342

2.277 R.W. Impey, P.A. Madden, I.R. McDonald: Chem. Phys. Lett. **88**, 589 (1982)

2.278 A. Rahman, F.H. Stillinger: J. Chem. Phys. **55**, 3336 (1971)

2.279 M. Mezei, D.L. Beveridge: J. Chem. Phys. **76**, 593 (1982)

2.280 O. Steinhauser: Mol. Phys. **45**, 335 (1982)

2.281 M.L. Klein, I.R. McDonald, B.J. Berne, M. Rao, D.L. Beveridge, P.K. Mehrotra: J. Chem. Phys. **71**, 3889 (1979)

2.282 F.H. Stillinger, A. Rahman: J. Chem. Phys. **61**, 4973 (1974)

2.283 A. Ben-Naim: In *Water and Aqueous Solutions* (Plenum, New York 1974)

2.284 D.W. Rebertus, B.J. Berne, D. Chandler: J. Chem. Phys. **70**, 3395 (1979)

2.285 J.S. Rowlinson: Trans. Faraday Soc. **47**, 120 (1951)

2.286 R.O. Watts: Chem. Phys. **57**, 185 (1981)

2.287 A. Ben-Naim, F.H. Stillinger: "Aspects of the statistical-mecanical theory of water" in *Structure and Transport Processes in Water and Aqueous Solutions*, ed. by R.A. Horne (Wiley-Interscience, New York 1972) Chap.8

2.288 A. Rahman, F.H. Stillinger: J. Am. Chem. Soc. **95**, 7943 (1973)

2.289 A. Geiger, F.H. Stillinger, A. Rahman: J. Chem. Phys. **70**, 4185 (1979)

2.290 F. Hirata, P.J. Rossky: J. Chem. Phys. **74**, 6867 (1981)

2.291 H. Popkie, H. Kistenmacher, E. Clementi: J. Chem. Phys. **59**, 1325 (1973)

2.292 O. Matsuoka, M. Yoshimine, E. Clementi: J. Chem. Phys. **64**, 1351 (1976)

2.293 W.L. Jorgensen: Chem. Phys. Lett. **92**, 405 (1982)

2.294 M. Mezei, S. Swaminathan, D.L. Beveridge: J. Am. Chem. Soc. **100**, 3255 (1978)

2.295 W.L. Jorgensen: J. Am. Chem. Soc. **101**, 2011 (1979)

2.296 I.R. McDonald, M.L. Klein: J. Chem. Phys. **68**, 4875 (1978)

2.297 M.D. Morse, S.A. Rice: J. Chem. Phys. **76**, 650 (1982)

2.298 F. Stillinger: Adv. Chem. Phys. **31**, 1 (1975)

2.299 P. Barnes: "Machine simulation of water", in *Progress in Liquid Physics*, ed. by C.A. Croxton (Wiley, Chichester 1978) p.391

2.300 D.W. Wood: "Computer simulation of water and aqueous solutions", in *Water, A. Comprehensive Treatise*, ed. by F. Franks (Plenum, New York 1979) Chap.6, pp.279-436

2.301 G. Nementhy, W.J. Peer, H.A. Scheraga: Ann. Rev. Biophys. Bioeng. **10**, 459 (1981)

2.302 D.L. Beveridge, M. Mezei, P.K. Mehrotra, R.T. Marchese, G. Ravishanker, T.R. Vasu, S. Swaminathan: "Monte Carlo computer simulation studies of the equilibrium properties and structure of liquid water", in *Molecular-Based Study and Prediction of Fluid Properties*, ed. by J.M. Haile and G.A. Mansoori (American Chemical Society, Washington, DC 1983)

2.303 M.R. Mruzik: Chem. Phys. Lett. **48**, 171 (1977)

2.304 F.F. Abraham: J. Chem. Phys. **61**, 1221 (1974)

2.305 W. Bol: Mol. Phys. **45**, 605 (1982)

2.306 K. Okazaki, S. Nosê, Y. Kataoka, T. Yamamoto: J. Chem. Phys. **75**, 5864 (1981)

2.307 F.H. Stillinger: Science **209**, 451 (1980)

2.308 A.H. Narten, H.A. Levy: J. Chem. Phys. **53**, 2263 (1971)

2.309 M. Mezei, D.L. Beveridge: J. Chem. Phys. **74**, 622 (1981)

2.310 M.D. Danford, H.A. Levy: J. Am. Chem. Soc. **84**, 3965 (1962)

2.311 H.S. Frank, W.Y. Wen: Discuss. Faraday Soc. **24**, 133 (1957)

2.312 L. Pauling: In *Hydrogen Bonding*, ed. by D. Hadzi and H.W. Thompson (Pergamon, London 1959) pp.1-5

2.313 G. Nemethy, H.A. Scheraga: J. Chem. Phys. **36**, 3382 (1962)

2.314 M.G. Sceats, M. Stavola, S.A. Rice: J. Chem. Phys. **70**, 3927 (1979)

2.315 G.M. Bell: J. Phys. C5, 889 (1972)

2.316 P.D. Fleming, G.H. Gibbs: J. Stat. Phys. **10**, 157, 351 (1974)

2.317 D.E. O'Reilly: Phys. Rev. A7, 1659 (1973)

2.318 O. Weres, S.A. Rice: J. Am. Chem. Soc. **94**, 8983 (1972)

2.319 A.H. Narten, M.D. Danford, H.A. Levy: Discuss. Faraday Soc. **43**, 97 (1967)

2.320 A.H. Narten, H.A. Levy: J. Chem. Phys. **55**, 2263 (1971)

2.321 F. Hadju, S. Lengyel, G. Pâlinkâs: J. Appl. Crystallogr. **9**, 134 (1976)

2.322 D.I. Page, J.G. Powles: Molec. Phys. **21**, 901 (1971)

2.323 J.G. Powles, J.C. Dore, D.I. Page: Mol. Phys. **24**, 1025 (1972)

2.324 A.H. Narten: J. Chem. Phys. **56**, 5681 (1972)
2.325 G. Walford, J.H. Clarke, J.C. Dore: Mol. Phys. **33**, 25 (1977)
2.326 G. Walford, J.C. Dore: Mol. Phys. **34**, 21 (1977)
2.327 I.P. Gibson, J.C. Dore, quoted by J.C. Dore: Faraday Discussions Chem. Soc.
      66, 82 (1978)
2.328 N. Ohtomo, K. Arakawa: Bull. Chem. Soc. Japan **51**, 1649 (1978)
2.329 N. Ohtomo, K. Arakawa: Bull. Chem. Soc. Japan **52**, 2755 (1979)
2.330 N. Ohtomo, K. Tokiwano, K. Arakawa: Bull. Chem. Soc. Japan **54**, 1802 (1981)
2.331 A.Y. Wu, E. Whalley, G. Dolling: Chem. Phys. Lett. **84**, 433 (1981)
2.332 A.Y. Wu, E. Whalley, G. Dolling: Mol. Phys. **47**, 603 (1982)
2.333 P.A. Egelstaff, J.A. Polo, J.H. Root, L.J. Hahn, S.H. Chen: Phys. Rev. Lett.
      47, 1733 (1981)
2.334 A.K. Soper, R.N. Silver: Phys. Rev. Lett. **49**, 471 (1982)
2.335 G.A. Gaballa, G.W. Neilson: Mol. Phys. **46**, 211 (1982)
2.336 E. Kálmán, G. Pálinkás, P. Kovács: Mol. Phys. **34**, 505 (1977)
2.337 A.H. Narten, W.E. Thiessen, L. Blum: Science **217**, 1033 (1982)
2.338 D. Hankins, J.W. Moskowitz, F.H. Stillinger: J. Chem. Phys. **53**, 4544 (1970) -
      erratum: J. Chem. Phys. **59**, 995 (1973)
2.339 H. Kistenmacher, G.C. Lie, H. Popkie, E. Clementi: J. Chem. Phys. **61**, 546
      (1974)
2.340 J.E. Del Bene, J.A. Pople: J. Chem. Phys. **58**, 3605 (1973)
2.341 B.R. Lentz, H.A. Scheraga: J. Chem. Phys. **58**, 5296 (1973)
2.342 E. Clementi, H. Kistenmacher, W. Kolos, S. Romano: Theor. Chim. Acta **55**, 287
      (1980)
2.343 W. Kolos: Theor. Chim. Acta **51**, 219 (1979)
2.344 E. Clementi, W. Kolos, G.C. Lie, G. Ranghino: Int. J. Quant. Chem. **17**, 377
      (1980)
2.345 P. Habitz, P. Bagus, P. Siegbahn, E. Clementi: Int. J. Quant. Chem. **23**, 1803 (1983)
2.346 P.A. Egelstaff, J.H. Root: Chem. Phys. Lett. **91**, 96 (1982)
2.347 A.J.C. Ladd: Mol. Phys. **33**, 1039 (1977)
2.348 G.N. Sarkisov, V.G. Dashevsky, G.G. Malenkov: Mol. Phys. **27**, 1249 (1974)
2.349 A. Geiger, A. Rahman, F.H. Stillinger: J. Chem. Phys. **70**, 263 (1979)
2.350 S. Okazaki, K. Nakanishi, H. Touhara, Y. Adachi: J. Chem. Phys. **71**, 2421 (1979)
2.351 S. Okazaki, K. Nakanishi, H. Touhara, N. Watanabe, Y. Adachi: J. Chem. Phys.
      74, 5863 (1981)
2.352 C. Pangali, M. Rao, B.J. Berne: J. Chem. Phys. **71**, 2975 (1979)
2.353 C. Pangali, M. Rao, B.J. Berne: J. Chem. Phys. **71**, 2982 (1979)
2.354 D.C. Rapaport, H.A. Scheraga: J. Phys. Chem. **86**, 873 (1982)
2.355 G. Alagona, A. Tani: J. Chem. Phys. **72**, 580 (1980)
2.356 V.G. Dashevsky, G.N. Sarkisov: Mol. Phys. **27**, 1271 (1974)
2.357 J.C. Owicki, H.A. Scheraga: J. Am. Chem. Soc. **99**, 7413 (1977)
2.358 S. Swaminathan, S.W. Harrison, D.L. Beveridge: J. Am. Chem. Soc. **100**, 5705
      (1978)
2.359 S. Swaminathan, D.L. Beveridge: J. Am. Chem. Soc. **101**, 5832 (1979)
2.360 G. Bolis, E. Clementi: J. Chem. Phys. Lett. **82**, 147 (1981)
2.361 G. Bolis, G. Corongiu, E. Clementi: Chem. Phys. Lett. **86**, 299 (1982)
2.362 P.J. Rossky, M. Karplus: J. Am. Chem. Soc. **101**, 1913 (1979)
2.363 P.K. Mehrotra, D.L. Beveridge: J. Am. Chem. Soc. **102**, 4287 (1980)
2.364 P.J. Rossky, M. Karplus, A. Rahman: Biopolymers **18**, 825 (1979)
2.365 L. Pratt, D. Chandler: J. Chem. Phys. **67**, 3683 (1977)
2.366 S. Goldman: J. Chem. Phys. **74**, 5851 (1981)
2.367 A. Geiger: Ber. Bunsenges. Phys. Chem. **85**, 32 (1981)
2.368 M. Mezei, D.L. Beveridge: J. Chem. Phys. **74**, 6902 (1981)
2.369 M. Rao, B.J. Berne: J. Phys. Chem. **85**, 1498 (1981)
2.370 J. Chandrasekar, W. Jorgensen: J. Chem. Phys. **77**, 5080 (1982)
2.371 K. Heinziger, P.C. Vogel: Z. Naturforsch. **29a**, 1164 (1974)
2.372 K. Heinziger, P.C. Vogel: Z. Naturforsch. **31a**, 463 (1976)
2.373 K. Heinziger: Z. Naturforsch. **31a**, 1073 (1976)
2.374 P.C. Vogel, K. Heinziger: Z. Naturforsch. **30a**, 789 (1975)
2.375 P.C. Vogel, K. Heinziger: Z. Naturforsch. **31a**, 476 (1976)
2.376 G. Pálinkás, W.O. Riede, K. Heinziger: Z. Naturforsch. **32a**, 1137 (1977)
2.377 P. Bopp, W. Dietz, K. Heinziger: Z. Naturforsch. **34a**, 1424 (1979)

2.378 Gy.I. Szász, K. Heinziger, G. Pálinkás: Chem. Phys. Lett. 78, 194 (1981)
2.379 Gy.I. Szász, K. Heinziger: Z. Naturforsch. 34a, 840 (1979)
2.380 W. Dietz, W.O. Riede, K. Heinziger: Z. Naturforsch. 37a, 1038 (1982)
2.381 Gy.I. Szász, W. Dietz, K. Heinziger, G. Pálinkás, T. Radnai: Chem. Phys. Lett. 92, 388 (1982)
2.382 R.O.Watts, E. Clementi, J. Fromm: J. Chem. Phys. 61, 2550 (1974)
2.383 E. Clementi, R. Barsotti: Theoret. Chim. Acta 43, 101 (1976)
2.384 E. Clementi, R. Barsotti: Chem. Phys. Lett. 59, 21 (1978)
2.385 E. Clementi, G. Corongiu, B. Jönsson, S. Romano: J. Chem. Phys. 72, 260 (1980)
2.386 S. Engström, B. Jönsson: Mol. Phys. 43, 1235 (1981)
2.387 J. Fromm, E. Clementi, R.O. Watts: J. Chem. Phys. 62, 1388 (1975)
2.388 R.O. Watts: Mol. Phys. 32, 659 (1976)
2.389 H. Kistenmacher, H. Popkie, E. Clementi: J. Chem. Phys. 61, 799 (1974)
2.390 C.L. Briant, J.J. Burton: J. Chem. Phys. 64, 2888 (1976)
2.391 M.R. Mruzik, F.F. Abraham, D.E. Schreiber, G.M. Pound: J. Chem. Phys. 64, 481 (1976)
2.392 A. Banerjee, R. Shepard, J. Simons: J. Chem. Phys. 73, 1814 (1980)
2.393 E. Clementi, G. Corongiu, F. Lelj: J. Chem. Phys. 70, 3726 (1979)
2.394 W.K. Lee, E.W. Prohofsky: J. Chem. Phys. 75, 3040 (1981)
2.395 Gy. I. Szász, K. Heinziger, W.O. Riede: Z. Naturforsch. 36a, 1067 (1981)
2.396 P. Bopp, K. Heinziger, G. Jancso: Z. Naturforsch. 32a, 620 (1977)
2.397 J.E. Enderby, G.W. Neilson: Rep. Progr. Phys. 44, 593 (1981)
2.398 G. Pálinkás, T. Radnai, Gy.I. Szász, K. Heinziger: J. Chem. Phys. 74, 3522 (1981)
2.399 T. Radnai, G. Pálinkás, Gy.I. Szász, K. Heinziger: Z. Naturforsch. 36a, 1076 (1981)
2.400 G. Pálinkás, T. Radnai, W. Dietz, Gy.I. Szász, K. Heinziger: Z. Naturforsch. 37a, 1049 (1982)
2.401 A.K. Soper, G.W. Neilson, J.E. Enderby, R.A. Howe: J. Phys. C10, 1793 (1977)
2.402 M. Neumann, F.J. Vesely, O. Steinhauser, P. Schuster: Mol. Phys. 35, 841 (1978)
2.403 M. Neumann, F.J. Vesely, O. Steinhauser, P. Schuster: Mol. Phys. 37, 1725 (1979)
2.404 B.J. Costa Cabral, D. Rinaldi, J.C. Rivail: Chem. Phys. Lett. 93, 157 (1982)
2.405 S. Romano, E. Clementi: Gazz. Chim. Ital. 108, 319 (1978)
2.406 S. Romano, E. Clementi: Int. J. Quantum Chem. 14, 839 (1978)
2.407 S. Romano, E. Clementi: Int. J. Quantum Chem. 17, 1007 (1980)
2.408 S. Romano: Int. J. Quantum Chem. 20, 921 (1981)
2.409 C.W. David: Chem. Phys. Lett. 78, 337 (1981)
2.410 J.M. Goodfellow: Proc. Nat. Acad. Sci. USA 79, 4977 (1982)
2.411 J.A. McCammon, B.R. Gelin, M. Karplus: Nature 267, 585 (1977)
2.412 J.A. McCammon, P.G. Wolynes, M. Karplus: Biochemistry 18, 927 (1979)
2.413 A.T. Hagler, J. Moult: Nature 272, 222 (1978)
2.414 E. Clementi, G. Corongiu, M. Gratarola, P. Habitz, C. Lupo, P. Otto, D. Vercauteren: Int. J. Quantum Chem.: Quantum Chemistry Symposium 16, 409 (1982)
2.415 W.K. Lee, E.W. Prohofsky: Chem. Phys. Lett. 85, 98 (1982)
2.416 E. Clementi, G. Corongiu: J. Chem. Phys. 72, 3979 (1980)
2.417 E. Clementi, G. Corongiu: Int. J. Quant. Chem. 16, 897 (1979)
2.418 E. Clementi, G. Corongiu: Biopolymers 18, 2431 (1979)
2.419 E. Clementi, G. Corongiu: Biopolymers 20, 551 (1980)
2.420 E. Corongiu, E. Clementi: Biopolymers 20, 2427 (1981)
2.421 E. Clementi, G. Corongiu: Biopolymers 21, 763 (1982)
2.422 E. Clementi, G. Corongiu: In *Biomolecular Stereodynamics*, ed. by R. Sarma (Adenine, New York 1981) pp.209-259
2.423 E. Clementi, G. Gorongiu: IBM Research Rep. POK-04 (Sept. 1981)
2.424 C.A. Croxton: *Statistical Mechanics of the Liquid Surface* (Wiley, New York 1980)
2.425 J.S. Rowlinson, B. Widom: *Molecular Theory of Capillarity* (Clarendon Press, Oxford 1982)
2.426 F.F. Abraham: Rep. Prog. Phys. 45, 1113 (1982)
2.427 C. Ebner, M.A. Lee, W.F. Saam: Phys. Rev. A21, 959 (1980)
2.428 M.H. Kalos, J.K. Percus, M. Rao: J. Stat. Phys. 17, 111 (1977)

2.429 M. Rao, B.J. Berne, J.K. Percus, M.H. Kalos: J. Chem. Phys. **71**, 3802 (1979)
2.430 M. Rao, B.J. Berne: Mol. Phys. **37**, 455 (1978)
2.431 F.F. Abraham: J. Chem. Phys. **72**, 1412 (1980)
2.432 S.M. Thompson, K.E. Gubbins: J. Chem. Phys. **70**, 4947 (1980)
2.433 S.M. Thompson, K.E. Gubbins: J. Chem. Phys. **74**, 6467 (1981)
2.434 T.A. Weber, E. Helfand: J. Chem. Phys. **72**, 4014 (1980)
2.435 C.Y. Lee, H.L. Scott: J. Chem. Phys. **73**, 4591 (1980)
2.436 H.L. Scott, C.Y. Lee: J. Chem. Phys. **73**, 5351 (1980)
2.437 J. Miyazaki, J.A. Barker, G.M. Pound: J. Chem. Phys. **64**, 3364 (1976)
2.438 J.P. Badiali, M.L. Rosinberg, D. Levesque, J.J. Weis: J. of Phys. C**16**, 2183 (1983)
2.439 M.P. d'Evelyn, S. Rice: Phys. Rev. Lett. **47**, 1844 (1981)
2.440 F.F. Abraham, Y. Singh: J. Chem. Phys. **67**, 2384 (1977)
2.441 F.F. Abraham, Y. Singh: J. Chem. Phys. **68**, 4767 (1978)
2.442 F.F. Abraham: J. Chem. Phys. **68**, 3713 (1978)
2.443 I.K. Snook, D. Henderson: J. Chem. Phys. **68**, 2134 (1978)
2.444 J.R. Grigera: J. Chem. Phys. **72**, 3439 (1980)
2.445 I.K. Snook, W.J. Van Megen: J. Chem. Phys. **70**, 3099 (1979)
2.446 S. Toxvaerd: J. Chem. Phys. **74**, 1998 (1981)
2.447 J.N. Cape: J. Chem. Soc. Faraday Trans II **78**, 317 (1982)
2.448 D.E. Sullivan, D. Levesque, J.J. Weis: J. Chem. Phys. **72**, 1170 (1980)
2.449 W.J. Van Megen, I.K. Snook: J. Chem. Soc. Faraday Trans II **75**, 1095 (1978)
2.450 I.K. Snook, W.J. Van Megen: J. Chem. Phys. **72**, 2907 (1980)
2.451 W.J. Van Megen, I.K. Snook: J. Chem. Phys. **74**, 1409 (1981)
2.452 I.K. Snook, W.J. Van Megen: J. Chem. Phys. **75**, 4738 (1981)
2.453 D.E. Sullivan, R. Barker, C.G. Gray, B. Streett, K.E. Gubbins: Mol. Phys. **44**, 597 (1981)
2.454 B. Jönsson: Chem. Phys. Lett. **82**, 520 (1981)
2.455 S. Toxvaerd, E. Praestgaard: J. Chem. Phys. **67**, 5291 (1977)
2.456 J.P. Hansen, L. Verlet: Phys. Rev. **184**, 151 (1969)
2.457 J.N. Cape, L.V. Woodcock: J. Chem. Phys. **73**, 2420 (1980)
2.458 A.J.C. Ladd, L.V. Woodcock: Mol. Phys. **36**, 611 (1978)
2.459 B. Borštnik, A. Ažman: Z. Naturforsch. **339**, 1557 (1978)
2.460 G.M. Torrie, J.P. Valleau: Chem. Phys. Lett. **65**, 343 (1979)
2.461 G.M. Torrie, J.P. Valleau: J. Chem. Phys. **73**, 5807 (1980)
2.462 G.M. Torrie, J.P. Valleau: J. Phys. Chem. **86**, 3251 (1982)
2.463 G.M. Torrie, J.P. Valleau, G.N. Patey: J. Chem. Phys. **76**, 4615 (1982)
2.464 C.W. Outhwaite, L.B. Bhuiyan, S. Levine: J. Chem. Soc. Faraday Trans II **76**, 1388 (1980)
2.465 W.J. Van Megen, I.K. Snook: J. Chem. Phys. **73**, 4656 (1980)
2.466 I.K. Snook, W.J. Van Megen: J. Chem. Phys. **75**, 4104 (1981)
2.467 B. Jönsson, H. Wennerström, B. Halle: J. Phys. Chem. **84**, 2179 (1980)

3. Monte Carlo Studies of Critical and Multicritical Phenomena

D. P. Landau[*]

With 10 Figures

In the late 1960s Monte Carlo studies of relatively simple lattice models began in earnest. Much of the work concentrated on contributing to our understanding of the effects of finite system size, finite sampling and metastabilities, boundary conditions, etc. In some cases these studies included complications in the system Hamiltonians although for the most part rather conventional systems were investigated. The first decade or so of this "serious" work was described in detail in [3.1]. Since then, there has been substantial growth in the use of Monte Carlo methods to study critical phenomena and a rapid expansion in the number of workers using this technique. Many recent studies involve rather complicated models and in many cases the ordered state was not even known prior to the simulation study. In addition, new methods such as Monte Carlo Renormalization Group (MCRG) techniques have been developed and used to augment, or even replace, standard Monte Carlo studies. (Since MCRG methods have been described in detail in [3.2], I need not describe them here. Another recently developed approach, the block distribution method, is described in Chap.1.) Results from Monte Carlo calculations are also used together with information obtained from other methods to understand various kinds of behavior. In this chapter recent studies of critical and multicritical phenomena primarily in models of adsorbed monolayers, surfaces, and bulk solids are reviewed.

In Sect.3.1 I review studies of phase diagrams and critical behavior in two-dimensional lattice gas and Ising models. Section 3.2 describes studies of surfaces of three-dimensional Ising models, and Sect.3.3 the bulk properties of three-dimensional Ising-binary alloy models. Section 3.4 considers Potts models, and Sect.3.5 continuous spin models. Dynamic critical phenomena are discussed in Sect.3.6 and work on other models which do not fit into any of the preceding categories is described in Sect.3.7.

3.1 Two-Dimensional Lattice-Gas Ising Models

3.1.1 Adsorbed Monolayers

The development of more sensitive diffraction methods has led to an increase in activity involving the study of adsorbed monolayers on crystalline substrates. Order-

---

[*]Work supported in part by NSF grant No. DMR-8300754.

disorder phenomena in adsorbed monolayers are often described by simple two-dimensional lattice gas models which are exactly equivalent to Ising models by a simple transformation [3.3,4]. Occupation variables c_i are assigned to each lattice site i such that $c_i = 1$ if the site is filled and $c_i = 0$ if it is empty. The Hamiltonian for this model is

$$\mathcal{H} - \mu N_a = -(\varepsilon + \mu) \sum_i c_i - \sum_{i \neq j} \phi_{ij} c_i c_j - \sum_{i \neq j \neq k} \phi_t c_i c_j c_k \quad , \tag{3.1}$$

where μ is the chemical potential, N_a is the total number of adatoms, ε is the binding energy, and ϕ_{ij} and ϕ_{ijk} represent two-body and three-body coupling, respectively. If we make the transformation to spin variables $S_i = \pm 1$ where

$$c_i = (1 - S_i)/2 \quad , \tag{3.2}$$

we obtain the Ising Hamiltonian:

$$\mathcal{H} = H \sum_i S_i - \sum_{i \neq j} J_{ij} S_i S_j - \sum_{i \neq j \neq k} J_t S_i S_j S_k \tag{3.3}$$

with

$$H = -\frac{1}{2}(\varepsilon + \mu) - \frac{1}{4} \sum_{j(\neq i)} \phi_{ij} - \frac{1}{8} \sum_{j \neq k(\neq i)} \phi_{ijk} \tag{3.4a}$$

$$J_{ij} = \frac{1}{4} \phi_{ij} + \frac{1}{8} \sum_{k(\neq i, j)} \phi_t \tag{3.4b}$$

$$J_t = -\frac{1}{8} \phi_t \quad . \tag{3.4c}$$

Binder and *Landau* [3.3] studied square lattices with nearest- (nn) and next-nearest-neighbor (nnn) coupling as a model of adsorption on (100) faces of cubic substrates. In a more detailed study [3.5] they obtained more accurate data and included a much more extensive analysis. They considered 40×40 lattices with periodic boundary conditions broken into four interpenetrating sublattices. Their results showed that mean-field theory yields *qualitatively incorrect* results when J_{nn} and J_{nnn} are both repulsive. For $R = J_{nnn}/J_{nn}$ between zero and 1/2 they found a degenerate phase in which empty rows alternate with rows with alternate sites occupied. Cooling the system from a high-temperature disordered state to $T = 0$ yielded a metastable domain state in which only the domain *corners* cost energy. A finite-size scaling analysis using data for lattices as large as 60×60 also showed that the critical behavior associated with the order-disorder transition of the (2×1) phase (which occurs for $R > 1/2$) is nonuniversal. To within experimental error, however, *Suzuki*'s "weak universality" [3.6] is valid. A careful study of the phase boundary for the $c(2 \times 2)$ state with $R = 0$ was made using lattices as large as 80×80. To within quite small error limits the results were identical to the "interface solution" of *Müller-Hartmann* and *Zittartz* [3.7]. The behavior of the model at low temperatures was related to hard-core lattice gases and transitions at $T = 0$ interpreted as generalized percolation transitions. Adsorption iso-

94

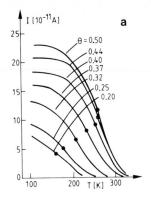

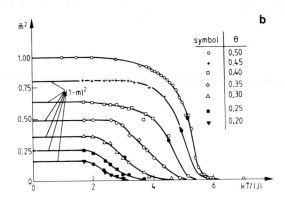

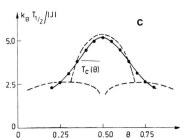

Fig.3.1a-c. Comparison of LEED data and Monte Carlo results for H on Pd(100). (a) Temperature dependence of LEED intensities at fixed coverage θ (the dots denote $T_{1/2}$); (b) temperature dependence of the order parameter of the c(2×2) structure obtained from Monte Carlo calculations (the dots denote $T_{1/2}$); (c) the phase diagram derived from $T_{1/2}$ (dots) and the correct one (dashed curve) [3.4]

therms were also obtained for a wide range of interaction parameters. These calculations were further extended by the inclusion of third nearest-neighbor (3nn) two-body interactions and three-body coupling J_t. Their calculations showed that new ordered phases could be stabilized by additional interactions and degenerate (4×2) and (4×4) ordered states were observed. A detailed comparison of data obtained for $J_t \neq 0$ was made with the experimental results of *Behm* et al. [3.8]. This comparison yielded estimates for the two-body and three-body coupling constants, but most important it demonstrated the ambiguity in the determination of phase boundaries and tricritical points for LEED data obtained at constant coverage (Fig.3.1). A more detailed study of the phase boundaries of the square lattice with nn, nnn, and 3nn interactions by *Landau* and *Binder* [3.9] has shown that the transitions from the (4×2) and (4×4) ordered states to the disordered state are first order.

Lattice gas models have also been used to understand adsorbed monolayers on other crystal faces. Several different groups have used Monte Carlo methods to investigate the order-disorder transition of the p(2×1) structure of oxygen on W(110) in terms of lattice gas models. *Williams* et al. [3.10,11] considered four different two-body interactions and from the order-parameter verus temperature results at fixed coverage they calculated LEED intensities which they compared with experimental results. (The simulations were actually done at fixed coverage using particle exchange, i.e., spin exchange, kinetics.) Interaction parameters were estimated but they were not unique since experimental results were "fitted" for only two coverages (25% and 53%). *Ching* et al. [3.12,13] included three-body as well as two-body interactions in their simulations. They also considered 75% coverage where a p(2×2) ordered state is stable at low temperatures. To reproduce the observed transitions they found it was necessary to have attractive interactions between nn sites as well as attractive and repulsive interactions involving more distant neighbors. The appearance of the p(2×2) phase was explained by significant three-body interactions. *Roelofs* et al. [3.14] studied a square-lattice gas model appropriate to the adsorption of O on Ni(111). They included repulsive nn interactions and attractive nnn coupling which was either zero or 30% of the nn coupling in magnitude. Calculations of the intensities and halfwidths of LEED intensities of the fractional order beams associated with the c(2×2) structure were used to estimate critical temperatures. *Binder* et al. [3.15] considered ordering anisotropic triangular (centered rectangular) lattices as well as square lattices. For the triangular lattice they considered near-neighbor two-body and three-body coupling. They determined a phase diagram which included (3×1) and (2×1) phases, providing a description of H on Fe(110). These calculations were extended by *Kinzel* and co-workers [3.16-18], who combined simulation results with transfer matrix scaling results to obtain phase diagrams and disorder lines. Monte Carlo calculations of the structure factor $S(q)$ were used to study incommensurate behavior in the (3×1) phase (Fig.3.2). Estimates for critical exponents for the (3×1) transition suggest that they depend upon the coupling ratios.

The lateral interactions in monolayers of Ag on W(110) were estimated by *Stoop* [3.19], who compared experimentally determined phase diagrams with Monte Carlo data on a centered rectangular lattice gas model with pairwise interactions extending out to fifth-nearest neighbors and including two different three-body interactions. (Monte Carlo data were used together with extensive ground-state calculations.) Best-fit estimates were obtained by assuming attractive nn, repulsive nnn and 3nn couplings, with attractive three-body interactions.

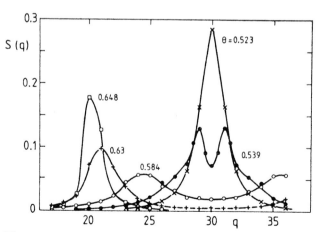

Fig.3.2. Structure factor $S(q)$ as a function of wave vector q (in units of $2\pi/60$) for coverage 1/2 $< \phi < 2/3$ at $kT/|J_2| = 1.4$. Data are from Monte Carlo calculations for systems of size 60 × 20. $R_1 = \frac{1}{2}$, $R_2 = \frac{1}{2}$, $R_t = -\frac{1}{4}$ [3.16]

The simple lattice-gas Ising model of (3.1,2) was modified by *Lee* and *Landau* [3.20] to include the possibility of two coadsorbed adatom species on (001) faces of cubic substrates. A simple modified BEG model was presented which was thought to be suitable for dissociative adsorption of CO on Mo(001) or W(001). Each site i on a square lattice was described by a site occupation variable P_i^λ which was 1 if the site was occupied by species λ and zero otherwise. A simple transformation yields an S = 1 Ising model

$$\mathcal{H} = K_{nn} \sum_{nn} s_{iz}^2 s_{jz}^2 + J_{nnn} \sum_{nnn} s_{iz} s_{jz} + \Delta \sum_i s_{iz}^2 \quad , \tag{3.5}$$

where K_{nn}, $J_{nnn} > 0$ in order to reproduce the experimentally observed c(2 × 2) phase. The resultant phase diagram also yielded an ordered (2 × 1) phase at high densities. A bicritical point was found on the c(2 × 2) to (2 × 1) phase boundary and a tricritical point on the c(2 × 2) to disordered boundary.

Several new studies were made of models which yield $\sqrt{3} \times \sqrt{3}$ ordered phases on the triangular lattice. (These models are given added importance by the large number of experimental studies of physisorption on graphite.) Although some of these results were reported solely in terms of Ising models, it is appropriate to discuss them here. *Wada* et al. [3.21] simulated triangular Ising models with repulsive nn coupling and with $R = J_{nnn}/J_{nn} = -0.1$. They found divergences in specific heat and suceptibility at $kT_f/J_{nn} = 0.40 \pm 0.05$ and a rounded specific heat maximum at higher temperatures. They also examined the time dependence of the sublattice magnetizations and concluded that the $\sqrt{3} \times \sqrt{3}$ low-temperature state undergoes a transition directly to the disordered state at T_f. *Fujiki* et al. [3.22] carried out simulations for the same model over a wide range of R. They determined not only bulk quantities but also sublattice occupations (magnetizations). From their data they deduced the existence of three "singular temperatures" which separate in turn the disordered phase, partially disordered phase, a three-sublattice ferromagnet and a two-sublattice ferromagnet ($\sqrt{3} \times \sqrt{3}$ structure). *Landau* [3.23] recently completed a detailed study of this model including a chemical potential (magnetic field) for R = -1, thus repeating and extending the simulations of *Mihura* and *Landau* [3.24] but using better statistics and finite-size analyses. These data for R = -1 confirmed the early finding [3.24] that the ($\sqrt{3} \times \sqrt{3}$) phases are separated from the disordered state by a phase boundary which is second order at high temperatures and which has tricritical points and first-order transitions at low temperatures (Fig.3.3). Both the critical and tricritical exponents are found to be consistent with those predicted for the three-state Potts model. The behavior at 50% coverage (i.e., zero field) was interpreted differently than in all previous work [3.21-23]. Finite-size scaling and topological analyses were used to show that there is a low-temperature ordered ($\sqrt{3} \times \sqrt{3}$) phase separated from the disordered state by an XY-like line of critical points which exists between upper and lower temperatures

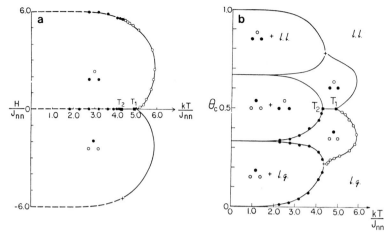

Fig.3.3a,b. Phase diagram for the triangular Ising lattice gas model with R = -1:
(a) Field-temperature plane. Open circles and solid curves show second-order
transitions, while closed circles and dashed lines indicate first-order transitions. (The solid line between T_1 and T_2 is XY-like.) Tricritical points are
marked by +; (b) Coverage-temperature plane. Open circles show critical coverages
(magnetizations) at second-order boundaries and closed circles show the boundaries
of coexistence regions. (Data points are shown only for $\phi \leq 0.5$.) $\ell.g.$ and $\ell.\ell.$
refer to lattice gas and lattice liquid regions respectively [3.23]

T_1 and T_2 respectively. Along this line between T_1 and T_2, vortex-like topological
excitations were identified (Fig.3.3) which are responsible for nonuniversal
(Kosterlitz-Thouless) critical behavior.

Chin and *Landau* [3.25] examined the phase transitions in the triangular lattice
gas Ising model with nn two- and three-body coupling. The results were compared and
contrasted with the phase diagrams obtained by *Schick* et al. [3.26] using real
space RG methods.

3.1.2 Ising Model Critical and Multicritical Behavior

Several different authors have used Monte Carlo renormalization groups to investigate two-dimensional Ising models. *Ma* [3.27] first introduced the combination of
real space renormalization group and Monte Carlo methods which he applied to the
Ising square lattice. However, his approach was cumbersome in that it required
knowing the renormalized Hamiltonian. *Friedman* and *Felsteiner* [3.28] used a different approach in which majority-rule functions and block-spin correlation functions
are compared. They obtained quite good results for the Ising square lattice although the convergence of the exponents with block size was slow. *D'Elia* and *Ceva*
[3.29] used essentially the same method to study critical exponents and fixed point
values in the Ising ferromagnet and the Baxter model, whereby critical exponent
results obtained using three different block-spin transformations were compared and
discussed. *Lewis* [3.30] used a slightly modified method to study the approach of
the critical temperature to the exact (infinite lattice) value.

Swendsen [3.31,32] introduced a new Monte Carlo renormalization group (MCRG) method which obtains exponent estimates by determining correlation functions for the original and transformed lattices and does not involve determination of the fixed-point Hamiltonian. This method was applied to Ising square lattice ferromagnets and antiferromagnets using a variety of block-spin transformations and scale factors. The eigenvalue estimates showed rapid convergence to the exact values and suggested that for the transformations used the original Hamiltonian was close to the fixed point. *Swendsen* and *Krinsky* [3.33] then applied the method to two Ising models with nonuniversal behavior. Ising spins $\sigma = \pm 1$ were placed on square lattices of size 100×100 and Hamiltonians for an nnn antiferromagnet

$$\mathcal{H} = -K_1 \sum_{nn} \sigma_i \sigma_j - K_2 \sum_{nnn} \sigma_i \sigma_k \tag{3.6}$$

and a Baxter model

$$\mathcal{H} = -K_2 \sum_{nnn} \sigma_i \sigma_j - K_4 \sum_{(ijkl)} \sigma_i \sigma_j \sigma_k \sigma_l \tag{3.7}$$

were studied. In both cases, ordering occurs on two interpenetrating lattices composed of nnn sites and it was necessary to use the $b = \sqrt{5}$ transformation introduced by *Van Leeuwen* [3.34] which rotates the transformed lattice with respect to the original lattice. Both models yielded eigenvalues which depended on the interactions used. The estimates for the Baxter-model eigenvalues y_T and y_H agreed well with exact values. The crossover eigenvalues y_S and y_M were found to have essentially the same values for both models.

Binder [3.35] introduced a new block-distribution method as an alternate MCRG approach. Whereas the method proposed by *Swendsen* [3.31] depends upon determining how the mean value of a number of correlations change with the application of the block-spin transformation, this method relies on estimating the manner in which higher-order moments (primarily the fourth-order cumulant) of the order parameter varies with either lattice size or with the size of subsystems within a larger lattice (Chap.1). For the Ising square lattice this method produced exponents which were quite close to the known values and an estimate for the fixed point value for fourth-order cumulant of $U^* \approx 0.52$ for subsystem blocks. *Landau* and *Swendsen* [3.36] applied an MCRG method to the Ising square lattice with anisotropic nearest-neighbor coupling. As the ratio of the interactions in the two different directions varied they found a marginal operator and a line of *Ising* fixed points.

Phase transitions in the zero-field Ising square lattice with nearest- and next-nearest-neighbor coupling were studied by *Landau* [3.37] using standard Monte Carlo simulations. Within experimental error the exponents for the simple antiferromagnetic state did not depend upon the ratio of couplings. Large changes in critical entropy and internal energy were observed as the interactions were varied. Tricritical behavior in the next-nearest-neighbor Ising antiferromagnet was studied using an MCRG technique [3.38,39]. For small magnetic fields two relevant eigenvalues were found identical to the zero-field thermal and magnetic eigenvalues [3.38]. An extended MCRG was developed and used to locate tricritical points in the nnn Ising antiferromagnet and the Blume-Capel model independent of eigenvalue convergence with RG iteration. Four relevant eigenvalues were found; the eigenvalue estimates were virtually identical for the two models: $y_1^e = 1.80 \pm 0.02$, $y_2^e = 0.84 \pm 0.05$, $y_1^o = 1.93 \pm 0.01$, $y_2^o = 1.13 \pm 0.02$. Additional Monte Carlo data showed apparent hysteresis at the phase boundary even just *above* the tricritical temperature. Results for different lattice sizes indicated that this behavior was a finite-size effect and could lead to a spurious estimate for T_t from standard Monte Carlo calculations.

A Monte Carlo renormalization group method was applied to the Baxter-Wu model (Ising spins on a triangular lattice with three-spin interactions) by *Novotny* et al. [3.40]. They found a thermal eigenvalue exponent y_T which agreed well with the exact result and magnetic and crossover exponents which agreed well with conjectured values. The convergence with RG iteration was rapid in contrast to that for the q = 4 Potts model, generally believed to be in the same universality class but exhibiting logarithmic corrections (Sect.3.4).

3.1.3 Models with Incommensurate Phases

Hornreich et al. [3.41] considered Ising square lattices with ferromagnetic nearest-neighbor interactions J_1 and antiferromagnetic interactions J_2 between spins separated by two lattice spacings along one axis (uniaxial case) or both axes (isotropic case). For the uniaxial case the data suggested that a Lifshitz point occurred at $J_2/J_1 = -0.30 \pm 0.05$. (The Lifshitz point divides the "λ line" into two segments — a transition to the ferromagnetic state occurs on one side and a transition to a sinusoidal phase occurs on the other). In the isotropic case the Lifshitz point appeared to move to T = 0 for $-0.45 > J_2/J_1 > -0.475$. *Selke* and *Fisher* [3.42,43] refined these calculations and extended them considerably. They studied long thin rectangular lattices in the uniaxial case (ANNNI model) so that the wave vector **q** of the modulated phase could assume very closely spaced values. In addition they pointed out the existence of (2,2) antiphase state at low temperatures for J_2/J_1 < 1/2. Near the sinusoidal-paramagnetic transition they found dislocation-like configurations which they interpreted as suggesting Kosterlitz-Thouless, XY-like behavior. For the isotropic model they identified a "chess-board-like" antiphase state. The phase boundaries for the isotropic case were also determined. *Selke* [3.44] later carried out yet more extensive Monte Carlo calculations on the ANNNI model. He examined M × N lattices with periodic boundary conditions where N varied from 5 to 30 and M was made very large ($176 \leq M \leq 352$) to allow careful analysis of variations of the **q** vector in the modulated phase. The size dependence of specific heat peaks and structure factors was carefully examined and attempts were made to extrapolate critical temperatures to infinite lattice size. This extrapolation suggests a rather different picture from that obtained directly from data for a single lattice size. Instead of appearing on the ferromagnetic-disordered phase boundary, in the infinite lattice the Lifshitz point appears to be on the sinusoidal-disordered boundary. Further evidence was presented for an XY-like transition bounding the modulated phase. *Barber* and *Selke* [3.45] used the block-distribution function method which analyzes only second moments to study the two-dimensional ANNNI model. An analysis of the incommensurate-commensurate transition suggests that the critical behavior is not exactly Kosterlitz-Thouless in nature.

Saito [3.46] used standard Monte Carlo methods to study different models which exhibit incommensurate phases, including the triangular Ising model with antiferromagnetic nn and nnn interactions. For small nnn coupling he found second-order transitions from a (2×2) ordered state to a disordered state with exponents consistent with those predicted for the four-state Potts model. For larger values of nnn coupling the transition becomes first order. When a magnetic field was included, *Saito* [3.47] found that for large J_{nnn} new $3 \times 3(3 \times 1)$ and (2×2) phases occur. The $3 \times 3(3 \times 1)$ phase appears to be separated from the disordered state by an incommensurate phase.

3.2 Surfaces and Interfaces

Binder and *Hohenberg* [3.48] first used the Monte Carlo method to study surface phase transitions by examining simple cubic Ising films with periodic boundary conditions in the x and y directions and free edges in the z direction. Interactions were restricted to nearest-neighbors, but the coupling J_s between sites in the surface layer·was allowed to be different from the ferromagnetic exchange J involving bulk spins. The behavior of the magnetization in the surface layer, m_1, was described by the usual critical form, i.e., $m_1 \propto (T-T_c)^{\beta_1}$, with $\beta_1^{eff} \approx 0.66$ for $J_s = (1 + \Delta)J$ with $\Delta = 0$ and $\beta_1^{eff} \approx 0.86$ for $\Delta = -0.5$. A multicritical point Δ_c corresponding to the simultaneous criticality of ferromagnetic surface and bulk phases is expected to be present for some positive value of Δ. To determine whether β_1^{eff} is strongly affected by crossover to the multicritical point, *Landau* and *Binder* [3.49] carried out a more complete study over a wide range of Δ. Since the limitation of both bulk and surface correlation lengths by finite lattice size will cause the results to differ from their corresponding infinite lattice values, it is important that the lattice be large in all directions. Therefore, *Landau* and *Binder* [3.49] used lattices as large as $55 \times 55 \times 40$. The surface layer magnetization data shown in Fig.3.4 are consistent with $\beta_1 \approx 0.78$ independent of Δ, but they also clearly show deviations from asymptotic behavior due to crossover for a wide range of Δ.

Abraham and *Smith* [3.50] proposed a similar model, which included impurities and modified couplings, for studying the surface of a solvent-solute system. They used a $10 \times 10 \times 16$ simple cubic lattice gas with the same boundary conditions as

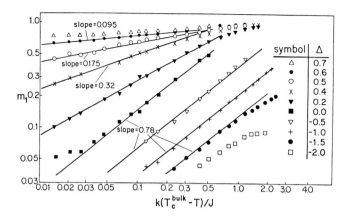

Fig.3.4. Critical behavior of the magnetization m_1 at the surface (note $k\bar{T}_c/J = 4.51$ for the bulk) for simple cubic lattices with nearest-neighbor coupling J in the bulk at $J_S = J(1+\Delta)$ in the surface plane [3.49]

in the magnetic surface model described earlier. Three-state site occupation variables were used to indicate each site occupied by a solute molecule, a solvent molecule, or empty. The interaction between two sites depended on the nature of the occupancy of each site but with the dominant interaction in the bulk being the attractive coupling between solvent molecules on nearest-neighbor sites. The results showed that either by increasing the solute-solute attractive coupling in the surface layer or by varying the chemical potential for solute molecules in the surface it was possible to produce a dramatic increase in the density of solute molecules at the surface. The surfactancy produced by enhancing interactions between solute molecules in the surface was interpreted as evidence for a phase transition. However, since the lattices studied were rather small and only 250 MCS/step were used for collecting data, conclusions must be drawn cautiously.

Interfacial properties of several different models have also been investigated. *Abraham* and *Smith* [3.51] studied interface roughening in a 65 × 65 two-dimensional spin-1/2 Ising model with periodic boundary conditions only in the horizontal direction; the bottom row was coupled to an extra row of spins permanently fixed in the up direction, the top row was connected to another extra row in which a long chain of spins (almost the entire row) was fixed to point down and the remaining spins also pointed up. Only nn coupling J was allowed except for sites in the top layer whose coupling to the "extra" layer was equal to aJ. As many as 2×10^4 MCS/site were used for computing averages. The magnetization profiles agreed well with exact results where they were available. Results for a > 1 showed clear binding of the interface to the edge. Data for pair correlations were also obtained; no theoretical predictions were available for comparison. Correlations across the interface showed anomalous long-range behavior. In addition, the autocorrelation functions decay very slowly with time for sites near the interface. *Selke* and *Pesch* [3.52] studied the interface free energy and the interface profile of the 3-state Potts model in two dimensions. They considered (N × M) square lattices (typically 40 × 42) with periodic boundary conditions in the horizontal direction and the top and bot-

tom rows held fixed in two different states. The interface free energy was extracted
from a comparison with the results for a corresponding lattice with no interface.
As many as 1.5×10^4 MCS/site were used for taking data near T_c. Their results suggest that in the interface, e.g., between regions of State 1 and State 2, droplets
of State 3 tend to occur at least near T_c. They also found that the interface free
energy σ tends towards zero as $(T_c-T)^\mu$ with $\mu = 0.85 \pm 0.15$, in agreement with the
theoretical value of $\mu = 5/6$ derived from scaling arguments. *Selke* and *Huse* [3.53]
studied the size dependence of the effective width W of the nonboundary states at
the interface for $N \times N$ q = 3 and q = 4 Potts models. They found that for 6 < N < 60 W
(or "net adsorption" as they call it) diverges as N^a where within the errors,
a = 1-β/ν. As the critical point is approached W diverges as $(T - T_c)^\omega$ with ω consistent with the scaling relation $\omega = \nu$-β. *Selke* and *Yeomans* [3.54] carried out
similar Monte Carlo studies of the two-dimensional Blume-Capel model (along with
interface free-energy calculations). In contrast to the Potts model the "net adsorption" W does not diverge as the second-order critical line is approached. However, W does appear to diverge weakly at the tricritical point. As the first-order
line is approached W diverges with exponent $\omega = 0.33 \pm 0.03$. *Binder* [3.55], on the
other hand, recently introduced a novel technique for determining interface free
energies between coexisting phases without having to compare results for two systems, i.e., one with an interface and one without an interface. Instead he determined the probability distribution of order parameters (appropriate for the phases
in question) in a single system with no stable interface. By then examining the
finite-size behavior of the minimum of the probability distribution the interface
free energy can be determined. This approach relies on the fact that interface
pairs will spontaneously form in finite systems due to thermal fluctuations. From
Monte Carlo data on Ising square lattices of size 20 × 20 and smaller he was able
to show that the method reproduced known results rather well. In three-dimensions
he studied simple cubic Ising models with linear dimensions L between L = 2 and
L = 12. The results showed that in three dimensions the interfacial free energy
goes to zero as $F_s/kT_c \sim 1.03(1-T/T_c)^2$. The Fisk-Widom universal amplitude ratio
was estimated for the first time $\beta^2 c \approx 0.092 \pm 0.05$. *Shugard* et al. [3.56,57] studied
surface roughening in several models by analyzing data for height-height correlation functions. Several different solid-on-solid models with cross section 60 × 60
and 120 × 120 were considered. Their data were consistent with the expected logarithmic divergence of the correlation function with distance for temperatures at
and above the roughening temperature T_R with an amplitude of $2/\pi^2$ as predicted by
Kosterlitz-Thouless theory. *Swendsen* [3.58] showed that finite-size effects can
play a subtle but important role in the analysis since variation of the correlation
function with distance becomes very slow for large distances. He also showed that
the finite-size dependence of the second moment of the height distribution profile could be used to verify the nature of the divergence and to determine the

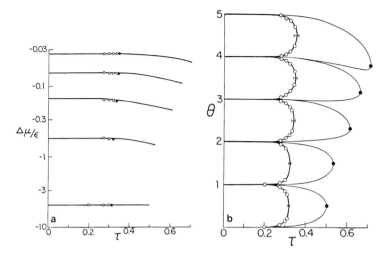

Fig.3.5a,b. Phase diagrams for the first five adsorbed layers of the de Oliveira-Griffiths model: (a) Chemical potential-temperature plane. Solid curves show mean-field results [3.59], open circles the location of first-order phase transitions, and closed circles the critical points as determined from Monte Carlo data. (b) Coverage-temperature pane. Light curves show mean-field results [3.59] for the boundaries to the coexistence region, the solid circles show the mean-field estimate for the critical point, and the open circles show the Monte Carlo results [3.61]

amplitude as well. The data showed that quite large lattices were needed to carry out this analysis at T_R.

Monte Carlo simulations have also been used to study multilayer adsorption on crystalline substrates. *De Oliveira* and *Griffiths* [3.59] proposed a generalized lattice gas model in which each site interacts with its nearest neighbors and with the substrate via a potential which falls off as the cube of the distance to the substrate. They found that for a strong substrate potential within the mean-field approximation, as the number of adsorbed atoms increases a series of first-order transitions occurs due to the formation of successive layers (complete wetting). *Ebner* [3.60] and *Kim* and *Landau* [3.61] studied this model on hcp lattices. Both studies confirmed the existence of first-order lines; however, as shown in Fig.3.5 the critical points associated with these lines appear to approach the roughening temperature T_R instead of the three-dimensional critical temperature predicted by mean-field theory. Similarly, the phase diagram in coverage-temperature space (Fig.3.5) showed distinct differences from the mean-field result. *Ebner* [3.60] also studied this model with a weaker substrate potential (including a short-range repulsive term) and observed the transition from complete wetting to partial wetting. This transition also occurred at a substantially lower temperature than was predicted by mean-field theory. Extrapolation of the transition temperature to zero chemical potential also yielded an estimate for the no-wetting to partial-wetting transition which appeared to be above the roughening temperature.

104

3.3 Three-Dimensional Binary-Alloy Ising Models

Monte Carlo calculations have been used by several different groups to study the critical behavior of simple cubic Ising models. *Rácz* and *Ruján* [3.62] used a Monte Carlo method to evaluate the Niemeijer-van Leeuwen cumulant expansion for the simple cubic Ising model. The accuracy of the results was quite modest; the authors suggest, however, that the limitations are primarily due to the use of the second-order cumulant expansion. *Blöte* and *Swendsen* [3.63] applied an MCRG method to the nn simple cubic model. Although the convergence with iteration was not quite as fast as in two dimensions, the best estimates obtained for exponents were consistent with those found by series and ε expansions. *Swendsen* [3.64] described a momentum space MCRG method which he applied to the simple cubic Ising ferromagnet. This method does not have the same limitations on initial lattice size as does the analogous real-space method using block-spin transformations. Using $17 \times 17 \times 17$ initial lattices he obtained exponent estimates which agreed with those obtained using other techniques. *Knak Jensen* and *Mouritsen* [3.65] made Monte Carlo studies of Ising ferromagnets on simple cubic and diamond lattices. Data were obtained for lattices containing as many as 108,000 sites and detailed analyses were made of the critical behavior of the internal energy using values for T_c which had previously been determined from series expansions. The analysis included the possibility of confluent singularities. If the leading confluent singularity is used in the fitting procedure the resultant amplitude ratio is consistent with series expansion results but not with the ε expansion prediction. If instead the amplitude of this term is set equal to zero and the next order confluent singularity is used in the fit, the resulting amplitude ratios agree well with ε expansion values. *Freedman* and *Baker* [3.66] used multispin coding to calculate the properties of an $N \times N \times N$ simple cubic Ising model. From their data they determined the renormalized coupling constant g_R as a function of correlation length ξ (and lattice size N). They found that g_R decreases systematically for increasing N instead of remaining constant as it should if hyperscaling is valid.

Tricritical behavior in a variety of layered Ising metamagnets and the nnn-Ising antiferromagnet in three dimensions was studied by *Herrmann* et al. [3.67,68] using MCRG methods. For ferromagnetic intrasublattice coupling, which was strong compared with the antiferromagnetic intersublattice coupling, mean-field tricritical exponents were found. For small intrasublattice couplings the observed exponent behavior was consistent with the breakup of the tricritical point into a double-critical end point and a critical end point as predicted by mean-field theory; however, no change could be detected in the phase diagram itself.

A number of studies have now been made of three-dimensional Ising models with multispin interactions. *Mouritsen* et al. [3.69] used Monte Carlo calculations to show that Ising models on fcc lattices with pure three-spin or pure four-spin interactions show only a single first-order phase transition. If both two- and four-

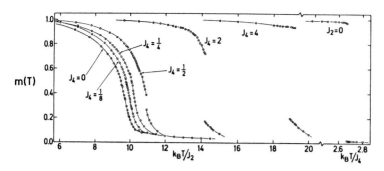

<u>Fig.3.6.</u> Ferromagnetic order parameter m(T) for the fcc Ising ferromagnet with various 2-spin coupling J_2 and 4-spin interactions J_4. Data are obtained from Monte Carlo Calculations of systems with N = 2000 spins, except in the case J_4 = 4 where N = 432 [3.70]

spin coupling are included, *Mouritsen* et al. [3.70] found that the transition becomes second-order as the two-spin interaction is increased. The tricritical coupling was estimated to be in the region $J_4/J_2 \approx 1/4$ - 1/2 (Fig.3.6). In the region of couplings for which the transition is second order, no change was seen in the critical exponents contrary to series expansion results. *Chakrabarti* et al. [3.71] studied the critical behavior of compressible Ising models with bilinear spin coupling by simulating an equivalent effective Ising Hamiltonian which includes a four-spin interaction. They too found first-order transitions for $J_4/J_2 = 0.25$ and 0.50. *Chakrabarti* [3.72] later found that addition of nonmagnetic impurities turned the first-order transition into a second-order transition.

Monte Carlo calculations have also been carried out for equivalent Ising models for binary AB alloys. *Binder* [3.73] determined the phase diagram for an Ising fcc antiferromagnet in a field by studying large (16,384 site) lattices. He found first-order transitions between A_3B, AB, and AB_3 phases and this disordered state. At the critical field between two ordered phases, the disordered phase is stable down to zero temperature due to frustration effects. These results show that all previous theoretical treatments of this model yielded incorrect results.

Phani et al. [3.74] studied an Ising binary-alloy model on an fcc lattice including both antiferromagnetic nn coupling J > 0 and ferromagnetic nnn interaction $-\alpha J$. Their data suggested that a second-order transition occurred for $\alpha \geq 1/4$ but that the transition was first order for smaller values of α. A more detailed investigation of this model, including the presence of a field (i.e., for concentration other than 50% A and 50% B atoms) was made by *Binder* et al. [3.75]. Both "spin-flip" and "spin-exchange" methods were used for lattices containing 2048 sites and 16,384 sites. (Finite-size effects had a negligible effect on the phase boundaries.) The results showed that phase diagrams derived using various approximate analytic methods are generally unreliable. Data for $\alpha = 1/4$ did suggest that the cluster variation method was fairly accurate for stoichiometric compositions but probably less

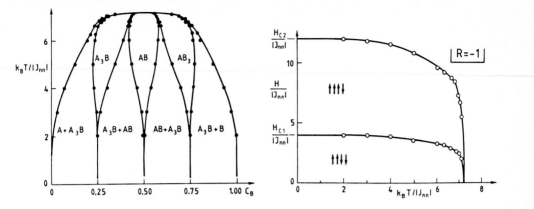

Fig.3.7. Phase diagram of the fcc Ising antiferromagnet with $R = J_{nnn}/J_{nn} = -1$. The left side shows the temperature dependence of the first-order transitions in a field and the right side shows the corresponding binary alloy phase diagram with $c_B = (1-m)/2$ [3.76]

reliable for other compositions. These calculations were extended still further by *Binder* [3.18,76] who estimated the location of first-order phase boundaries (particularly when substantial hysteresis occurs) by analysis of the free energy and entropy (Sect.1.2.2). This approach allowed the extraction of accurate phase boundaries such as the one shown in Fig.3.7 for $J_{nnn}/J_{nn} = -1$.

Phase transitions in the ammonium halides were studied by *Piragene* and *Schneider* [3.77] using simple cubic Ising models with complicated first-, second-, and third-nearest-neighbor coupling. Lattices as large as $8 \times 8 \times 8$ with periodic boundary conditions were used and bulk properties and phase boundaries were determined for different interactions. Second-order transitions were found from the γ and δ phases ("antiferromagnetic" and "ferromagnetic") to the β(disordered) phase and the phase boundaries were compared with the results of mean-field theory and the cluster approximation method.

The effect of competing interactions in an Ising model on the bcc lattice was examined by *Banavar* et al. [3.78] who combined renormalization group ϵ expansion methods and normal Monte Carlo methods to determine the phase diagram. They used ferromagnetic nn and antiferromagnetic nnn interactions near the ratio for which mean-field theory predicts a bicritical point for ferromagnetic and Type II anti-ferromagnetic [AF(2)] ordered phases. The results suggested that as a function of coupling the entire AF(2) phase boundary becomes first order and that the second-order ferromagnetic phase boundary intersects it at a critical end point.

Jain and *Landau* [3.79] studied the $S = 1$ Blume-Capel model on an fcc lattice using a standard Monte Carlo method. The Hamiltonian for this model is ($S = 1$)

$$\mathcal{H} = J \sum_{nn} S_{iz}S_{jz} + \Delta \sum_i S_{iz}^2 + H \sum_i S_{iz} \quad . \tag{3.8}$$

The phase boundary was traced out in the Δ - T plane and a tricritical point was found at $kT_t/12J = 0.256 \pm 0.004$, $\Delta_t/12J = 0.471 \pm 0.004$, in good agreement with estimates from series expansions. The tricritical exponents were consistent with classical, mean-field predictions. The tricritical "wings" were traced out in Δ - T - H space and found to have critical exponents consistent with three-dimensional Ising exponents.

Crossover from one-dimensional to three-dimensional behavior was investigated by *Graim* and *Landau* [3.80] who considered coupled Ising chains. They studied finite-size effects as a function of sample shape and included ratios of interchain coupling $J_\perp$ to interchain coupling $J_{||}$ as small as $\Delta = J_\perp/J_{||} = 0.003$. Both standard Monte Carlo sampling as well as the "n-fold way" method of *Bortz* et al. [3.81] were used. For small Δ the system behaved like independent Ising chains at high temperature before undergoing a transition to an ordered state at low temperature. They found that asymptotic $(\Delta \rightarrow 0)$ variation of T_c was valid over a surprisingly wide range of $\Delta(\Delta \lesssim 0.2)$. *Maier* et al. [3.82] proposed a model for layered squaric acid with an ice rule interaction mechanism and showed that it was equivalent to a system of Ising chains with weak interchain coupling. Monte Carlo calculations were made to determine the susceptibility and order-parameter behavior and estimate interchain and intrachain interaction parameters.

Three-dimensional lattice gas models were also used to investigate hydrogen in metal behavior. *Dietrich* and *Wagner* [3.83] studied the behavior of a simple cubic model with long-range, oscillatory interactions as a model for the incoherent α-β phase transition in PdH_x. The resultant phase diagram is compared with experimental results. *Bond* and *Ross* [3.84] studied fcc lattices appropriate to the α' Pd-D system. Repulsive nn interactions and attractive nnn coupling were used and data were obtained using both canonical and grand-canonical ensemble simulations. The general agreement between the simulation and experimental results is good up to a deuterium concentration of 68%.

A simple cubic model with long-range coupling was studied by *Kretschmer* and *Binder* [3.85]. They included nn exchange J and a dipolar interaction of strength μ^2/a^3. Several methods of treating the long-range interactions were considered. Depending on the ratio of exchange to dipolar coupling, the model may show ferromagnetic, antiferromagnetic, and possibly helical phases. *Pawley* and *Tibballs* [3.86] used a pseudo-Ising model to study the KDP-type structure. Phase transitions for weak ordering fields agree with predictions. *Saito* [3.46] used standard Monte Carlo calculations to investigate the behavior of three-dimensional stacking of triangular Ising layers with antiferromagnetic nn interaction J and nnn coupling RJ. For $R \gtrsim 2.0$ he finds an incommensurate phase with a second-order transition to the disordered state. The transition from the disordered to commensurate state (for $R \lesssim 2.0$) is first order as is the commensurate-incommensurate state. The character of the multicritical point at $R \approx 2.0$ was not studied.

Selke's study of the D = 3 ANNNI model [3.87] has been extended by *Selke* and *Fisher* [3.88,89] to include an analysis of the temperature dependence of the wave vector. The Monte Carlo data together with series expansion and mean-field calculations yield a phase diagram which includes a T = 0 multiphase point for $-J_2/J_1 = 0.5$ where an infinite set of antiphase structures of type $<2^n 3>$ become degenerate. These phases are "buried" in an incommensurate phase; the low-temperature ordered phases for $-J_2/J_1 < 0.5$ are ferromagnetic (with a Lifshitz point at $-J_2/J_1 \sim 0.27$) and <2> antiphase for $-J_2/J_1 > 0.5$.

3.4 Potts Models

Binder [3.90] used standard Monte Carlo methods to study two-dimensional q-state Potts models

$$\mathcal{H} = -J \sum_{nn} \delta_{s_i s_j} \quad , \quad s_i = 1, 2, \ldots, q \tag{3.9}$$

on lattices as large as 200 × 200 with periodic boundary conditions. For q = 3 he found exponents $\alpha \approx 0.4$, $\beta \approx 0,1$, $\gamma \approx 1.45$. For q = 5,6 the transition is found to be weakly first order. Pronounced hysteresis, which is strongly dependent on the number of Monte Carlo steps used, is observed. Even with 10^4 MCS/site the discontinuity observed in the internal energy is much greater than the known exact values. For q = 5,6 the data strongly suggest "pseudocritical" behavior with singularities superimposed on the discontinuities. *Swendsen* [3.31] used a Monte Carlo renormalization group (MCRG) method to study the two-dimensional q = 3 Potts model. *Rebbi* and *Swendsen* [3.91] extended this study by using multispin coding together with a MCRG method to include two-dimensional Potts model with $q \leq 7$. Using b = 2 transformations they were able to obtain as many as five RG iterations beginning with a 96 × 128 lattice. Their results were consistent with nearly singular behavior for q > 4 and suggested the appearance of logarithmic corrections for q = 4.

The nature of the phase transition in the 3-state Potts model in three dimensions has been studied by several different groups. *Knak Jensen* and *Mouritsen* [3.92] used a standard Monte Carlo method to study 30 × 30 × 30 simple cubic lattices. They found a first-order transition with a jump of about 8% in the internal energy and 40% in the order parameter. *Herrmann* [3.93] examined both q = 3 and q = 4 Potts models in three dimensions. He too deduced first-order transitions superimposed on divergences with the jump in internal energy at the transition being much greater for q = 4 than q = 3. *Blöte* and *Swendsen* used an MCRG method [3.94,95] to study q = 3 Potts models in three and four dimensions. The MCRG flows strongly suggested first-order behavior but did not show evidence for a discontinuity fixed point. More recently, *Fucito* and *Vulpiani* [3.96] studied the three-dimensional q = 3 Potts model with nearest- and next-nearest-neighbor coupling (J_{nn} and J_{nnn} respectively), using standard Monte Carlo calculations. Using a modified "mixed-phase" initialization they too found strong evidence for a first-order transition when both nearest- and next-nearest-neighbor coupling are ferromagnetic ($J_{nn} = J_{nnn}$). When the next-nearest-neighbor interaction is antiferromagnetic ($J_{nnn} = -0.2J_{nn}$) they found that the transition is second order. A $q = 3$ Potts model on a square lattice with antiferromagnetic interactions in one direction and ferromagnetic coupling in the other was studied by *Kinzel* et al. [3.97]. Using a combination of Migdal RG calculations and standard Monte Carlo simulations they found rounded specific heat peaks which they concluded were not at the critical temperature.

Selke and *Yeomans* [3.98] determined the phase diagram of the two-dimensional three-state chiral Potts model described by the Hamiltonian

$$\mathcal{H} = -J_x \sum_{(ij)} \cos\left[\frac{2\pi}{3}(n_i - n_j + \Delta)\right] - J_y \sum_{ik} \cos\left[\frac{2\pi}{3}(n_i - n_j)\right] \quad , \qquad (3.10)$$

where n_i, $n_j = 0,1$, or 2 and the sums are over nearest neighbors in the x and y directions, respectively. Using 90×30 lattices they estimated that a Lifshitz point occurs at $\Delta \sim 0.4$. Crossover behavior near the Lifshitz point leads to a significant change in the effective value of the specific heat exponent α but only a minor change in the order parameter exponent β. In the modulated phase kinks, "dislocations" and wall motion are important.

Saito [3.99] used the standard Monte Carlo technique to study the phase diagram of the three-state Potts model on a triangular lattice with both two- and three-body interactions. Data were obtained for lattices as large as 120×120 with as many as 16,000 MCS/site retained for averages. He found the ferromagnetic-disordered state transition changes from second to first order when the two-site coupling is sufficiently antiferromagnetic. The antiferromagnetic-disordered transition as well as the ferromagnetic-antiferromagnetic coexistence line are first order.

Grest and *Banavar* [3.100] studied q-state Potts models on a square lattice with antiferromagnetic nearest- and ferromagnetic next-nearest-neighbor interactions. For $q = 3$ with $J_{nnn} = 0$ they found a phase transition to a "broken-sublattice symmetry" state at finite temperature. With weak nnn coupling the $q = 3$ Potts model appears to have an antiferromagnetic state at low temperatures and then undergoes successive transitions to a broken-sublattice symmetry state and then to a disordered state as the temperature is increased. For $q \geq 4$ with ferromagnetic nnn coupling the system orders in a highly degenerate antiferromagnetic state and appears to undergo two first-order transitions before becoming completely disordered. *Banavar* et al. [3.101] studied three-dimensional $q = 3$ and $q = 4$ antiferromagnetic Potts models. Their data indicate ordering at low temperatures; quenching to low temperatures produces a glassy "plastic crystal" phase for $q = 4$.

Swendsen et al. [3.102] used an MCRG method to study the two-dimensional $q = 4$ Potts model. They introduced a plaquette interaction F which is nonzero only when all four bonds around a unit square connect unlike states. This latter interaction is used to control the chemical potential of "effective vacancies"; actual vacancies were never introduced. Their data yield an estimate for $\nu^{-1} = 1.49 \pm 0.01$ and confirm previous conjectures regarding universality, a marginal direction, and logarithmic corrections.

Berker and *Andelman* [3.103] used a Monte Carlo method to study q-state Potts models in two dimensions with q as large as 20. They also included in the Hamiltonian a plaquette interaction F which is nonzero only when all four bonds around a unit square (plaquette) connect unlike states. By varying F and q they studied the "condensation" of "effective vacancies".

Novotny [3.104] transformed the arbitrary (noninteger) q-state planar Potts model into an ice-type model whose weights depend upon q. These models were simulated using a Monte Carlo method. He found good agreement with known results for the ferromagnetic Potts model, and for the antiferromagnetic case found that the position and magnitude of the specific heat peak was lattice-size independent for $q \geq 0$, thus suggesting that no continuous transition occurs.

3.5 Continuous Spin Models

Monte Carlo results now exist for evaluating the suitability of the Kosterlitz-Thouless theory of the phase transition in two-dimensional XY-like models. *Miyashita* et al. [3.105] simulated a plane-rotator model on 50×50 square lattices. To decrease computing time, however, they allowed the spin-vectors to assume only one of a large but discrete number of directions. This method introduces a small effective anisotropy into the model, but no effects of this anisotropy were observed. Pictures of typical spin configurations show vortex pairs superimposed on a background of spin waves. They estimated $kT_c \sim 1.15$ J. *Tobochnik* and *Chester* [3.106] extended the investigation to lattices as large as 90×90 and allowed spin directions to vary continuously. From an analysis of the susceptibility divergence in terms of the Kosterlitz-Thouless exponential form $\chi T \propto \exp(bt^{-\nu})$, they estimated $kT_c/J \approx 0.89$ with an error of a few percent. They found that the specific heat peak is nondivergent and that it occurs well above T_c. "Snapshots" of vortex-core configurations (Fig.3.8) show that the bound vortex pairs to indeed begin to unbind at T_c, whereas the temperature dependence of the overall density of vortex pairs shows no special behavior at T_c. (For more details on simulations of Kosterlitz-Thouless transitions, see also Chap.7.) *Van Himbergen* and *Chakravarty* [3.107] also studied the two-dimensional plane-rotator model (although they call it an XY model) with better statistics (10^4 MCS/site) than previous workers. They also determined the helicity modulus from the internal energy difference between antiperiodic and periodic boundary conditions. The data showed a sharp increase in the derivative of the helicity modulus at $kT/J \sim 1$ and the magnitude of the peak increased rapidly with increasing lattice size. *Betsuyaku* [3.108] studied the plane-rotator model on 66×66 square

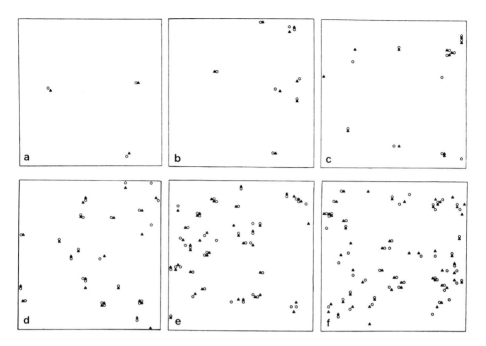

Fig.3.8a-f. Positions of vortex cores in the 60×60 plane-rotator model for typical configurations: o positive vortices, ▲ negative vortices: (a) $kT/J = 0.8$; (b) $kT/J = 0.85$; (c) $kT/J = 0.90$; (d) $kT/J = 0.95$; (e) $kT/J = 1.00$; (f) $kT/J = 1.05$ [3.106]

lattices using a special pseudoblock-spin transformation together with the Monte Carlo simulation. The $N \times N$ lattice was divided up into square blocks of side ℓ, where $1 \leq \ell \leq (N/2)$. After a configuration was generated using a standard importance sampling Monte Carlo scheme, block spins were assigned by randomly picking a spin from within each block. "Block-pair" correlations between nearest-neighbor blocks were then determined as a function of ℓ. The block-pair correlation decreases as ℓ^{-m} at low temperatures for a very wide range of ℓ. Above $kT/J \sim 0.90$ the correlation no longer decreases according to a power law and an exponential decay sets in above $kT/J \sim 1.02$.

Landau [3.109] studied the XY model with fourth-order anisotropy $h_4 \sum_i \cos(4\theta_i)$ added, using the Monte Carlo block-distribution method proposed by Binder. He simulated $L \times L$ square lattices, with $L < 20$, with 6×10^4 MCS retained for computing cumulant averages $U_L = 1 - \langle m^4 \rangle_L / (3\langle m^2 \rangle_L^2)$. For $h_4 = 0$ the data were consistent with the Kosterlitz-Thouless description of a phase transition with $\nu = \infty$ with $kT_c/J = 0.73 \pm 0.03$. The fixed-point value U^* varied between 0.652 and 2/3. As h_4 was increased, T_c varied only slowly but ν decreased rapidly.

Other generalizations of the plane-rotator model were studied by *Swendsen* [3.110] using standard Monte Carlo methods. A "cross-product" operator $H_x = K_x \sum (1 - S_i \cdot S_k)(1 - S_j \cdot S_\ell)$ was added where the sum is over elementary squares, or

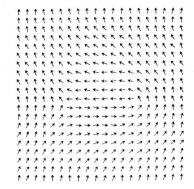

Fig.3.9. Spin configuration in the plane-rotator with nn coupling K_1 = 20.0 and plaquette coupling K_x = 9.5. The vortex-antivortex pair have charge onehalf [3.110]

plaquettes, and each scalar product involves spins at opposite corners of the plaquette. For $K_x > 0$ Swendsen found that fractionally charged vortices appeared, such as those shown in Fig.3.9. When antiferromagnetic third-nearest-neighbor coupling is added, pronounced hysteresis appears as a function of K_x, indicating a first-order transition but showing no trace of the expected intermediate state. Interpretation of the data is made difficult by slowly diffusing domain walls, long relaxation times, and excitations involving large length scales.

Kawabata and *Bishop* [3.111] studied the two-dimensional Heisenberg model with nearest-neighbor coupling in search of long-lived, metastable "nontopological" defects. Lattices as large as 100×100 were considered and metastable "vortex-like" defects were found.

The phase diagrams and critical behavior of a two-dimensional anisotropic Heisenberg antiferromagnet was investigated by *Landau* and *Binder* [3.112]. Classical spins were placed on a square lattice with periodic boundary conditions and Hamiltonian

$$\mathcal{H} = J \sum_{nn} [(1 - \Delta)(S_{ix}S_{jx} + S_{iy}S_{jy}) + S_{iz}S_{jz}] + H_\| \sum_i S_{iz} \quad . \tag{3.11}$$

For small values of $H_\|$ the system ordered in a simple antiferromagnetic state and underwent an Ising-like transition to the paramagnetic state. As the field was increased at low temperatures the system underwent a transition to the spin-flop state. (Since this state has an effective spin dimensionality of two it is expected to be XY-like.) Examination of the projections of the spins onto the xy plane did indeed show antiferromagnetic vortices bound together in vortex-anti-vortex pairs at low temperatures. At the critical temperatures these pairs began to unbind and the high-temperature susceptibility showed an exponential Kosterlitz-Thouless divergence. For all fields the vortex-pair density increased with temperature as exp(-2μ/kT) where 2μ is the pair creation energy. As the field increased and the spins tended to tip towards the z direction, the pair creation energy decreased. From these data it was not possible to determine if the bicritical umbilicus terminated at T = 0 or at some finite temperature. They also studied the isotropic

($\Delta = 0$) Heisenberg antiferromagnet for which T_c is believed to be zero for $H_{||} = 0$. Transition temperatures for $H_{||} > 0$ were estimated from the vortex-pair unbinding; the resultant phase diagram showed an umbilicus, but for $H/J = 0.01$ the critical temperature was already 80% of its maximum value.

Tobochnik [3.113] used a Monte Carlo renormalization group method to investigate the phase transitions in q-state clock (vector Potts) models with $q = 4,5,6$, whereby 16×16 and 32×32 lattices were simulated using up to 10^5 MCS/site. Block spins were determined by vectorially summing the spins in 2×2 blocks, i.e., carrying out a $b = 2$ block spin renormalization group transformation. Thermodynamic averages on the original and transformed lattices are compared and T_c is determined by finding the temperature at which the renormalization group transformation does not alter the thermodynamic average. For $q = 5$ and 6 three phases were seen: an Ising-like ordered phase, an XY-like phase, and a disordered phase. Critical temperatures were estimated as $kT_1/J \sim 0.68$ and $kT_2/J \simeq 1.15$ for $q = 6$ and $kT_1/J \sim 0.93$ and $kT_2/J \sim 1.08$ for $q = 5$. These values must be treated as suspect, however, since for $q = 4$ a single transition with $kT_c/J = 1.21 \pm 0.02$ was found, whereas a simple transformation shows that the $q = 4$ model is equivalent to an Ising model with $J' = J/2$ and hence $kT_c/J = 1.135$.

Several different Monte Carlo studies have also been made of models believed to have Lifshitz points. *Selke* [3.114,115] first studied a plane rotator on a simple cubic lattice with ferromagnetic nearest-neighbor interactions and antiferromagnetic coupling between spins separated by two lattice spacings along one axis of the system. Lattices of size $16 \times 16 \times 16$ were studied and close to the Lifshitz point the order parameter was depressed from the nearest-neighbor model estimate of $\beta = 0.32 \pm 0.02$ to $\beta_{LP} = 0.20 \pm 0.02$. *Selke* [3.116] also used an analysis of the finite lattice behavior of the magnetization to confirm this estimate. In addition, an analysis of the high-temperature susceptibility yielded a value of $\gamma_{LP} = 1.5 \pm 0.1$ as compared with the ordinary XY value of $\gamma = 1.31$.

3.6 Dynamic Critical Behavior

The time-dependent behavior of many-body systems may be determined by equations of motion which must be integrated over time and which cannot be studied by Monte Carlo methods. There are a large number of kinetic models, however, describable by a master equation, whose time-dependent behavior near phase transitions is both interesting and accessible by Monte Carlo methods. This "dynamic" critical behavior can best be described by time-displaced correlation functions, e.g., $\phi_M = \langle M(0)M(t)\rangle - \langle M\rangle^2$. As the critical temperature is approached the relaxation time $\tau = \int_0^\infty \phi_M(t')dt'$ diverges as $\tau \propto (T - T_c)^{-\Delta}$, where Δ is a dynamic critical exponent. *Suzuki* [3.117] predicted that the ratio Δ/ν, where ν is the usual correlation length exponent, is universal. Previous work [3.118] in two dimensions has been carried out on the Ising square lattice. *Katz* et al. [3.119] studied the Baxter-Wu model (triangular Ising model with three-spin interactions) on 52×52 lattices using 12,000 MCS/site to determine averages. For $|1 - T/T_c| \lesssim 0.1$ the divergence of the susceptibility and relaxation times were described by effective exponents $\gamma = 1.0$ and $\Delta = 1.3$. Since $\nu = 1$ and $\Delta \sim 2$ for the square lattice and $\nu = 2/3$ for the Baxter-Wu model, the *Suzuki* prediction is that $\Delta = 4/3$. The agreement is excellent.

Tobochnik et al. [3.120] devised a dynamic MCRG method which they then applied to the Ising square lattice. Spin configurations are first generated by an ordinary Monte Carlo method, then block spins are constructed using the majority rule on $b \times b$ blocks of spins. This process may be repeated on the block-spin lattice to produce a lattice reduced in scale by a factor of b^2 from the original lattice and so forth. Time-dependent correlation functions were constructed for each renormalized lattice as well as the original lattice. The correlation functions for lattices differing by m transformations are then compared to find the temperatures and times for which the two match. From dynamic scaling the two times at which the correlations match are related by $t' = tb^{mz}$ where z is the dynamic critical exponent. Using this technique with 8×8 and 16×16 lattices they found $z = 2.22 \pm 0.13$. *Yalabik* and *Gunton* [3.121] used this same approach to study the kinetic behavior of the Ising square lattice using spin-exchange instead of spin-flip transitions (for 16×16 and 8×8 lattices) and found $z = 3.80$ as compared with the theoretical value of $z = 3.75$. For a three-dimensional spin-flip model on $8 \times 8 \times 8$ and $16 \times 16 \times 16$ lattices they obtained $z \approx 2.08$, which is in reasonable agreement with an expansion estimate of $z = 1.99$. *Katz* et al. [3.122] applied the dynamic Monte Carlo renormalization group method to the Ising square lattice with spin-flip kinetics using larger lattices than those considered by *Tobochnik* et al. [3.120]. They also tried to match correlations at more different times and found some ambiguity in determining z, which they attributed in part to inaccuracy in the data and finite-size effects. *Angles D'Auriac* et al. [3.123] used the "n-fold" way method of Bortz et al. to test the dynamic finite-size scaling hypothesis. Square lattices between 2×2 and 15×15 were studied. They estimate the critical exponent $z = 2.0 \pm 0.1$ ($\tau = AL^z$). *Kalle* and *Winkelmann* [3.123] extended and improved previous multispin coding techniques and studied $600 \times 600 \times 600$ simple cubic Ising models. At $T/T_c = 1.4$ they found that the magnetization of an initially completely magnetized system decays as $\exp(-t/\tau)$ with $\tau = 2.9$.

Novotny and *Landau* [3.125,126] used standard spin-flip Monte Carle methods to study dynamic "impure critical behavior" in the Baxter-Wu model with quenched site impurities on 48×48 lattices. Even small impurity concentrations led to a dramatically increased critical exponent $z \approx 2.0$; but the data showed two distinct relaxation times. *Serinko* et al. [3.127] investigated the tricritical dynamic behavior in 80×80 Ising square lattices using a spin-flip Monte Carlo method. They found a dynamic tricritical exponent ≈ 1.1.

3.7 Other Models

A variety of models other than those discussed in the previous sections have also been studied using Monte Carlo methods.

3.7.1 Miscellaneous Magnetic Models

Ising models in dimension d = 4,5 have been studied using standard Monte Carlo methods and in d = 4 by MCRG. *Mouritsen* and *Knak Jensen* analyzed the critical behavior of the order parameter for the d = 4 hypercubic Ising ferromagnet including logarithmic corrections [3.128] and later corrections to scaling [3.129] to fit data farther from T_c. They concluded that the Monte Carlo data were consistent with the behavior predicted by renormalization group theory. For the d = 5 hypercubic Ising ferromagnet very close to T_c, the order parameter could be fitted by mean-field behavior with no logarithmic corrections. *Blöte* and *Swendsen* [3.130] investigated the d = 4 Ising model using an MCRG method. They found rapid convergence to classical exponent values (logarithmic correction terms are expected to produce slow convergence) with no indication of a marginal operator.

Ditzian et al. [3.131] used a combination of Monte Carlo, series expansions, mean-field, and renormalization group theory to investigate the d = 3 Ashkin-Teller model. They found substantial differences from the corresponding d = 2 phase diagram including a new phase in which the symmetry between the two spin variables is broken.

Sablik et al. [3.132] studied the "cubic model" on an fcc lattice using a standard spin-flip method. The model has a Hamiltonian

$$\mathscr{H} = -J \sum_{nn} \sigma_i \sigma_j \delta_{\alpha_i \alpha_j} \quad , \tag{3.12}$$

where $\sigma = \pm 1$ and $\alpha = x$, y or z. They found a first-order transition at $kT_c/J = 3.45 \pm 0.01$.

Weinkauf and *Zittartz* [3.133] studied phase transitions in a three-dimensional Ising gauge model with nearest-neighbor ferromagnetic two-spin coupling, in addition to the plaquette interaction. Substantial difficulties were encountered in achieving thermal equilibrium so "special" procedures were used in the study. The results were consistent with the proposed qualitative picture of the phase diagram; the observed metastability at the transition between the ferromagnetic and the gauge phases was interpreted as indicating a first-order transition.

Monte Carlo methods have also been used to study the effects of random, quenched impurities on critical behavior. *Landau* [3.134] used a standard Monte Carlo method to study the impure simple cubic Ising model. The transition remained relatively sharp with increasing impurity content, although specific heat peaks did decrease in maximum value. No clear evidence for modified critical behavior was found, although for this model *Harris*'s prediction [3.135] is that the "impure critical re-

116

gion" should be quite narrow: $\Delta T_c \sim x^{1/\alpha}$ where x is the fraction of sites occupied by impurities. *Novotny* and *Landau* [3.136,137] took advantage of the large value of α in the Baxter-Wu model to search for modified critical behavior in a much wider region. Long time-scale fluctuations ($\tau_0 \sim 5000$-$10{,}000$ MCS) in domain walls between equivalent ground states complicated interpretation of the data and the presence of impurities magnified the problem by "pinning" domain walls. Long runs using a number of different starting configurations and impurity distributions were therefore used. They found that even a small fraction of impurities (0.056) depressed the specific heat peak dramatically. Using finite-size scaling analyses they suggested that the pure lattice critical exponents ($\alpha = 2/3$, $\nu = 2/3$, $\gamma \approx 1.17$) changed with the addition of impurities to ($\alpha \lesssim 0$, $\nu = 1.00 \pm 0.07$, $\gamma = 1.95 \pm 0.08$). An MCRG study [3.138] of the "impure critical behavior" gives exponents consistent with these values.

3.7.2 Superconductors

Several studies of superconductor models have now been reported. *Kawabata* [3.139] studied the free vortex density as a function of applied magnetic field in a two-dimensional lattice model with interactions which fall off logarithmically with distance. The results were in qualitative agreement with the Doniach-Huberman theory. *Dasgupta* and *Halperin* [3.140] studied a three-dimensional lattice superconductor model which included both angular (phase variables) and directed-link variables. Monte Carlo data, together with duality arguments, suggest that the transition is equivalent to an inverted XY transition and not first-order transition as predicted. *Ebner* and *Stroud* [3.141] simulated a three-dimensional inhomogeneous superconductor made up of superconducting grains coupled together by Josephson tunneling. Calculations were carried out on $5 \times 5 \times 5$ and $10 \times 10 \times 10$

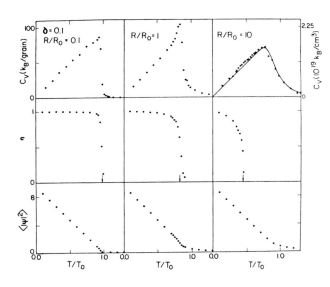

Fig.3.10. Specific heat C_V, phase-order parameter η, and mean square energy-gap parameter $\langle |\psi|^2 \rangle$ for an ordered simple cubic lattice of coupled superconducting grains. The size parameter $\delta = 0.1$ for all three coupling strengths (normal state tunneling resistance) $R/R_0 = 0.1$, 1, 10. Arrows denote estimated positions of the phase-ordering transition. The solid curve (upper right) is for isolated particles ($R = \infty$) [3.141]

simple cubic arrays. Different values were considered for the grain-size parameter δ, the normal state tunneling resistance, and introduced disorder in several possible ways. They found that the specific heat peak (single-grain transition) did not necessarily occur at the bulk transition temperature (Fig.3.10). With random single-grain transition temperatures it was even possible to observe a resistive transition temperature which lies above the specific heat peak.

3.7.3 Interacting Electric Multipoles

Extensive studies have been made of (relatively small) lattices with sites interacting via electrostatic multipole interactions. Since this work was not reviewed in [3.1], the early work is also reviewed here. *Mandell* [3.142] studied the behavior of 32 classical free-rotator quadrupoles on an fcc lattice with periodic boundary conditions. The system was found to have a first-order transition at $kT/\Gamma = 11 \pm 1$ where Γ is the quadrupolar coupling constant. Data were presented for the energy, specific heat, and correlation functions. *Mandell* [3.143] also studied fcc and hcp lattice models with both nearest- and next-nearest-neighbor quadrupolar interactions. He found good agreement with the transition temperature, latent heat, and susceptibilities of N_2. *O'Shea* [3.144] simulated an fcc lattice of classical, tetrahedral octopoles with nearest-neighbor octopole-octopole interactions. Transitions were found between a low-temperature ordered phase, an intermediate phase which is a mixture of hindered and freely rotating octopoles and a high temperature rotationally disordered state. The Monte Carlo results give a similar phase diagram but with quantitatively different critical temperatures from those given by mean-field theory. *O'Shea* and *Klein* [3.145] studied a system of classical octopoles on triangular lattices (6×6 and 12×12) with periodic boundary conditions. The orientational behavior of this system seems to be in some way similar to that of the 3-d system. A high-temperature rotationally disordered phase is separated from a low-temperature ordered phase by an intermediate, partially ordered phase which shows no sublattice structure. *O'Shea* and *Klein* [3.146] also studied a sytem of classical quadrupoles on a triangular lattice. They examined small (6×6) systems with zero substrate potential and with a substrate potential which confined the quadrupoles to lie in the plane parallel to the substrate. The freely rotating case shows a transition from an ordered "zig-zag" state to a low-temperature "pinwheel" structure which is missing in the strong substrate case. Orientational phases in bilayer lattice models appropriate to methane on graphite were simulated by *Maki* and *O'Shea* [3.147]. Octopole-octopole interactions were noted between molecules on a lattice composed of sites on two adjacent (111) planes of an fcc solid. In the free bilayer case a low-temperature ferrorotational ordered state is found at low temperature and is separated from the high-temperature disordered state by a partially disordered state. In the tripod bilayer system (lower layer molecules confined to stable tripod orientations) the behavior of the two layers is almost independent. Orienta-

tional phases of a quadrupolar bilayer were investigated by *O'Shea* and *Klein*
[3.148]. Both nearest- and next-nearest-neighbor coupling were included. For the
nearest-neighbor free bilayer several low-temperature structures were observed,
including a complicated eight-sublattice structure. The free nnn bilayer shows
simpler behavior. The in-plane bilayer showed a "herringbone" transition at a temperature quite close to the monolayer critical temperature. The relevance of these
results to nitrogen physisorbed on graphite was discussed. *Klenin* and *Pate* [3.149]
studied rather larger lattices (up to 2000 sites) of quadrupolar monolayers. They
found that for the zero substrate case multiple phase transitions do occur but that
defect formation (isolated pinwheels) prevents observation of the highly anisotropic intermediate phase.

3.7.4 Liquid Crystals

A number of Monte Carlo studies have also been carried out on systems of classical
molecules used to model liquid crystals. *Vieillard-Baron* investigated two-dimensional systems of hard ellipses [3.150] and hard spherocylinders [3.151] to study
solid-nematic and nematic-liquid phase transitions. *Jansen* et al. [3.152] studied
the behavior of a lattice version of the Maier-Saupe model

$$\mathcal{H} = -\frac{1}{2} \varepsilon \sum_{nn} P_2(\cos\theta_{ij}) \tag{3.13}$$

on $20 \times 20 \times 20$ lattices, improving the accuracy of the earlier work by *Lebwohl* and
Lasher [3.153]. They found a first-order transition at $kT_c/\varepsilon = 0.894 \pm 0.001$ with a
jump in order parameter of $S = 0.333 \pm 0.009$. *Luckhurst* and *Romano* [3.154] simulated
biaxial particles (noncylindrically symmetric molecules) on an fcc lattice interacting via anisotropic nn interactions. They found a first-order isotropic-(uniaxial)-
nematic phase and a second-order uniaxial-biaxial transition. *Denham* et al. [3.155]
simulated triangular systems of cylindrically symmetric particles (up to 3600 particles) interacting via a weak anisotropic potential. The results suggest a second-
order (or higher) transition from an orientationally disordered to a partially
ordered phase. *Luckhurst* and *Romano* [3.155] also carried out computer simulations
on systems of particles not confined to a lattice. Particles interacted via Maier-
Saupe anisotropic potential which had a Lennard-Jones-type distance dependence.
The results were compared to those obtained for the lattice model as well as those
derived from mean-field theory. *Luckhurst* et al. [3.157] studied the effect of an
external field on a simple cubic system of cylindrically symmetric particles interacting with a nn-$P_2(\cos\theta_{ij})$ potential. Orientational order and internal energy were
determined for different field strengths. Comparisons were made with predictions
from mean-field theory.

3.8 Conclusion and Outlook

It is now clear that Monte Carlo simulations of critical behavior have come of age. They can certainly play an important role in both the study of complex models and the comparison between theory and experiment. It is, however, important that practioners in this area remember that care must be taken to insure that effects of finite sampling time, finite-system size, etc., are carefully examined so that the conclusions extracted from the data are correct! The reader is therefore encouraged to review the work and pitfalls presented in [3.1] and to pay particular attention to the guidelines discussed by *Binder* in the first chapter of [3.158].

References

3.1   D.P. Landau: "Phase Diagrams of Mixtures and Magnetic Systems" and "Applications in Surface Physics", in *Monte Carlo Methods in Statistical Physics*, ed. by K. Binder, Topics Current Phys, Vol.7 (Springer, Berlin, Heidelberg, New York 1979) Chaps.3 and 9
3.2   R.H. Swendsen: "Monte Carlo Renormalization", in *Real-Space Renormalization*, ed. by T.W. Burkhardt, J.M.J. van Leeuwen, Topics Current Phys., Vol.30 (Springer, Berlin, Heidelberg, New York 1982) Chap.3
3.3   K. Binder, D.P. Landau: Surf. Sci. **61**, 577 (1976)
3.4   K. Binder, D.P. Landau: Surf. Sci. **108**, 503 (1981)
3.5   K. Binder, D.P. Landau: Phys. Rev. **B21**, 1921 (1980)
3.6   M. Suzuki: Prog. Theor. Phys. **51**, 1992 (1974)
3.7   E. Müller-Hartmann, J. Zittartz: Z. Phys. **B27**, 261 (1977)
3.8   R.J. Behm, K. Christmann, G. Ertl: Surf. Sci. **99**, 320 (1980)
3.9   D.P. Landau, K. Binder: Phys. Rev. **B31**, 5946 (1985)
3.10 E.D. Williams, S.L. Cunningham, W.H. Weinberg: J. Vac Sci. Technol **15**, 417 (1978)
3.11 E.D. Williams, S.L. Cunningham, W.H. Weinberg: J. Chem. Phys. **68**, 4688 (1978)
3.12 W.Y. Ching, D.L. Huber, M. Fishkis, M.G. Lagally: J. Vac. Sci. Technol. **15**, 653 (1978)
3.13 W.C. Ching, D.L. Huber, M.G. Lagally, G.-C. Wang: Surf. Sci. **77**, 550 (1978)
3.14 L.D. Roelofs, R.L. Park, T.L. Einstein: J. Vac. Sci. Technol. **16**, 478 (1979)
3.15 K. Binder, W. Kinzel, D.P. Landau: Surf. Sci. **117**, 232 (1982)
3.16 W. Kinzel, W. Selke, K. Binder: Surf. Sci. **121**, 13 (1982)
3.17 W. Selke, K. Binder, W. Kinzel: Surf. Sci. **125**, 74 (1983)
3.18 K. Binder, W. Kinzel, W. Selke: J. Magn. and Magn. Mater **31-34**, 1445 (1983)
3.19 L.C.A. Stoop: To be published
3.20 H.H. Lee, D.P. Landau: Phys. Rev. **B20**, 2893 (1979)
3.21 K. Wada, T. Tsukada, T. Ishikawa: J. Phys. Soc. Japan **51**, 1311 (1982)
3.22 S. Fujiki, S. Shutch, Y. Abe, S. Katsura: J. Phys. Soc. Japan **52**, 1531 (1983)
3.23 D.P. Landau: Phys. Rev. **B27**, 5604 (1983)
3.24 B. Mihura, D.P. Landau: Phys. Rev. Lett. **38**, 977 (1977)
3.25 K.K. Chin, D.P. Landau: To be published. See also:
       K.K. Chin: "Monte Carlo Study of a Triangular Ising-Lattice Gas Model";
       M.S. Thesis, University of Georgia (1982)
3.26 M. Schick, J.S. Walker, M. Wortis: Phys. Rev. **B16**, 2205 (1977)
3.27 S.-K. Ma: Phys. Rev. Lett. **37**, 461 (1976)
3.28 Z. Friedman, J. Felsteiner: Phys. Rev. **B15**, 5317 (1977)
3.29 S. D'Elia, H. Ceva: Phys. Rev. **B26**, 5207 (1982)
3.30 A.L. Lewis: Phys. Rev. **B16**, 1249 (1977)
3.31 R.H. Swendsen: Phys. Rev. Lett. **42**, 859 (1979)
3.32 R.H. Swendsen: Phys. Rev. **B20**, 2080 (1979)
3.33 R.H. Swendsen, S. Krinsky: Phys. Rev. Lett. **43**, 177 (1979)
3.34 J.M.J. van Leeuwen: Phys. Rev. Lett. **34**, 1056 (1976)

3.35 K. Binder: Phys. Rev. Lett. **47**, 693 (1981); and Z. Phys. B**43**, 119 (1981)
3.36 D.P. Landau, R.H. Swendsen: Phys. Rev. B**30**, 2787 (1984)
3.37 D.P. Landau: Phys. Rev. B**21**, 1285 (1980)
3.38 D.P. Landau, J. Tombrello, R.H. Swendsen: In *Ordering in Two-Dimensions*, ed. by S.K. Sinha (North Holland, Amsterdam 1980) p.351
3.39 D.P. Landau, R.H. Swendsen: Phys. Rev. Lett. **46**, 1437 (1981)
3.40 M.A. Novotny, D.P. Landau, R.H. Swendsen: Phys. Rev. B**26**, 330 (1980)
3.41 R.M. Hornreich, R. Liebman, H.G. Schuster, W. Selke: Z. Phys. B**35**, 91 (1979)
3.42 W. Selke, M.E. Fisher: J. Mag. Magn. Mater. **15-18**, 403 (1980)
3.43 W. Selke, M.E. Fisher: Z. Physik B**40**, 71 (1980)
3.44 W. Selke: Z. Phys. B**43**, 335 (1981)
3.45 M. Barber, W. Selke: J. Phys. A**15**, L617 (1982)
3.46 Y. Saito: Phys. Rev. B**24**, 6652 (1981)
3.47 Y. Saito: J. Magn. and Magn. Mater. **31-34**, 1049 (1983)
3.48 K. Binder, P.C. Hohenberg: Phys. Rev. B**9**, 2194 (1974)
3.49 K. Binder, D.P. Landau: Phys. Rev. Lett. **52**, 318 (1984); see also in: *Phase Transitions and Critical Phenomena*, Vol.8, ed. by C. Domb and J.L. Lebowitz (Academic, London 1983) p.1
3.50 D.B. Abraham, E.R. Smith: J. Phys. A**14**, L193 (1981)
3.51 D.B. Abraham, E.R. Smith: J. Phys. A**14**, 2059 (1981)
3.52 W. Selke, W. Pesch: Z. Phys. B**47**, 335 (1982)
3.53 W. Selke, D.A. Huse: Z. für Physik B**50**, 113 (1983)
3.54 W. Selke, J. Yeomans: J. Phys. A**16**, 2789 (1983)
3.55 K. Binder: Phys. Rev. A**25**, 1699 (1982)
3.56 W.J. Shugard, J.D. Weeks, G.H. Gilmer: Phys. Rev. B**25**, 2022 (1982)
3.57 W.J. Shugard, J.D. Weeks, G.H. Gilmer: Phys. Rev. Lett. **41**, 1399 (1982)
3.58 R.H. Swendsen: Phys. Rev. B**25**, 2019 (1982)
3.59 M.J. de Oliveira, R.B. Griffiths: Surf. Sci. **71**, 687 (1978)
3.60 C. Ebner: Phys. Rev. **23**, 1925 (1981)
3.61 I.M. Kim, D.P. Landau: Surf. Sci. **110**, 415 (1981)
3.62 Z. Rácz, R. Ruján: Z. Phys. B**28**, 287 (1977)
3.63 H.W.J. Blöte, R.H. Swendsen: Phys. Rev. B**20**, 2077 (1979)
3.64 R.H. Swendsen: Phys. Rev. Lett. **47**, 1159 (1981)
3.65 S.J. Knak Jensen, O.G. Mouritsen: J. Phys. A**15**, 2631 (1982)
3.66 B.A. Freedman, G.A. Baker: J. Phys. A**15**, L715 (1982)
3.67 H.J. Herrmann, E.B. Rasmussen, D.P. Landau: J. Appl. Phys. **53**, 7994 (1982)
3.68 H.J. Herrmann, D.P. Landau: To be published
3.69 O.G. Mouritsen, S.J. Knak Jensen, B. Frank: Phys. Rev. B**23**, 967 (1981)
3.70 O.G. Mouritsen, S.J. Knak Jensen, B. Frank: Phys. Rev. B**24**, 347 (1981)
3.71 B.K. Chakrabarti, N. Bhattacharyya, S.K. Sinha: J. Phys. C**15**, L777 (1982)
3.72 B.K. Chakrabarti: J. Phys. C**15**, L1195 (1982)
3.73 K. Binder: Phys. Rev. Lett. **45**, 811 (1980)
3.74 M.K. Phani, J.L. Lebowitz, M.H. Kalos, C.C. Tsai: Phys. Rev. Lett. **42**, 577 (1979)
3.75 K. Binder, J.L. Lebowitz, M.K. Phani, M. Kalos: Acta Metallurgica **29**, 1655 (1981); U. Gahn: J. Phys. Chem. Solids **43**, 977 (1981)
3.76 K. Binder: Z. Phys. B**45**, 61 (1981)
3.77 T.V. Piragene, V.E. Shneider: Sov. Phys. Solid State **21**, 1742 (1979)
3.78 J.R. Banavar, D. Jasnow, D.P. Landau: Phys. Rev. B**20**, 3820 (1979)
3.79 A.K. Jain, D.P. Landau: Phys. Rev. B**22**, 445 (1980)
3.80 T. Graim, D.P. Landau: Phys. Rev. B**24**, 5156 (1981)
3.81 A.B. Bortz, M.H. Kalos, J.L. Lebowitz: J. Comput. Phys. **17**, 10 (1975)
3.82 H.D. Maier, H.E. Müser, J. Petersson: Z. Phys. B**46**, 251 (1982)
3.83 S. Dietrich, H. Wagner: Z. Phys. B**36**, 121 (1979)
3.84 R.A. Bond, D.K. Ross: J. Phys. F**12**, 597 (1982)
3.85 R. Kretschmer, K. Binder: Z. Phys. B**34**, 375 (1979)
3.86 G.S. Pawley, J.E. Tibballs: Ferroelectrics Lett. **44**, 33 (1982)
3.87 W. Selke: Z. Phys. B**29**, 133 (1978)
3.88 W. Selke: Sol. State Comm. **27**, 1417 (1978)
3.89 W. Selke, M.E. Fisher: Phys. Rev. B**20**, 257 (1979)

3.90 K. Binder: J. Stat. Phys. **24**, 69 (1981)
3.91 C. Rebbi, R.H. Swendsen: Phys. Rev. B**21**, 4094 (1980)
3.92 S.J. Knak Jensen, O.G. Mouritsen: Phys. Rev. Lett. **43**, 1736 (1979)
3.93 H.J. Herrmann: Z. Phys. B**35**, 171 (1979)
3.94 H.W.J. Blöte, R.H. Swendsen: J. Appl. Phys. **50**, 7382 (1979)
3.95 H.W.J. Blöte, R.H. Swendsen: J. Magn. Mag. Mat. **15-18**, 399 (1980)
3.96 F. Fucito, A. Vulpiani: Physics Lett. **89**A, 33 (1982)
3.97 W. Kinzel, W. Selke, F.Y. Wu: J. Phys. A**14**, L399 (1981)
3.98 W. Selke, J.M. Yeomans: Z. Phys. B**46**, 311 (1982)
3.99 Y. Saito: J. Phys. A**15**, 1885 (1982)
3.100 G.S. Grest, J.R. Banavar: Phys. Rev. Lett. **46**, 1458 (1981)
3.101 J.R. Banavar, G.S. Grest, D. Jasnow: Phys. Rev. Lett. **45**, 1424 (1980)
3.102 R.H. Swendsen, D. Andelman, A.N. Berker: Phys. Rev. B**24**, 6732 (1981)
3.103 A.N. Berker, D. Andelman: J. Appl. Phys. **53**, 7923 (1982)
3.104 N. Novotny: J. Appl. Phys. **53**, 7997 (1982)
3.105 S. Miyashita, H. Nishimori, A. Kuroda, M. Suzuki: Prog. Theor. Phys. **60**, 1669 (1978)
3.106 J. Tobochnik, G.V. Chester: Phys. Rev. B**20**, 3761 (1979)
3.107 J.E. Van Himbergen, S. Chakravarty: Phys. Rev. B**23**, 359 (1981)
3.108 H. Betsuyaku: Physica **106**A, 311 (1981)
3.109 D.P. Landau: J. Mag. Mag. Mat. **31-34**, 1115 (1983)
3.110 R.H. Swendsen: Phys. Rev. Lett. **49**, 1302 (1982)
3.111 C. Kawabata, A.R. Bishop: Solid State Commun. **33**, 453 (1980)
3.112 D.P. Landau, K. Binder: Phys. Rev. B**24**, 1391 (1981)
3.113 J. Tobochnik: To be published. The published version of this manuscript includes changes, see Phys. Rev. B**26**, 6201 (1982), and figures in this published version were corrected in Phys. Rev. B**27**, 6972 (1983). Correct results for the q=6 model can be found in M.S.S. Challa and D.P. Landau, Phys. Rev. B**33**, 437 (1986)
3.114 W. Selke: J. Magn. Mag. Mat. **9**, 7 (1978)
3.115 W. Selke: Sol. State Commun. **27**, 1417 (1978)
3.116 W. Selke: J. Phys. C**13**, L261 (1980)
3.117 M. Suzuki: *Proc. Int. Conf. on Frontiers of Theoretical Physics*, ed. by F.C. Auluck, L.S. Kothari, V.S. Nanda (Macmillan Co. of India, Delhi)
3.118 See, e.g., E. Stoll, K. Binder, T. Schneider: Phys. Rev. B**8**, 3266 (1973)
3.119 A.L. Katz, D.P. Landau, J.D. Gunton: J. Phys. A**12**, L299 (1979)
3.120 J. Tobochnik, S. Sarker, R. Cordery: Phys. Rev. Lett. **46**, 1417 (1981)
3.121 M.C. Yalabik, J.D. Gunton: Phys. Rev. B**25**, 534 (1982)
3.122 S.L. Katz, J.D. Gunton, C.P. Liu: Phys. Rev. B**25**, 6008 (1982)
3.123 J.C. Angles D'Auriac, R. Maynard, R. Rammal: To be published
3.124 C. Kalle, V. Winkelmann: J. Statist. Phys. (to be published)
3.125 M.A. Novotny, D.P. Landau: Bull. Am. Phys. Soc. **24**, 361 (1979)
3.126 M.A. Notvotny, D.P. Landau: To be published. The published manuscript, Phys. Rev. B**32**, 5874 (1985) showed that the data were consistent with a value of $Z\approx2.75$ over a wide range of $(T-T_c)$
3.127 R. Serinko, H.J. Herrmann, D.P. Landau: To be published
3.128 O.G. Mouritsen, S.J. Knak Jensen: Phys. Rev. B**19**, 3663 (1979)
3.129 O.G. Mouritsen, S.J. Knak Jensen: J. Phys. A**12**, L339 (1979)
3.130 H.W.J. Blöte, R.H. Swendsen: Phys. Rev. B**22**, 4481 (1980)
3.131 R.V. Ditzian, J.R. Banavar, G.S. Grest, L.P. Kadanoff: Phys. Rev. B**22**, 2542 (1980)
3.132 M.J. Sablik, M. Cook, N. Vuong, L. Phani, M. Zawonski, M. Phani: J. Appl. Phys. **50**, 7385 (1979)
3.133 A. Weinkauf, J. Zittartz: Z. Phys. B**45**, 223 (1982)
3.134 D.P. Landau: Phys. Rev. B**22**, 2450 (1980)
3.135 A.B. Harris: J. Phys. C**7**, 1671 (1974)
3.136 M.A. Novotny, D.P. Landau: J. Magn. Mag. Mat. **15-18**, 247 (1980)
3.137 M.A. Novotny, D.P. Landau: Phys. Rev. B**24**, 1468 (1981)
3.138 M.A. Novotny, D.P. Landau: To be published
3.139 C. Kawabata: Physics Lett. **77**A, 181 (1980)
3.140 C. Dasgupta, B.I. Halperin: Phys. Rev. Lett. **47**, 1556 (1981)

3.141 C. Ebner, D. Stroud: Phys. Rev. B**25**, 5711 (1982)
3.142 M.J. Mandell: J. Chem. Phys. **60**, 1432 (1974)
3.143 M.J. Mandell: J. Chem. Phys. **60**, 4880 (1974)
3.144 S.F. O'Shea: J. Chem. Phys. **68**, 5435 (1978)
3.145 S.F. O'Shea, M.L. Klein: J. Chem. Phys. **71**, 2399 (1979)
3.146 S.F. O'Shea, M.L. Klein: Chem. Phys. Lett. **66**, 381 (1979)
3.147 K. Maki, S.F. O'Shea: J. Chem. Phys. **73**, 3358 (1980)
3.148 S.F. O'Shea, M.L. Klein: Phys. Rev. B**25**, 5882 (1982)
3.149 M.A. Klenin, S.F. Pate: Phys. Rev. B**26**, 3969 (1982)
3.150 J. Vieillard-Baron: J. Chem. Phys. **56**, 4729 (1972)
3.151 J. Vieillard-Baron: Mol. Phys. **28**, 809 (1974)
3.152 H.J.F. Jansen, G. Vertogen, J.G.J. Ypma: Mol. Cryst. Liq. Cryst. **38**, 87 (1977)
3.153 P.A. Lebwohl, G. Lasher: Phys. Rev. A**6**, 426 (1977)
3.154 G.R. Luckhurst, S. Romano: Mol. Phys. **40**, 129 (1980)
3.155 J.Y. Denham, G.R. Luckhurst, C. Zannoni, J.W. Lewis: Mol. Cryst. Liq. Cryst. **60**, 185 (1980)
3.156 G.R. Luckhurst, S. Romano: Proc. R. Soc. Lond. A**373**, 111 (1980)
3.157 G.R. Luckhurst, P. Simpson, C. Zannoni: Chem. Phys. Lett. **78**, 429 (1981)
3.158 K. Binder: "Introduction: Theory and 'Technical' Aspects of Monte Carlo Simulations", in *Monte Carlo Methods in Statistical Physics*, ed. by K. Binder, Topics Current Phys., Vol.7 (Springer, Berlin, Heidelberg, New York 1979) Chap.1

4. Few- and Many-Fermion Problems

K.E.Schmidt[+] and M.H.Kalos[*]

With 7 Figures

The success of Monte Carlo methods such as the Green's function Monte Carlo method (GFMC) [4.1-8] in calculating the ground-state properties of many-body systems has led to their application to Fermi systems. For boson systems, ground-state properties may be calculated with only statistical errors. These Bose calculations have been done for large numbers of particles (N > 100) and in three spatial dimensions. Although some exact fermion calculations have been done [4.9-14], these calculations have generally been in one spatial dimension, where the fermion system can be mapped onto an equivalent Bose system, or with only a few particles (N ~ 3), where the difficulties associated with fermion systems can be controlled.

Monte Carlo calculations of fermion states are less straightforward than those for bosons, a problem which may be traced to the necessary antisymmetry of their wave functions. Consequently, we are forced to deal with wave functions which are not positive definite and are therefore not directly interpretable as probability densities. Since the wave function may be negative, technical complications connected with cancellation of positive and negative terms arise. We refer collectively to these difficulties, peculiar to Monte Carlo methods, as the "fermion problem". Closely associated with this problem is the fact that GFMC methods normally solve for the lowest energy state of a many-body Hamiltonian.

Another, purely technical, difficulty would remain even if the above problems were solved. The simplest nontrivial antisymmetric function is a determinant. For many-body systems, the evaluation of this determinant is a significant or dominant part of the calculational effort, a fact that controls the number of particles that can be modeled on a given computer. However, for nonrelativistic many-body systems, this limit is not serious and calculations with N > 100 are possible.

An exposition of the technical barriers to the solution of fermion problems is given in the following sections. Section 4.1 reviews the GFMC method including recent reformulations. The GFMC method is developed in imaginary time formalism and related to the time-integrated form as reviewed by *Ceperley* and *Kalos* [4.1]. The explicit use of imaginary time can be technically useful and helps relate the

[*]This work was supported by the U.S. Department of Energy under Grant AC 02-79ER10353.
[+]Work supported under the auspices of the U.S. Department of Energy.

Green's function methods to path integral methods. Section 4.2 discusses the short-time approximation sometimes used and its relationship to exact methods. The special difficulties associated with fermion Monte Carlo methods along with the method of transient estimation are explained in detail in Sect.4.3. The fixed-node approximation is discussed in Sect.4.4. In Sect.4.5, we discuss an exact method for few-fermion systems. Some speculations on possible treatments in the general case are given in Sect.4.6. We shall deal almost exclusively with the fermion ground state. Some interesting recent applications to quantum spin systems [4.9] will therefore not be discussed. Extension to finite temperatures [4.10,14,15] is generally straightforward; however, some of our conclusions for the ground state must be generalized with care when applied at finite temperatures.

4.1 Review of the GFMC Method

The GFMC method was described in [4.1] and elsewhere [4.2-6]. In this section we briefly review the GFMC method considering the Green's function as the imaginary time development operator. The connection between this formalism and the standard GFMC described in [4.1] will be shown.

We begin with the nonrelativistic N-body Hamiltonian in three spatial dimensions:

$$H = \frac{-\hbar^2}{2m} \sum_i \nabla_i^2 + V(R) \quad . \tag{4.1}$$

Here, $V(R)$ is the potential given by

$$V(R) = \sum_i V_1(r_i) + \sum_{i<j} V_2(r_{ij}) + \dots \quad , \tag{4.2}$$

where V_1, V_2 are one- and two-body potentials, and R represents the 3N coordinates of the N particles. It is convenient to define the 3N-dimensional Laplacian as

$$\nabla_R^2 = \sum_i \nabla_i^2 \tag{4.3}$$

and $\hbar^2/2m = 1$.

In the following we assume all particles identical. The generalization to other situations is straightforward. We shall deal exclusively with coordinate-space calculations which have the advantage that the Hamiltonian is local in the coordinate representation. However, we are aware that the nonlocality of other bases may help in resolving the fermion problem in some cases.

The Schrödinger equation can be written as

$$H\Psi = i \frac{\partial \Psi}{\partial t} \quad . \tag{4.4}$$

If we replace i times t by τ, an imaginary time variable, the Schrödinger equation has the structure of a diffusion equation [4.16-23],

$$H\Psi = \frac{-\partial \Psi}{\partial \tau} \quad . \tag{4.5}$$

Formally, the solution is given by

$$\Psi(R,\tau) = \exp(-H\tau)\Psi(R,0) \quad . \tag{4.6}$$

If $\Psi(R,0)$ is expanded in the eigenstates of H,

$$H\psi_n = E_n\psi_n \; , \; E_0 < E_1 < E_2 \cdots \; , \tag{4.7}$$

$$\Psi(R,0) = \sum_n a_n\psi_n(R) \quad , \tag{4.8}$$

then $\Psi(R,\tau)$ is given by

$$\Psi(R,\tau) = \sum_n a_n \, e^{-E_n\tau}\psi_n(R) \quad , \tag{4.9}$$

and for large τ:

$$\Psi(R,\tau \to \infty) = a_0 \, e^{-E_0\tau}[\psi_0(R) + O(e^{-(E_1-E_0)\tau})]$$

$$= a_0 \, e^{-E_0\tau}\psi_0(R) \quad . \tag{4.10}$$

In (4.10) $\psi_0(R)$ is that eigenstate of H with the lowest eigenvalue E_0 and nonzero overlap with the function $\Psi(R,0)$. It should be noted that if $\Psi(R,0)$ is chosen to be antisymmetric, only those a_n corresponding to antisymmetric ψ_n will be nonzero, and the large τ solution (4.10) will necessarily be antisymmetric.

The connection between the long-time behavior of a diffusion equation and the Schrödinger equation was apparently first suggested by *Fermi* (cf. *Metropolis* and *Ulam* [4.16]). They also suggested that the diffusion process could be simulated by Monte Carlo methods.

The GFMC method is simply one way of simulating the diffusion equation (4.5) on a computer. Several points about the structure of (4.5) should be kept in mind: (i) Ψ and not ψ^2 corresponds to the density of the diffusing objects. (ii) These objects are not the physical particles that make up the many-body system. Rather, they are pointers or markers that move in a 3N-dimensional space. The position of one of these objects — henceforth called "walkers" — in the 3N-dimensional space gives the positions of all the real particles in three-dimensional space. (iii) The potential $V(R)$ is an external influence upon these walkers and gives the creation or absorption probabilities per unit time in our diffusion process. The walkers do not interact with each other, i.e., the Schrödinger equation is linear. (iv) The classical diffusion equation models the behavior of diffusing particles by a continuum density. We shall model a continuum wave function as the average density of an ensemble of our abstract walkers.

Table 4.1 compares solving a classical diffusion problem and solving the Schrödinger equation in imaginary time. In a classical problem, say, of perfume diffusing from a bottle to fill a room, a straightforward approach might be to write

Table 4.1. Comparison between solving a classical diffusion problem, and the Schrödinger equation in imaginary time

Deterministic *Dynamics*	Langevin *Dynamics*	Differential *Equation*
Newtonian dynamics of particles and medium →	Medium replaced by a random force and absorption. Fluctuations present →	Diffusion equation (Continuum model with no fluctuations)
	Continuum replaced by a set of walkers under the influence of a random force and absorption. Fluctuations present (GFMC) ←	Schrödinger equation (Continuum model with no fluctuations)

down Newton's equations for the perfume molecules and the molecules of the medium through which they diffuse. If all the interactions as well as the initial conditions were known, the density of diffusing particles would be determined by solving Newton's equations as shown in the first column of Table 4.1.

These equations may be approximated by replacing the medium by a random force and an absorption term (if for example the room had activated charcoal walls), and by assuming that collisions among the perfume molecules are unimportant. This would lead to Langevin dynamics for the perfume molecules as shown in the second column of Table 4.1. Finally, if a macroscopic description is desired, we can go to a continuum limit and recover the diffusion equation as shown in the third column. An important difference between the diffusion equation and Langevin dynamics is that the former contains no density fluctuations at equilibrium, while the latter will always contain density fluctuations for a finite number of particles.

The GFMC method can be thought of as taking the reverse path from the classical case. Here we start with a continuum wave function with no fluctuations, i.e., the wave function has a unique value at a given value of **R**. In the GFMC method, we replace this continuum by a set of walkers moving under the influence of a random force and absorption. Just as in the classical case, this necessarily introduces statistical fluctuations into the numerical solution. These fluctuations can be reduced by simply taking a large ensemble of walkers.

To develop the GFMC method formally, we first identify the imaginary time development operator in coordinate space with the Green's function

$$<R_2|e^{-H\tau}|R_1> = G(R_2, R_1; \tau)$$

$$= \sum_n \psi_n(R_2)\psi_n^*(R_1) \; e^{-E_n\tau} \quad . \tag{4.11}$$

The Green's function at τ is the solution at τ for the diffusion from a point source at $\tau = 0$

$$H_{R_2} G(R_2,R_1;\tau) = \frac{-\partial G}{\partial \tau}(R_2,R_1;\tau) \quad , \tag{4.12}$$

$$G(R_2,R_1;0) = \delta(R_2 - R_1) \quad . \tag{4.13}$$

The wave function is then given by

$$\Psi(R_2,\tau_2) = \int dR_1 G(R_2,R_1;\tau_2 - \tau_1)\Psi(R_1,\tau_1) \quad . \tag{4.14}$$

If we can obtain an expression for G that can be used in (4.14), then we can solve Ψ at arbitrary τ.

Before obtaining the necessary equations for G, the calculation of the ground-state energy must be considered. There are two related ways to calculate this energy. First, it may be calculated using (4.10). The normalization of the wave function will be exponential in τ for large τ. This normalization is proportional to the number of walkers present, so that a "growth" estimate of E may be inferred from the growth or decline in the population size. The second method is called a "mixed" estimate. It can be derived from $E(\tau)$, the expectation value of H at time τ:

$$E(\tau) = \frac{\int dR \ \Psi^*(R,\tau)H\Psi(R,\tau)}{\int dR \ \Psi^2(R,\tau)} \quad . \tag{4.15}$$

We can substitute for $\Psi(R,\tau)$ using (4.6) and get

$$E(\tau) = \frac{\int dR \ \Psi^*(R,0) \ e^{-H\tau} \ H \ e^{-H\tau} \ \Psi(R,0)}{\int dR \ \Psi^*(R,0) \ e^{-H\tau} \ e^{-H\tau} \ \Psi(R,0)} \quad . \tag{4.16}$$

Using the property that H and $e^{-H\tau}$ are Hermitian,

$$E(\tau) = \frac{\int dR \ H\Psi^*(R,0)\Psi(R,2\tau)}{\int dR \ \Psi^*(R,0)\Psi(R,2\tau)} \tag{4.17}$$

$$\equiv E_{mixed}(2\tau) \quad .$$

The mixed estimate is particularly useful since $\Psi(R,\tau)$ is given only by the positions of the walkers. That is, $\Psi(R,\tau)$ is approximated by a sum of delta functions at each of the positions of the walkers. However, the initial wave function $\Psi(R,0)$ is usually an analytic trial function. The value of $H\Psi^*(R,0)$ is therefore easier to calculate than the value of $H\Psi^*(R,\tau)$. It is obvious from (4.16) that $E_{mixed}(\tau)$ is an upper bound to the ground-state energy for all τ and equal to the ground-state energy for large τ [4.25].

The growth estimate is the logarithmic derivative of the normalization with respect to τ:

$$E_{growth}(\tau) = \frac{(-\partial/\partial \tau) \int dR \ \Psi^*(R,0) \ \Psi(R,\tau)}{\int dR \ \Psi^*(R,0)\Psi(R,\tau)} \quad . \tag{4.18}$$

129

Using (4.6), it is easy to show the equality of the growth and mixed estimates. They are not statistically independent calculations of energy.

It should be noted that a constant energy may be added to the Hamiltonian without changing the physics. In particular, it is useful to add a constant such that the value of E_0 is approximately zero, so that the normalization will remain approximately constant.

We have now shown that given G, we can calculate the ground-state energy. But G is not known in general. Therefore, an expansion of G about some set of known Green's functions is used, which can be carried out very much as described in [4.1]. A simple formal example follows. We take a known Green's function to be the Green's function G_U for the Hamiltonian H_U. Following (4.12) we get

$$H_U \exp(-H_U \tau_1) = \frac{-\partial}{\partial \tau_1} \exp(-H_U \tau_1) \quad , \tag{4.19}$$

$$H \exp[-H(\tau_2 - \tau_1)] = \frac{\partial}{\partial \tau_1} \exp[-H(\tau_2 - \tau_1)] \quad . \tag{4.20}$$

We multiply (4.19) on the left by $\exp[-H(\tau_2 - \tau_1)]$ and (4.20) on the right by $\exp(-H_U \tau_1)$, substract the two resulting equations, and integrate over τ_1 from 0 to τ_2. This gives

$$\exp(-H\tau_2) = \exp(-H_U \tau_2) + \int_0^{\tau_2} d\tau_1 \exp[-H(\tau_2 - \tau_1)](H_U - H) \exp(-H_U \tau_1) \quad , \tag{4.21}$$

which is an expansion of $\exp(-H\tau)$ in terms of known quantities.

In the more general case, the Green's function G_U is defined to be zero outside some domain $D(R)$, which introduces surface terms into the equation. Furthermore, the base Green's functions will generally be different for different domains, and the domains can be chosen as a function of position as explained in [4.1,6]. A more general expansion equivalent to that used in [4.1] is

$$G(R_2, R_1; \tau) = G_{U(R_1)}(R_2, R_1; \tau)$$

$$+ \int_{D(R_1)} dR_3 \int_0^\tau d\tau' \, G(R_2, R_3; \tau - \tau')[U(R_1) - V(R_3)]G_{U(R_1)}(R_3, R_1; \tau')$$

$$+ \int_{S(R_1)} dR_3 \int_0^\tau d\tau' \, G(R_2, R_3; \tau - \tau')[-\nabla_N G_{U(R_1)}(R_3, R_1; \tau')] \quad . \tag{4.22}$$

Here the integrals are respectively over the volume and surface of the domain $D(R_1)$. The Hamiltonian H_U is taken to be

$$H_U = -\nabla_R^2 + U \quad , \tag{4.23}$$

where U is a constant inside the domain. Both U and the domain D are functions of the initial point R_1. The ∇_N term in (4.22) indicates the normal gradient to the surface.

More general expansions can be constructed by making U a function of coordinates and of τ, possibly by establishing a good approximation to G and calculating the necessary U to give this G_U. Some work in this direction has been carried out by *Ceperley* [4.25] for Coulomb systems.

Equations (4.14,22) completely determine the wave function from the known potentials and the G_U's. The Monte Carlo method is most easily applied if the U's and domains D are chosen such that $U(R_1) - V(R_3) > 0$ for all R_3 in (4.22). The terms in (4.22) are then positive definite and can be used to define probabilities as in [4.1].

Importance sampling techniques are very important when applying the Monte Carlo method. Since these methods are not specifically needed in understanding the GFMC method as applied to fermions, we refer the interested reader to [4.1,6].

The imaginary-time formalism is very simply related to the standard formalism given in [4.1]. If we are interested only in behavior for large times, we can propagate the wave function any amount at each step. In particular, a time-integrated Green's function G can be defined as

$$G(R_2,R_1) = \int_0^\infty G(R_2,R_1;\tau)d\tau \quad , \tag{4.24}$$

which can formally be written as

$$\int_0^\infty d\tau \, e^{-H\tau} = H^{-1} \quad . \tag{4.25}$$

Clearly the eigenvalues of H must all be positive for (4.24,25) to be well defined. This can always be arranged by adding a constant to H as explained earlier. An integral equation for $\psi(R)$ can then be written as

$$\psi_0 = E_0 H^{-1} \psi_0 \quad . \tag{4.26}$$

and an iterative equation is given by

$$\psi^{n+1}(R) = E_T \int dR' \, G(R,R')\psi^n(R') \quad , \tag{4.27}$$

where E_T is a trial value for E_0. Equation (4.27) can converge to $\psi_0(R)$ for large n by applying the same techniques as in (4.10) and the preceding equations. Further, by integrating (4.22) over τ_2 from 0 to ∞ the expansion of G given in [4.1] is recovered.

The standard GFMC equations are thus seen to be the time-integrated form of the time-dependent equations given above. This same duality appears in field theoretic methods where formalisms for the Green's function [4.26] or transfer matrix [4.27] (time-dependent form) and resolvent operators [4.28] (time-integrated form) appear. For ground-state calculations the choice between the two formalisms is only a matter of taste, convenience, and possibly computing efficiency. In fact many of the calculations using the time-integrated form actually calculate the time-dependent quantities and integrate using the Monte Carlo method [4.6]. A time-dependent formalism may have some advantages for calculating time-dependent correlations. Since the diffusion equation for the Green's function (4.12) is identical with the Bloch equation [4.15,29] or the density matrix when τ is identified with $(kT)^{-1}$ (where T is the temperature and k is Boltzmann's constant), the time-dependent formalism can be thought of as generating the $T \to 0$ limit of the Bloch equation.

To get useful information at finite temperature, a trace over the density matrix is required. This corresponds to taking periodic boundary conditions in the time variable, rather than simply weighting with an antisymmetric trial function at the end points. Either the trace must be taken only over fermion (i.e., antisymmetric) configurations, or the density matrix must be antisymmetrized [4.9,10,14]. This leads to some additional complications in calculations at finite temperatures.

4.2 The Short Time Approximation

A popular approach to solving (4.14,22) uses the short-time approximation. To extract this approximation from (4.22), the following steps are taken: (1) the domains D(R) are taken to be all space, so that the surface integral vanishes; (2) $U(R_1)$ is taken to be $V(R_1)$; (3) τ is made small; (4) the volume integral in (4.22) is neglected. It should be pointed out that only the last step necessarily introduces an approximation, although Steps (1) and (2) will usually destroy the positive definite character of the equation and make Monte Carlo simulation more difficult.

With these steps taken, (4.22) becomes

$$G(R_2,R_1;\tau) = G_{U(R_1)}(R_2,R_1;\tau) \quad . \tag{4.28}$$

The Green's function G_U for an infinite domain with $U = V(R_1)$ is given by

$$G_{U(R_1)}(R_2,R_1;\tau) = \frac{1}{(4\pi\tau)^{3N/2}} \exp[-(R_2 - R_1)^2/4\tau - V(R_1)\tau] \quad . \tag{4.29}$$

If τ is small, the Gaussian term in (4.29) restricts the magnitude $|R_2 - R_1|$ to values of order $\tau^{\frac{1}{2}}$. Expanding in $V(R_3)$ in (4.22) about R_1 gives

$$V(R_3) = V(R_1) + \nabla_{R_1} V(R_1) \cdot (R_3 - R_1) + \ldots \tag{4.30}$$

$$= V(R_1) + O(\tau^{\frac{1}{2}}) \quad .$$

Thus the short time approximation using (4.28) gives an error $O(\tau^{\frac{1}{2}})$ for most potentials, and will give the correct answer only in the limit $\tau \to 0$. The error in the time step can be made smaller by a more judicious choice of U in (4.28). However, neglecting the volume integral term in (4.22) will always require the limit $\tau \to 0$.

A more usual method for deriving the short time approximation employs the Trotter formula for G

$$G(R_2,R_1;\tau) = <R_2|e^{-H\tau}|R_1> \quad , \tag{4.31}$$

which can be written as

$$<R_2|e^{-H\tau}|R_1> = <R_2|e^{-(p^2+V)\tau}|R_1> \tag{4.32}$$

$$= <R_2|e^{-p^2\tau} e^{-V\tau} \exp\{0([p^2,V]\tau^2)\}|R_1> \quad . \tag{4.33}$$

If the commutator terms are ignored, then (4.28,29) are recovered. Examination of the commutator terms shows the same structure as (4.30) and the error term is $0(\tau^{\frac{1}{2}})$ here as well.

The main virtue of the short-time approximation is its simplicity both conceptually and in application. A price is paid for this simplicity in that the extrapolation to $\tau \to 0$ is difficult and may cost more computer time than doing the calculation with the full expansion for G. It would appear to be fruitful to use correlated sampling or reweighting techniques [4.1] to calculate this time step error.

The short-time approximation gives the most straightforward relation between path integrals in imaginary time and GFMC. This is most easily seen by iterating (4.14) many times. The wave function at large imaginary time T becomes

$$\Psi(R,T) = \int \prod_{c=1}^{\ell-1} dR_i \ G(R,R_1;T/\ell)G(R_1,R_2;T/\ell) \cdots$$

$$G(R_{\ell-1},R_\ell;T/\ell) \ \Psi(R_\ell,0) \quad . \tag{4.34}$$

In the limit $\ell \to \infty$, (4.34) becomes a path integral from R_ℓ to R [4.29]. The Green's function gives the weighting factors of the action, and the integrals over the R_i sum over all paths.

Within the short-time approximation, (4.34) gives an expression for $\Psi(R,T)$ that is simply a many-dimensional integral over a product of analytically known functions. This integral is often evaluated using the Monte Carlo method together with the sampling algorithm of *Metropolis* et al. [4.30]. Naive application of this procedure will generally be slower than the methods given in [4.1,2,6].

As might be expected, an approximation like the short-time approximation would be defined for the standard GFMC method. To restrict the integral in (4.25) to small times, a large constant E_c could be added to the Hamiltonian. The same steps given above could then be applied to the time-integrated Green's function equation yielding a good approximation in the limit $E_c \to \infty$.

4.3 The Fermion Problem and the Method of Transient Estimation

It was mentioned in the introduction that the root cause of the difficulties associated with Monte Carlo calculations for fermion systems is that the GFMC equations cannot be made positive definite. To gain some insight into the origin and effects of this complication, let us look at a projection method which is often called "transient estimation".

One may regain a set of equations which will contain only positive definite quantities as follows. Divide the initial wave function explicitly into positive and negative contributions:

$$\Psi(R,0) = \psi_T^A(R) = \psi^+(R,0) - \psi^-(R,0) \quad . \tag{4.35}$$

The antisymmetric function $\psi_T^A(R)$ is taken to be the starting function. Clearly there are infinitely many ways to split $\psi_T^A(R)$ into positive and negative pieces. The obvious choice is to take

$$\psi^+(R,0) = \frac{1}{2}\left[\psi_T^A(R) + |\psi_T^A(R)|\right] \quad , \tag{4.36}$$

and

$$\psi^-(R,0) = -\frac{1}{2}\left[\psi_T^A(R) - |\psi_T^A(R)|\right] \quad . \tag{4.37}$$

Both $\psi^+(R,0)$ and $\psi^-(R,0)$ are positive definite and can be evolved in imaginary time using (4.14). The Green's function equation is unchanged; all needed quantities are positive definite and can be interpreted as probabilities. The Monte Carlo method may then be applied as before. The analysis below (4.10) indicates that only the eigenstate of H with the lowest energy that has nonzero overlap with $\psi_T^A(R)$ will survive. Since $\psi_T^A(R)$ is explicitly antisymmetric, it is orthogonal to all states with energy lower than the Fermi ground state. Therefore, at large τ, $\Psi(R,\tau)$ defined by

$$\Psi(R,\tau) = \psi^+(R,\tau) - \psi^-(R,\tau) \tag{4.38}$$

will converge to the fermion ground state.

A difficulty arises since $\psi^+(R,0)$ and $\psi^-(R,0)$ separately have nonzero overlap with the Bose ground state (as well as with other states of mixed symmetry). Thus ψ^+ and ψ^- will converge independently to the Bose ground state. Schematically ψ^+ and ψ^- can be written as

$$\psi^\pm(R,\tau \to \infty) = \exp(-E_F\tau)\{\pm a_F\psi_F(R) + a_B\psi_B(R) \exp[-(E_B - E_F)\tau]\} \quad , \tag{4.39}$$

where the subscript F indicates the Fermi state and the subscript B indicates the Bose state (or any state of mixed symmetry). The mixed estimate for the energy is

$$E_{mixed}(\tau) = \frac{\int dR \ \psi_T^A(R)H\Psi(R,\tau)}{\int dR \ \psi_T^A(R)\Psi(R,\tau)} \quad , \tag{4.40}$$

which converges to E_F as $\tau \to \infty$. The Bose terms in (4.39) do not contribute to the mixed estimate (4.40). They do however, contribute to the variance which increases exponentially relative to the average value of the energy. The reason for this can

134

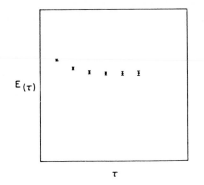 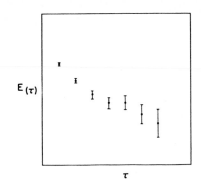

Fig.4.1. Schematic representation of a calculation using transient estimation where
the energy has converged before the error
bars have grown too large

Fig.4.2. Schematic representation of a
calculation using transient estimation.
The error bars have grown so large that
they mask convergence

be seen by examining (4.39). Since $E_B < E_F$, the Bose term grows exponentially compared to the Fermi term. The exact cancellation seen in (4.40) takes place only on
the average. The fluctuations in the Bose state eventually grow large enough to
overwhelm any antisymmetric signal in a background of symmetric noise. This is the
major difficulty encountered for these nonrelativistic Fermi systems.

It should be noted that this difficulty is completely disguised by the equations
as originally written, since the Monte Carlo method is unable to subtract one distribution from another. For example, it appears at first glance that the above
problem with (4.39) could be eliminated if only the distributions were made antisymmetric since this would eliminate the a_B term. This is not the case except in
one spatial dimension, as will be seen later in this section.

If the method of transient estimation is applied and the energy constant of the
Hamiltonian adjusted so that the average normalization of ψ^+ or ψ^- is maintained,
the calculation of the ground-state energy might behave as shown schematically in
Figs.4.1,2. In Fig.4.1 we show what might be termed a best case. Here it is apparent
that the energy has converged to the fermion value before the exponentially growing variance has taken over. Figure 4.2 shows the opposite and, unfortunately, more
normal case where the variance grows and obscures the convergence of the energy to
the fermion value.

The desired result of Fig.4.1 can occur under various circumstances. If, for
example, the starting wave function $\psi_T^A(R)$ is very close to the correct Fermi state,
then we might expect this behavior. Since the energy convergence depends mostly on
the low lying fermion excited states, which die out like $\exp[-(E_{Fex} - E_F)\tau]$, where
E_{Fex} is the energy of these states, and the variance grows like $\exp[-(E_F - E_B)\tau]$,
we might expect a result like Fig.4.1 if

$$\frac{E_F - E_B}{E_{Fex} - E_F} \ll 1 \quad . \tag{4.41}$$

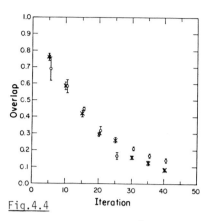

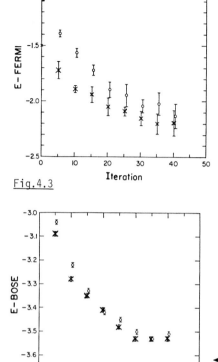

Fig.4.3

Fig.4.4

Fig.4.3. Energy of the ³He wave function calculated every five iterations of (4.27). The
circles are results from a Slater-Jastrow
wave function. The crosses are obtained using
a wave function including backflow and triplet
correlations. A slight lateral displacement
of one set of results has been made for additional clarity

Fig.4.4. Overlap of fermion wave functions
with trial wave functions. Symbol identification as in Fig.4.3

◀ Fig.4.5. Energy of the boson component of $\psi^{\pm}$
for the two choices of initial wave function.
The correct boson energy is -3.54 ± 0.01 K;
symbol identification as in Fig.4.3

This kind of energy level spacing, however, is extremely unlikely.

Results of a calculation of the energy of liquid ³He [4.24] are shown in Figs.
4.3-5 for two different starting wave functions. These calculations used the time-
integrated equations so the abscissa is the number of iterations of (4.27) rather
than the value of τ. Figure 4.3 shows the mixed energy. The growth of the variance
is not so noticeable since it wastes computer time to calculate the energy accurately for the initial iterations. Since most of the computer time was spent calculating the mixed estimate for the energy and not in iterating (4.27), far fewer
configurations were used to calculate the energy for the smaller number of iterations. Figures 4.4,5 show explicitly the convergence of ψ^+ and ψ^- separately to
the Bose ground state. Figure 4.4 in essence shows the amount of fermion signal remaining in the Bose noise. Figure 4.5 shows that ψ^+ or ψ^- separately predict the
correct answer for the Bose energy after 40 iterations.

A useful result can still be obtained from these calculations. Equations
(4.15-17) show that the energies given in Fig.4.3 are upper bounds to the true
Fermi ground-state energy at every point. By averaging the last three entries the
result for liquid ³He that $E_F < -2.20 \pm 0.05$ K at the experimental equilibrium density

was obtained. This is the best upper bound to the ground-state energy of liquid ^{3}He found to date.

As we mentioned earlier, the complication that arises in fermion Monte Carlo methods is not generally caused by a lack of antisymmetry in the Green's function. If the Green's function were sufficiently long range, i.e., if we knew the Green's function analytically for large τ, then antisymmetrizing it would resolve the fermion problem. This should become more apparent in Sect.4.5 where antisymmetrizing the Green's function is a subset of an exact solution for few-fermion systems. Antisymmetrizing the Green's function will, however, eliminate sign cancellation difficulties in treating fermion systems in one spatial dimension [4.14].

In one spatial dimension the nodal structure of the Fermi ground state is completely determined by the requirement of antisymmetry. This is simply because particles can be interchanged only by passing them through one another. Since the particles cannot be in the same states, the wave function must go to zero when the particles are at the same point. If the Green's function is antisymmetrized, the nodal structure of $G(R',R;\tau)$ in the R' variable is completely independent of R. The nodes in the R' variable occur when two particles lie at the same point. This Green's function contains the correct nodal structure of the fermion ground state. If we look at all paths that arrive at a point R, we see that they all have the same sign independent of the point of origin. Thus, there is no sign cancellation difficulty, and fermion systems in one spatial dimension may be calculated with no special Monte Carlo difficulty.

It is easy to see that the same technique will not work in more than one spatial dimension, because exchanges in more than one dimension can (and usually do) take place without the particles passing through each other. The nodal structure is not wholly determined by the requirement of antisymmetry. To understand this in more detail we look at the case of two particles in two dimensions interacting with a central force. Antisymmetry means the wave function changes sign under the interchange $r_{12} \rightarrow -r_{12}$. In Fig.4.6 a possible nodal structure for the wave function is shown. Since the potential is central, the Green's function must have nodes as shown in Fig.4.7 for two different initial points. The points in the neighborhood of the point x in Fig.4.7 would be reached with a net negative sign in Fig.4.7a but a net positive sign in Fig.4.7b. These paths tend to cancel in the average but not in the variance, so the difficulty with signs is not eliminated.

The fact that antisymmetrizing the Green's function cannot solve the fermion problem in more than one spatial dimension can be understood if we take $\tau \rightarrow 0$. The Green's function will be nonzero for R only in a neighborhood of radius $\sim \tau^{\frac{1}{2}}$ about the point R'. Clearly, however, antisymmetrizing the Green's function cannot help unless an antisymmetric image of the point R is also within $\tau^{\frac{1}{2}}$ of R'. In Fig.4.7 this happens only if r_{12} is within $\tau^{\frac{1}{2}}$ of the origin. For small τ this neighborhood becomes negligibly small.

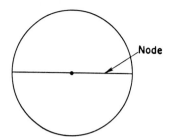

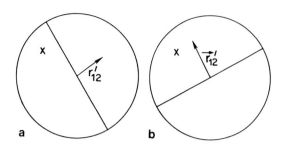

Fig.4.6. The r_{12} plane for two fermions in two dimensions. The horizontal line gives a possible nodal structure

Fig.4.7a,b. The r_{12} plane showing the node of the antisymmetric Green's function for two different values of r_{12} [(a,b) respectively]. Note that the paths leading to point x have different signs even though they both started on the same side of the node in Fig.4.6

4.4 The Fixed Node Approximation

In most cases the nodes of the Fermi ground state are not known. When they are, fermion problems can be solved by calculating the wave function with the boundary condition that it be zero whenever the true wave function is negative [4.18,22,23]. By antisymmetrizing the resulting wave function, the full fermion ground state is generated.

An approximate solution can be obtained by assuming an approximate nodal surface and calculating the wave function and energy as discussed above. The simplest way of defining some fixed node is to take it as the manifold of zeros of some trial wave function. The boundary condition is that the calculated wave function is zero whenever the trial wave function is negative. The calculated wave function may then be antisymmetrized to obtain appropriate ground-state expectation values.

In practice, this final antisymmetrization is carried out automatically. For example, the ground-state energy is calculated from the mixed estimator

$$E_{mixed}(\tau) = \frac{\int dR[H\psi_T^A(R)]\psi_{FN}(R)}{\int dR\ \psi_T^A(R)\psi_{FN}(R)} \quad , \tag{4.42}$$

where, as before, the superscript A on ψ_T^A indicates a fully antisymmetric trial function, and ψ_{FN} is the fixed node solution. Since the Hamiltonian is a symmetric operator under particle interchange, any components of ψ_{FN} which are not antisymmetric will give zero contribution to the matrix elements of (4.42). Since ψ_{FN} is zero under odd permutations there is no increase in variance. This can also be seen by realizing that ψ_{FN} is an eigenstate of H (although when the assumed fixed node is not the correct one, the normal derivatives of ψ are not continuous at zero). The fixed-node solution is an upper bound to the ground-state energy [4.22].

The fixed-node approximation is easy to implement. All that is necessary is to add to the Hamiltonian an infinitely repulsive potential in regions where the trial function is negative. This automatically enforces the boundary condition. In the short time approximation, the fixed-node approximation can be implemented by simply discarding any walkers that move into a region where $\psi_T^A < 0$. In the exact GFMC with domains, the domains are chosen so as not to overlap such a region. This is done by estimating an upper bound to the distance to the node, and setting a domain radius to be a fraction of that distance.

As explained in the previous section, the nodal structure of the Fermi ground state in one spatial dimension is known exactly. Therefore the fixed-node approximation becomes exact in one spatial dimension. It is equivalent to calculating a system of Bose particles in one dimension with a hard core that keeps particles from occupying the same position.

Another, perhaps less drastic, approximation can be made by replacing the infinitely repulsive potential used in the fixed node approximation with a softer potential. A similar kind of approximation can be made by allowing the walkers to remain in a region of negative $\psi_T^A(R)$ for a finite amount of imaginary time τ before discarding them [4.20].

The fixed-node approximation has been applied to the problems of the electron gas [4.21], the atomic and molecular systems of Be and LiH [4.22,23], and other systems [4.18]. The results for the electronic systems are quite good and give results competitive with more traditional methods. Presumably this is because the Hartree-Fock solutions generally used to determine the nodal structure give physically reasonable results for the soft Coulomb potential. In liquid ^{3}He, where the hard-core interaction makes the Hartee-Fock solution completely useless, less satisfactory results are obtained. *Ceperley* [4.31] has obtained an energy of -2.06 ± 0.05 K for liquid ^{3}He using the nodes of a Slater determinant of plane waves.

An interesting combination of the method of transient estimation and the fixed-node approximation was used by *Ceperley* and *Alder* [4.21] in their calculation on electron gas. The fixed-node solution was first calculated, and then the configurations were diffused without the fixed-node approximation. The resulting energy as a function of the imaginary time τ seems to be stable, indicating that the fixed-node solution was nearly correct. It should be pointed out that this technique cannot guarantee an upper bound to the ground-state energy, although both the method of transient estimation and the fixed-node approximation separately give upper bounds to the energy for all τ.

4.5 An Exact Solution for Few-Fermion Systems

A true solution to the fermion problem would avoid the difficulties associated with the method of transient estimation without introducing approximations such as the fixed-node approximation. Such a solution is possible for few-particle systems. What is needed is a method that controls the exponential growth of the symmetric components of ψ^+ and ψ^-.

In the Monte Carlo method, ψ^+ and ψ^- can be represented by a sum of delta functions at each of the positions of the walkers

$$\Psi^+(R,\tau) = \sum_{i+} \delta(R - R_{i+}) \quad , \tag{4.43}$$

$$\Psi^-(R,\tau) = \sum_{i-} \delta(R - R_{i}-) \quad . \tag{4.44}$$

Substituting these equations into (4.14) yields

$$\Psi(R,\tau + \Delta\tau) = \sum_{i+} G(R,R_{i+};\Delta\tau) - \sum_{i-} G(R,R_{i}-;\Delta\tau) \quad , \tag{4.45}$$

and analytic forms of G are to be subtracted (rather than sampled). If a position R_{i+} is sufficiently close to a position $R_{i}-$, then a substantial fraction of the Green's functions for these positions will cancel. We identify the new ψ^+ and ψ^- at imaginary time $\tau + \Delta\tau$ as

$$\Psi^\pm(R,\tau + \Delta\tau) = \pm \frac{1}{2} [\Psi(R,\tau + \Delta\tau) \pm |\Psi(R,\tau + \Delta\tau)|] \quad , \tag{4.46}$$

with $\Psi(R,\tau + \Delta\tau)$ described by (4.45). For a sufficient density of configurations the cancellation inherent in (4.45) stabilizes the amount of boson solution present in $\psi^\pm$, and the growth of variance will be controlled. In practice, sampling a new set of configurations from (4.46) is too difficult. Instead of taking all the configurations together, only pairs of configurations are taken together. The walkers are chosen in pairs so that one walker from ψ^+ and one from ψ^- are close together in configuration space. The resulting distributions $\psi^\pm(R,\tau + \Delta\tau)$ are sampled from the positive and negative parts of a Green's function for advancing a pair of walkers,

$$G^\pm(R;R_{i+},R_{i}-;\Delta\tau) = G(R,R_{i+};\Delta\tau) - G(R,R_{i}-;\Delta\tau) \quad . \tag{4.47}$$

This method has been used successfully [4.32] on a model three-body system and yields stable estimates for the Fermi ground-state expectation value.

The method is limited to few-body systems since the points R_{i+} and $R_{i}-$ must be close enough in phase space to allow sufficient cancellation. The use of various symmetries of the problem can help to bring points closer together, but what is required is a density of points in configuration space sufficient to allow cancellation of a large part of the symmetric (or not antisymmetric) components of ψ^+ and ψ^-. Clearly this will be more difficult to achieve as the number of particles grows.

4.6 Speculations and Conclusions

Several ideas suggest themselves for extension of the methods outlined previously. A method which systematically improves the fixed-node approximation would be very useful. For example, the net flux of diffusing walkers through an element of the nodal surface can be used to identify the direction in which an element of the nodal surface should be moved [4.33]. It may be possible to perform calculations with trial functions with different nodal surfaces and from these calculations derive a combination of trial functions with a better nodal structure [4.34].

A figure of merit which measures the quality of a given nodal structure might be derived from the property that the gradient of the potential evaluated at the true nodal surface must have no component perpendicular to the surface. This is why the net flux of diffusing walkers is zero. If a method of parametrizing the points on a nodal surface were found, it would be possible to select a best trial function by requiring that this property hold at a selected set of nodal surface points. Naturally, since the nodal surface for an N-body system is a 3N-dimensional manifold, an accurate specification is no trivial task, except possibly as the zeros of a trial function with some adjustable parameters. Even then, the true nodal surface will not generally be included among those generated by possible values of those parameters.

When a nodal surface is the zero of an antisymmetric trial function, $\psi_T^A(R,B)$ with parameters B, one measure of its quality is the degree to which ψ_T^A matches the Monte Carlo density which represents the converging wave function $\Psi(R,\tau)$. A simple and effective measure of such a match is the overlap between the two functions

$$X(B) = \frac{\int \psi_T^A(R,B)\Psi(R,\tau)\ dR}{\left\{\int |\psi_T^A(R,B)|^2\ dR \int |\Psi(R,\tau)|^2\ dR\right\}^{\frac{1}{2}}} \ . \tag{4.48}$$

Computing relative values of $X(B)$ is straightforward, and reweighting may be used to find values of B that maximize X. One can imagine cycling among three stages of calculations: (1) a Monte Carlo stage in which the nodes are determined by some B; (2) relaxation of the fixed node constraint during an interval of τ; and (3) selecting B to maximize the overlap X. If $\psi_T^A(R,B)$ contains the correct ground state for some B, then this procedure will converge to it without exponential decay of signal to noise ratio. It is not clear how well it will do (from the point of view of bias) when the trial function does not give the ground state. It will be possible to assure an upper bound.

The method in Sect.4.5 can be thought of as a way of filtering out some symmetric components from $\psi^\pm$. A more efficient filtering method taking advantage of the global distribution of configurations would be useful. It is possible to regard this as a problem in optimization, namely, to remove that set of configurations

which, over the course of a large Monte Carlo calculation, most accurately represents a symmetric distribution (and whose removal will therefore have little effect
on any antisymmetric expectations) [4.34]. Results obtained in a test problem are
encouraging, but the energy estimates are somewhat biased in a way that is not yet
properly understood.

In conclusion, we have presented some of the methods that have been used in Monte
Carlo calculations of the properties of fermion systems. None of the methods presented is completely satisfactory. For problems in one spatial dimension a variety
of means are available that effectively map the system onto a one-dimensional Bose
system. Most of the problems we are interested in are in more than one spatial dimension. For few-body systems the method given in Sect.4.5 can be used. For many-
particle systems the fixed-node approximation and the method of transient estimation are useful. Although neither of these latter methods is esthetically pleas-
ing, useful results can be obtained.

References

4.1 D.M. Ceperley, M.H. Kalos: In *Monte Carlo Methods in Statistical Physics*, ed.
 by K. Binder, Topics Current Phys., Vol.7 (Springer, Berlin, Heidelberg, New
 York 1979) Chap.4
4.2 M.H. Kalos: Phys. Rev. **128**, 1891 (1962)
4.3 M.H. Kalos: J. Comp. Phys. **1**, 127 (1966)
4.4 M.H. Kalos: Nucl. Phys. A**126**, 609 (1969)
4.5 M.H. Kalos: Phys. Rev. A**2**, 250 (1970)
4.6 M.H. Kalos, D. Levesque, L. Verlet: Phys. Rev. A**9**, 2178 (1974)
4.7 P.A. Whitlock, D.M. Ceperley, G.V. Chester, M.H. Kalos: Phys. Rev. B**19**, 5598
 (1979)
4.8 M.H. Kalos, M.A. Lee, P.A. Whitlock, G.V. Chester: Phys. Rev. B**24**, 115 (1981)
4.9 J.J. Cullen, D.P. Landau: Phys. Rev. B**27**, 297 (1983);
 A. Wiesler: Phys. Lett. **89**A, 359 (1982);
 J.W. Lyklema: Phys. Rev. Lett. **49**, 88 (1982)
4.10 H. De Raedt, A. Lagendijk: Phys. Rev. Lett. **46**, 77 (1981);
 H. De Raedt, A. Lagendijk: J. Stat. Phys. **27**, 731 (1982)
4.11 F. Fucito, E. Marinari, G. Parisi, C. Rebbi: Nucl. Phys. B**180** [F52], 369 (1981)
4.12 D.J. Scalapino, R.L. Sugar: Phys. Rev. Lett. **46**, 519 (1981)
4.13 D.J. Scalapino, R.L. Sugar: Phys. Rev. B**24**, 4295 (1981)
4.14 J.E. Hirsch, D.J. Scalapino, R.L. Sugar, R. Blankebecler: Phys. Rev. Lett.
 47, 1628 (1981)
4.15 P.A. Whitlock, M.H. Kalos: J. Comp. Phys. **30**, 361 (1979)
4.16 N. Metropolis, S. Ulam: J. Am. Stat. Ass. **44**, 335 (1949)
4.17 J.B. Anderson: J. Chem. Phys. **63**, 1499 (1975)
4.18 J.B. Anderson: J. Chem. Phys. **65**, 4121 (1976)
4.19 J.B. Anderson: J. Quantum Chem. **15**, 109 (1979)
4.20 J.B. Anderson: J. Chem. Phys. **73**, 3897 (1980)
4.21 D. Ceperley, B. Alder: Phys. Rev. Lett. **45**, 566 (1980)
4.22 J.W. Moskowitz, K.E. Schmidt, M.A. Lee, M.H. Kalos: J. Chem. Phys. **77**, 349
 (1982)
4.23 P.J. Reynolds, D.M. Ceperley, B.J. Alder, W.A. Lester: J. Chem. Phys. **77**,
 5593 (1982)
4.24 M.A. Lee, K.E. Schmidt, M.H. Kalos, G.V. Chester: Phys. Rev. Lett. **46**, 728
 (1981)
4.25 D.M. Ceperley: J. Comp. Phys. **51**, 404 (1983)

4.26 A.L. Fetter, J.D. Waleska: *Quantum Theory of Many-Particle Systems* (McGraw-Hill, New York 1971)
4.27 J.B. Kogut: Rev. Mod. Phys. **51**, 659 (1979)
4.28 N.M. Hugenholtz: Rept. Prog. Phys. **28**, 201 (1965)
4.29 See, e.g., R.P. Feynman: *Statistical Mechanics* (Benjamin, New York 1972)
4.30 N. Metropolis, A.W. Rosenbluth, H. Rosenbluth, A. Teller, E. Teller: J. Chem. Phys. **21**, 1087 (1953)
4.31 D.M. Ceperley: Private communication
4.32 D. Arnow, M.H. Kalos, M.A. Lee, K.E. Schmidt: J. Chem. Phys. **77**, 1 (1982)
4.33 D.M. Ceperley: In Proc. NATO ARW on Monte Carlo Methods in Quantum Problems, ed. by M.H. Kalos (D. Reidel, Dordrecht 1984)
4.34 M.H. Kalos (ed.): In Proc. NATO ARW on Monte Carlo Methods in Quantum Problems (D. Reidel, Dordrecht 1984)

5. Simulations of Polymer Models

A. Baumgärtner

With 21 Figures

Presented here is a review of Monte Carlo calculations for polymeric systems aimed at organizing the important results of polymer simulations. Topics considered include the variants of chain configuration generating technique, athermal and thermodynamic properties of single chains and chains at finite concentrations, and the dynamics of polymer melts.

This review is primarily concerned with the most recent results. Even with this restriction, the subject is large so that the scope of this chapter unfortunately necessitated some omissions.

5.1 Background

The linear polymer consisting of many concatenated atoms provides the basis for a class of materials with unmatched diversity of properties. Textile fibers, plastics, rubber, cellulose, proteins from living organisms, and the polynucleotides that store and transfer genetic information are examples.

Polymers are large only in the sense that they consist of thousands of atoms. Their length typically ranges from 10^{-5} to 10^{-2} cm. They are, however, no wider than 10^{-7} cm. A typical number for the degree of polymerization, i.e., the number of monomeric constitutents, is about 10^4.

Polymers can be described by *chemical* formulas, which accurately give the patterns of chemical bonds and the atoms connected by them. However, the most interesting aspects of polymeric substances, namely their unique mechanical and physical properties, are largely related to their *geometric* properties.

As a simple example, consider a hydrocarbon chain such as polyethylene $(-CH_2)_N$ with N monomers [5.1]. The angle θ between successive C-C bonds is essentially fixed, but rotations of about $\phi = 0°$, $+120°$, $-120°$ around the neighboring C-C bond are permitted (Fig.5.1). The energy between successive groups depends on the rotational angle ϕ as shown in Fig.5.1. There are three minima, corresponding to three principal conformations called *trans*, *gauche*(+) and *gauche*(-). Other configurations are strongly disfavored because those cause the preceding and following CH_2 groups to overlap each other (Fig.5.1). The rotation potential is usually given by

$$V(\phi) = V_0[x(1 - \cos\phi) + (1 - x)(1 - \cos3\phi)] \quad , \tag{5.1}$$

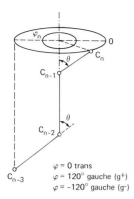

$\varphi = 0$ trans
$\varphi = 120°$ gauche (g+)
$\varphi = -120°$ gauche (g-)

Fig.5.1. The trans-*gauche* configuration of the hydrocarbon
chain and its rotation potential

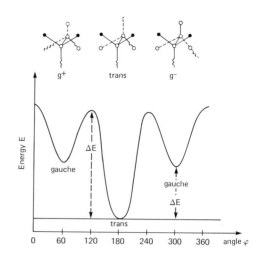

where V_0 and x are some empirical parameters of the order of 4 kJmol^{-1} and 0.1 respectively [5.1].

Rotations about the many bonds of the chain can lead to a multitude of different configurations in three-dimensional space, and the chain appears often as a *random coil*, rather than fully stretched. Consider, for example, a chain of N bonds with three rotational possibilities available to each bond, the total number of configurations for the macromolecule as a whole being 3^N. So it turns out that statistical mechanics is a useful and highly appropriate tool for even a single polymer chain. This latter case would apply, for example, to polymers in solution so dilute that they are separated from one another, Sect.5.3.1.

Only a small portion of the space pervaded by a chain in a random configuration is actually occupied by it (typically 1%). The remainder may be filled by solvent molecules. If the solvent is removed, *intermolecular forces* require that the space vacated somehow be filled. Three principle possibilities exist: (1) The polymer may collapse to a globular, dense configuration which is realized only in globular proteins with highly specialized chemical structure that predispose them to adopt compact configurations (Sect.5.3.2). (2) The chains may remain in a random structure with each of them interpenetrating the domains pervaded by other chains, thereby becoming highly entangled. This occurs under suitable conditions in nearly all polymers (Sect.5.3.3). (3) The polymers may associate in bundles in which neighboring chains are parallel to one another, exhibiting a highly ordered crystalline state (Sect.5.3.4).

Polymers may also occur as viscous liquids, rubbery solids, glasses, or semi-crystalline solids. Polymers of the same chemical composition may occur in any of these states, depending on temperature and their chain length.

Much effort has been devoted to deduce the physical properties of polymers from their chemical formulas. However, the aim of this chapter is to demonstrate how some of the basic properties of polymers can be understood (using here the Monte Carlo

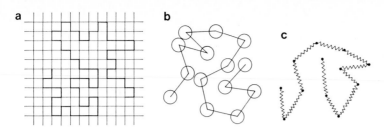

Fig.5.2. Three basic polymer models: (a) the lattice model (e.g., the square lattice model); (b) the bead-rod model (e.g., the pearl necklace model); (c) the bead-spring model

method as a special tool) within a more *global* point of view, where the chemical details of the chain structure are omitted as much as possible in order to extract simple, universal features which remain true for a large class of macromolecules. Much progress has been made for both static [5.2,3] and dynamic [5.3-5] phenomena in this respect within the last forty years. Especially recently *scaling concepts* have proven to be a highly appropriate tool [5.3].

Consequently some very useful *polymer models* have been proposed and investigated. The three most standard models are:

1) *Lattice models* where a flexible polymer chain is replaced by a self-avoiding random walk on a periodic lattice, as shown in Fig.5.2a for the square lattice. The walk is a succession of N steps subject to the condition that no lattice site may be visited more than once in the walk, so taking into account the "excluded volume effect". Although the mathematical problem of calculating the properties of self-avoiding walks is formidable, the model is well suited to computer enumerations, and has been the subject of many investigations [5.6]. For sufficiently short chains (up to $N \simeq 20$) such enumerations provide exact results which are then used to conjecture the pattern of asymptotic behavior ($N \to \infty$). These results are in good agreement with Monte Carlo calculation on lattice polymers (up to $N \simeq 600$) initiated and developed largely by *Wall* and his collaborators [5.7], and by *Rosenbluth* and *Rosenbluth* [5.8] (for later improvement of this method, see, e.g., [5.9,10]).

2) The *bead-rod model* (or freely jointed chain [5.1,11] consists of N successively connected hard rods of equal length and vanishing diameter (Fig.5.2b). The angles between neighboring segments are not restricted. One way to take into account the excluded volume is to put hard spheres centered at each connection point on the chain ("pearl necklace model") [5.12-15] or alternatively a soft, but purely repulsive potential [5.16]

$$V(r) = \begin{cases} 4\varepsilon[(\sigma/r)^{12} - (\sigma/r)^6 + 1/4] & ; \quad r \le 2^{1/6}\sigma \\ 0 & \quad r > 2^{1/6}\sigma \end{cases} \tag{5.2}$$

where ε and σ are some parameters and $r = |\mathbf{r}_i - \mathbf{r}_j|$ is the distance between the beads i and j. But also soft-core potentials with short-range attractive tails of

147

Lennard-Jones type

$$V(r) = 4\varepsilon[(\sigma/r)^{12} - (\sigma/r)^6]$$ (5.3)

have been successfully applied to this model [5.17-20].

3) The *bead-spring model* chain contains N beads connected by harmonic springs (Fig.5.2c). To produce finite extendibility of the spring a modified harmonic potential between the beads has to be introduced [5.21]:

$$v(r) = \begin{cases} -V_0 \ln[1 - (r/r_0)^2] & ; \quad r_0 \geq r \\ 0 & ; \quad r_0 < r \end{cases}$$ (5.4)

where r_0 and V_0 are some constants. The excluded volume is usually taken into account by a Lennard-Jones potential of type (5.2) [5.21-25].

5.2 Variants of the Monte Carlo Sampling Techniques

In this section the four most important techniques to generate random self-avoiding configurations of polymer chain models are described. Two of the techniques ("kink jump" and "Brownian dynamics") are of special interest because they provide additional information on the dynamic behavior of polymer chains in cases were hydrodynamic interactions are negligible, which is realized, e.g., in polymer melts (Sect.5.4.2).

1) The *"random walk"* generation technique. By this method one uses random numbers to generate random walks of equal step length on a lattice or in the continuum. To obtain unbiased results, the walk is stopped as soon as the self-avoiding walk condition is violated. Then a new chain can be generated. Obviously, this method has the following advantages: (i) it provides chain configurations over a broad range of total step length N; (ii) since the different configurations are statistically independent of each other, the accuracy of the averages can be estimated by standard statistical analysis; (iii) from recording the fraction W(N) of successful attempts to construct configurations one obtains the entropy S of the chain in the athermal case by

$$(S - S_0)/k_B = \ln W(N) \quad ,$$ (5.5)

where S_0 is the entropy in the absence of excluded volume; for the thermal case see Sect.5.3.1. The disadvantage of the simple sampling method is that it becomes very ineffective to generate very long chains. Since the entropy is to leading order [5.26] $(S - S_0) = -\lambda N + (\gamma - 1)\ln N + O(N-1)$, where $\gamma \simeq 7/6$ and $\gamma \simeq 4/3$ in three and two space dimensions respectively, the fraction W of successful attempts decreases exponentially fast $W(N) \propto \exp(-\lambda N)$. To overcome this "attrition effect" several "enrichment" techniques have been developed. One of the most effective techniques is the *Rosenbluth* method [5.8] enabling a walk to be continued in the presence of
148

closure. One returns to the last step before the self-interaction to choose a step from the remaining choices. If successful, the walk continues. This step must be weighted by an appropriate factor so that all the walks generated have the same a priori probability of occurrence. This technique can also be applied to the thermal case where the walks have to be weighted according to the Boltzmann distribution [5.8-10].

2) The "*reptation*" method is one of the most efficient generation techniques [5.27] and is applicable to any linear polymer chain model. Starting from an arbitrary configuration, one first selects one of the ends of the chain at random and then removes this end link of the chain and adds it to the other end, specifying randomly the orientation of the link. The resulting state is accepted as a new configuration, if the excluded-volume condition is not violated. (For the thermal case the Metropolis criterion has to be considered, Chap.1 and Sect.5.3.2.) This mechanism, which corresponds to a "slithering snake"-like motion of the chain along itself, provides an approach towards equilibrium [5.28].

3) The stochastic "*kink-jump*" method was originally proposed for lattice polymers [5.29], but it is also applicable to bead-rod models [5.12]. The stochastic motion of the chain results from successive *local* jumps randomly distributed along the chain; each of them affects a very small part of the chain. The simplest dynamics of local character for the bead-rod model consist in trying rotations of two links around the axis joining their end points through an angle ϕ chosen randomly from the interval $[-\Delta\phi,+\Delta\phi]$ (compare Fig.5.2b). The parameter $\Delta\phi$ is arbitrary, and it is convenient to choose it so that about one-half of the attempted rotations are successful. If an end point of the chain is chosen the link is rotated to a new position by specifying two randomly chosen angles (ϕ,θ) for the case $d = 3$, with $\cos\theta$ being equally distributed in the interval $-1 < \cos\theta \leq 1$. For the case $d = 2$ it is only the angle ϕ which can then be randomly chosen $[-\pi < \phi < \pi]$. Thus one starts a suitable initial configuration of the chain, and then chooses randomly one of the beads for rotation. The transition probability $W_i(\mathbf{r}_i \rightarrow \mathbf{r}_i')$ for this rotation is calculated, and compared to a random number $0 < \eta < 1$. If $W_i > \eta$, the rotation is actually performed and the new configuration is accepted; otherwise this rotation is rejected, and the old configuration is counted once more in averaging. In this manner, a sequence of configuration is generated. Since the system tends to equilibrium by construction (compare Chap.1 for general considerations) there is a correspondence between the time lapse and the number of configurations. So that the time unit does not depend on the chain length N, this unit is defined as a sequence in which, on the average, any bead has the possibility to move once. This Monte Carlo step per particle contains a sequence of N chain configurations. Thus for describing the evolution of the system we may use a parameter t, called time, which takes on the sequential values $t_k = k/N$, $k = 1,2...$. The dynamics of the "kink-jump" procedure is that of the Rouse model [5.10,12,29-32], where hydrodynamic interactions are neglected. So, the "kink-jump" dynamics does not provide any information

about the real dynamics of dilute polymer solutions (Zimm model [5.33]), where hydrodynamic effects are important. However, the "kink-jump" model provides a powerful tool to investigate the dynamics of polymer melts (Sect.5.4.2). As already discussed in [Ref.5.34, Chap.1], one may define time-dependent quantities as, e.g., the correlation function between the beads i and j

$$g(t) = <[\mathbf{r}_i(0) - \mathbf{r}_j(t)]^2> \quad . \tag{5.6}$$

The above-described "kink-jump" procedure for continuum chains has to be modified for applications to lattice polymer models: (a) To preserve the underlying lattice structure in general more than two links have to be moved simultaneously. For the diamond lattice three or four successive links can be moved keeping the end points of this chain fraction fixed [5.35,36]. (b) If there are k choices to perform an n-link motion (i.e., 2 choices of 4-bond motion on the diamond lattice [5.36]), all possible choices of motions can be combined. But in order to fulfill the "detailed balance" condition [Ref.5.34, Chap.1], the types of motion during the Monte Carlo procedure have to be selected randomly.

4) The "*Brownian-dynamics*" method [5.37] provides the static and dynamic properties of the "bead-spring" model without hydrodynamic effects. Following *Kirkwood* [5.38], *Rouse* [5.30] and *Zwanzig* [5.39], it is assumed that neglecting inertial terms (high viscosity limit) and hydrodynamic forces the time evolution of the polymer probability density f(R,t) at time t is given by a Smoluchowski equation [5.37, 40]

$$\partial f(R,t)/\partial t = D \sum_{j=1}^{N} \nabla_j[\nabla_j f(R,t) + \beta f(R,t)\nabla_j U(R)] \quad , \tag{5.7}$$

where the coordinates of the beads are denoted by $R = \{\mathbf{r}_i, \ 1 \leq i \leq N\}$. Here $U(\mathbf{R})$ is the total potential energy of the chain consisting of a Lennard-Jones-type potential (5.2) and a modified harmonic potential for the springs (5.4), and D is the diffusion constant. The probability density approaches the equilibrium distribution of the chain $f(\mathbf{R},t) \rightarrow \exp[-\beta U(\mathbf{R})]/Z$ as $t \rightarrow \infty$. The solution of the diffusion equation (5.7) is simulated by a special Monte Carlo scheme [5.22]. The two most essential steps are:

(1) The new position $\mathbf{r}_i$ for a randomly selected particle i is sampled from the equation

$$\mathbf{r}_i(t + \tau) = \mathbf{r}_i(t) - \tau\beta D\nabla_i U + \chi_i \quad , \tag{5.8}$$

where $<\chi_i\chi_j> = 6D\tau$.

(2) The move is accepted with probability q(R,R') where

$$q(R,R') = \min\left(1, \{\exp[-\beta U(R')]G(R',R,\tau)\}/\{\exp[-\beta U(R)]G(R,R',\tau)\}\right) \tag{5.9}$$

which is constructed so that $q(R,R')G(R,R',\tau)$ satisfies the detailed balance condition. The probability density $G(R,R',\tau)$ for the polymer to diffuse from point R to R' in time τ is given by

$$G(R,R') = (2\pi\tau D)^{-3N/2} \exp\{-[R - R' + \tau D\nabla U(R')]^2/4D\tau\} \quad . \tag{5.10}$$

It is possible to generalize the above method to include a hydrodynamical interaction matrix (Oseen tensor), but this has not yet been done [5.22].

5.3 Equilibrium Configurations

5.3.1 Asymptotic Properties of Single Chains in Good Solvents

The effective intramolecular forces between monomers are usually assumed to be of van der Waals consisting of short-range repulsion and longer range attraction (i.e., Lennard-Jones potentials). The energy parameter which scales the temperature, ε/k_BT, measures the energy of the monomer-monomer interaction relative to the energies of monomer-solvent and solvent-solvent interactions. At sufficiently high temperatures the repulsive forces dominate, leading to a swollen polymer coil. The configurational properties of this state are discussed below. At lower temperatures the attractive forces become increasingly important (the solvent becomes poorer) until the coil collapses to a dense globule, Sect.5.3.2.

The global properties of a polymer chain are described by the mean square end-to-end distance $<R_N^2>$ and the mean square radius of gyration $<S_N^2>$:

$$<R_N^2> = <(r_1 - r_{N+1})^2> \tag{5.11}$$

$$<S_N^2> = \sum_{i=1}^{N} \sum_{j=i+1}^{N+1} <(r_i - r_j)^2>/(N + 1)^2 \quad . \tag{5.12}$$

Many analytical (mean-field approximations [5.2,41], Lagrangian field theory [5.42, 43]) as well as numerical calculations (exact enumerations [5.6], Monte Carlo simulations [5.7,12-16]) support a power law in the limit $N \to \infty$

$$<R_N^2> \propto <S_N^2> \propto N^{2\nu} \quad , \tag{5.13}$$

where $\nu \simeq 0.588$ for dimensions $d = 3$ and $\nu \simeq 3/4$ for $d = 2$. The exponent ν is universal (i.e., independent of the underlying model) and is related to the correlation length exponent in the $n = 0$ vector model [5.42,44].

A *finite-size scaling analysis* [5.45] of Monte-Carlo-generated end-to-end distances for the "pearl necklace" model [5.14] presented in Fig.5.3 demonstrates very convincingly the general scaling form

$$<R_N>^{\frac{1}{2}} = \ell N^{\frac{1}{2}} f(N\delta^{d/\phi}) \quad , \tag{5.14}$$

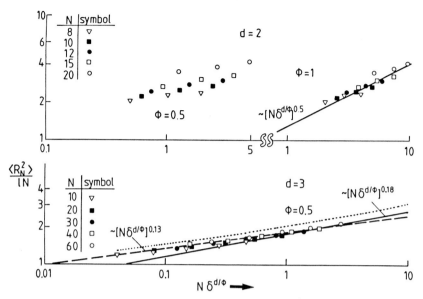

<u>Fig.5.3.</u> Log-log plot of the "expansion factor" $\langle R^2 \rangle / \ell N$ of the polymer chain versus scaled chain length for various values of N and excluded volume parameter δ. Data for dimensionality d = 2 and d = 3 are shown

where ϕ is the appropriate crossover exponent, $\delta = h/\ell$ is the ratio between the hard-sphere diameter h and rod length ℓ, and f(x) has the asymptotic behavior (d < 4)

$$f(x) = Ax^{(\nu - \frac{1}{2})} \quad , \quad x \to \infty \quad , \tag{5.15}$$

where A is a nonuniversal amplitude. For d > 4 the chain behaves asymptotically ideally [f(x) → A, x → ∞], while for d = 4 logarithmic corrections are important [f(x) → A(ln x)$^{\frac{1}{4}}$, x → ∞] [5.46].

However, this numerical analysis is based on data for fairly short chains (N ≤ 30 for d = 3; N ≤ 60 for d = 3), requiring then correction terms for the leading asymptotic behavior of very long chains (5.15). The values for the crossover exponent ϕ from Gaussian to nontrivial behavior has been adopted from [5.45]: $\phi = 1/2$ (d = 3) and $\phi = 1$ (d = 2). For d = 3, the slope of the straight line is found to be about 0.13 which would imply an "effective exponent" $\nu_{eff} = 0.56$. This estimate is in fact fairly close to 0.59. But one must note that the data do not even fall in the regime x ≫ 1 where the simple power law (5.15) should be valid: hence the accuracy of such direct Monte Carlo estimates for critical exponents is rather doubtful. For comparison, in Fig.5.3 the closed-form approximation for the crossover function for the "expansion factor" $\langle R_N^2 \rangle / \ell N$ discussed by *Domb* and *Barrett* [5.47] is shown by the dotted line. This approximation predicts somewhat too large values. The full line in Fig.5.3 represents the "true" asymptotic behavior with $\nu = 0.59$.

One of the most useful and accurate numerical methods to estimate ν is the
Monte Carlo renormalization group method. This combination of real space renormalization with Monte Carlo methods has become a very powerful tool in studies of critical phenomena [5.48-51] and has also been developed and applied to polymers
[5.14,52]. The method is based on *De Gennes'* suggestion of renormalization along
the chemical sequence of the chain [5.3,53-55]. As usual in the renormalization
group analysis of critical phenomena [5.56], one considers iterative elimination
of degrees of freedom. Starting with N_0 links of length ℓ_0 and sphere diameter h_0,
the chain is mapped onto another chain consisting of $N_1 = N_0/s$ so-called block links
of length ℓ_1 with sphere diameter h_1 (Fig.5.4). This is analogous to the real-
space renormalization group transformation of spin systems [5.57]. (A review on
the application of real-space renormalization methods to linear polymers, branched
polymers and gels is given in [5.58].) The length ℓ_1 of the block link is identified as the root mean square internal distance between points s units apart:
$\ell_1 = <(r_n - r_{n+s})^2>^{\frac{1}{2}}$. The renormalized hard-sphere diameter h_1 is calculated using
the condition that the renormalization transformation leaves the end-to-end dis-
tance invariant [5.14]: $R(\ell_0,h_0,N_0) = R(\ell_1,h_1,N_0/s)$. In practice, h_1 is estimated
using some interpolation technique between various Monte-Carlo-generated end-to-
end distances $R(\ell,h,N)$ [5.14,52]. Suppose we iterate the MC renormalization transformation k times with $k \to \infty$, one approaches a nontrivial fix point: $\delta_k \to \delta_{k+1}$
$\to \ldots \to \delta^* > 0$, where $\delta_0 = \delta_1 = \delta^*$. Since for k large enough s^k is a large number
already, then (5.14,15) also apply to the distance ℓ_k,

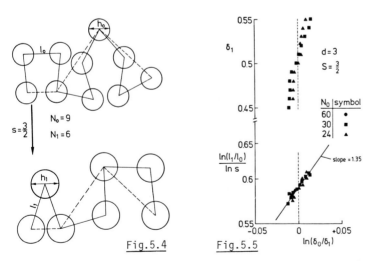

Fig.5.4   Fig.5.5

Fig.5.4. Monte Carlo renormalization transformation for scale factor s = 3/2 .
The excluded volume parameter $\delta_0 = h_0/\ell_0$ changes to $\delta_1 = h_1/\ell_1$

Fig.5.5. Renormalized excluded volume parameter δ_1 (*upper part*) and change of
the length scale $\ln(\ell_1/\ell_0)$ (*lower part*) versus $\ln(\delta_0/\delta_1)$ for various chain
lengths N_0 and scale factor s = 3/2

$$\ell_k = \ell_0(s^k)^{\frac{1}{2}}f(s^k\delta^{d/\phi}) = A\ell_0 s^{\nu k}\delta^{d(\nu-\frac{1}{2})/\phi} \, , \quad \text{for} \quad \delta_0 = \delta^* \, . \tag{5.16}$$

Hence we find for the change of the length scale at the fixed point

$$\nu = \ln(\ell_{k+1}/\ell_k)/\ln s \, , \quad \text{for} \quad \delta_k = \delta_{k+1} \, . \tag{5.17}$$

Thus the ratio $\ln(\ell_1/\ell_0)/\ln s$ is plotted in Fig.5.5, since from the value of this quantity at the fixed point one can read off the exponent ν directly. The estimated values are $\nu = 0.58 \pm 0.01$ for $d = 3$, and $\nu = 0.74 \pm 0.01$ for $d = 2$.

The *internal correlations* between monomers are of fundamental importance, and can be described by the *structure factor* and by the *probability distribution* of internal distances. The structure factor, which can be measured using neutron scattering techniques [5.59], is defined as

$$S(\mathbf{q}) = \left\langle N^{-2} \left| \sum_{k=1}^{N} \exp(i\mathbf{q} \cdot \mathbf{r_k}) \right|^2 \right\rangle \, , \tag{5.18}$$

where $\mathbf{q}$ is the wave vector. Furthermore $S(\mathbf{q})$ has the properties [5.41,59,60]

$$S(\mathbf{q}) \begin{cases} = 1 - \langle s_N^2 \rangle q^2/3 + O(q^4) & \text{for} \quad \langle s_N^2 \rangle \ll q^{-2} \\ \propto q^{-1/\nu}N & \text{for} \quad \langle s_N^2 \rangle > q^{-2} > \ell^2 \\ \propto 1/N & \text{for} \quad \langle s_N^2 \rangle \gg q^{-2} \, . \end{cases} \tag{5.19}$$

For intermediate q the excluded volume interaction causes the scattering function to behave as $q^{-1/\nu}$, whereas for large q the monomers are uncorrelated and $S(q) \to 1/N$. This behavior has also been found in Monte Carlo simulations on the bead-rod model [5.12,17] and on the bead-spring model [5.22,37].

The probability distribution of internal distances has not yet been investigated experimentally, but exact enumerations [5.61-67] and Monte Carlo simulations [5.7,13, 68] have provided quite accurate results supporting our understanding of the universal internal properties of a polymer chain and providing a powerful test of analytical predictions [5.43,63,69,70]. The distribution function $P_{ij}(r)$ of the distance r between two elements of the chain i and j behaves as

$$P_{ij}(r) = |i - j|^{-3\nu}f(r/|i - j|^\nu) \begin{cases} f(x) \sim x^{\theta_s} & \text{for} \quad x \ll 1 \\ f(x) \sim \exp(-x^{\delta_s}) & \text{for} \quad x \gg 1 \, . \end{cases} \tag{5.20}$$

The exponents of the end-to-end distribution (case $s = 0$) $P_{1,N}(r)$ are given by the relations $\delta_0 = 1/(1 - \nu)$ [5.69] and $\theta_0 = (\gamma - 1)/\nu$ [5.43], where $\gamma \simeq 1.16$ is the polymer analog of the susceptibility exponent. Recently *Des Cloiseaux* [5.70] suggested that besides the end-to-end correlation two other general cases exists: $s = 1$ corresponds to the correlation between an end point and an interior point ($i = 1$,

$N \gg |i-j| \gg 1$); $s = 2$ corresponds to the correlation between two interior points ($i \gg 1, |i-j| \gg 1, N-j \gg 1$). The exponents θ_s and δ_s have been estimated using exact enumerations, Monte Carlo and ε-expansion techniques based on a Lagrangian field theory for the $n = 0$ vector model [5.70]. A detailed comparison between the various results is given in [5.13]. Recent Monte Carlo calculations [5.13,68] suggest the values: $\theta_1 = 0.55 \pm 0.06$, $\theta_2 = 0.9 \pm 0.1$, $\delta_1 = 2.6 \pm 0.15$, $\delta_2 = 2.48 \pm 0.05$. The agreement between $\theta_0 = 0.270 \pm 0.006$, $\delta_0 = 2.44 \pm 0.05$ and the above exact values 0.271 and 2.43, respectively, is quite remarkable and supports the reliability of the Monte Carlo calculations. In the light of the various numerical investigations it has been suggested that the probability distribution (5.20) is given in general for the three principal cases by [5.13,63,67,68,71]:

$$P_{ij}(x) = (1/B)x^{\theta_s} \exp(-x^{\delta_s}) \quad , \tag{5.21a}$$

where

$$x = r/G_s|i-j|^{\nu} \tag{5.21b}$$

$$1/B = [\delta_s/\Gamma((3+\theta_s)/\delta_s)](G_s|i-j|^{\nu})^{-3} \tag{5.21c}$$

and $\Gamma(z)$ is the gamma function. Here B is a normalization constant which gives $\int_0^{\infty} P(x)x^2 \, dx = 1$. The constant G_s is model dependent. In Figs.5.6,7 the short-range and the long-range parts respectively of P_{ij} are obtained from Monte Carlo simulations of the pearl necklace model with $N = 161$ beads [5.13]. The distribution function has been constructed by setting up histograms for each pair (i,j) with mesh

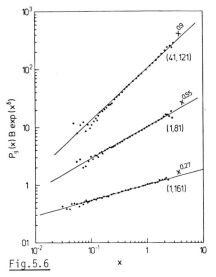

Fig.5.6

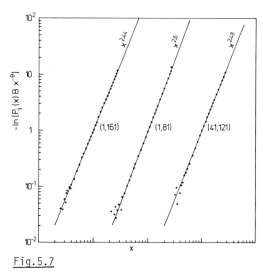

Fig.5.7

<u>Fig.5.6.</u> Log-log plot of $P_{ij}(x)B\exp(x^{\delta})$ versus x for the three principal pairs (i,j)

<u>Fig.5.7.</u> Log-log plot of $-\ln[P_{ij}(x)Bx^{-\theta}]$ versus x for the three principal pairs (i,j)

size 0.1ℓ. Only the three principal cases are shown. The power-law behavior can be followed over the whole region of x. This supports conjecture (5.21). Analyzing the pair correlations for nonprincipal cases, some conjectures on the crossover between the principal cases have been given in [5.13]. The estimated exponents for s = 2 (two interior points) are recovered by other Monte Carlo calculations on ring polymers where correlations with end points (s = 0 and s = 1) are excluded [5.68].

An interesting problem where scaling arguments have also been successfully applied to give a conclusive interpretation of Monte Carlo data is the *crossover between dimensionalities*, i.e., the asymptotic behavior of a polymer chain confined within a slab of width $D \gg \ell$ [5.72]. The Monte-Carlo-generated data confirmed the scaling law proposed by *De Gennes* and *Daoud* [5.73]:

$$R(N,D) = R_3(N)f(D/R_3(N)) \tag{5.22a}$$

with

$$f(x) \rightarrow \begin{cases} x^{-\frac{1}{4}} & \text{for} \quad x \ll 1 \\ \\ 1 & \text{for} \quad x \gg 1 \ , \end{cases} \tag{5.22b}$$

where $R_3(N) \equiv R(N,\infty) \propto N^{0.59}$ is the end-to-end distance without confinement in three dimensions.

Another important effect on the configurational properties of polymer chains is due to ionizable monomers. These *polyelectrolytes* are common in nature and in industrial applications requiring polymer-water systems. The Coulomb repulsion between the ionized groups is strong and complex; scaling properties of polyectrolytes are not fully understood [5.3]. Recently the first Monte Carlo investigations [5.74] on the configurational properties of charged cubic lattice polymers with free counterions have been reported. These calculations are restricted to short chains ($N \leq 16$) and should therefore be treated with caution: further work is required.

Monte Carlo methods often form a highly appropriate tool to study the athermal configurational properties of special polymers: *ring-polymers* [5.68,75] which occur in some natural macromolecules (e.g., DNA [5.76], *branched polymers* [5.77,78] which occur, e.g., at the sol-gel transition [5.79], *block copolymers* [5.80,81] and *proteins* [5.82-85].

5.3.2 Phase Transitions of Single Chains

Collapse Transition

One of the most interesting problems concerning the thermodynamic properties of a polymer chain in very dilute solution is the collapse transition. The intramolecular forces are usually assumed to be van der Waals, consisting of strong, short-ranged repulsive and weak long-ranged attractive interactions. By suitable changes in temperature or in solvent composition, the chain crosses over from an extended swollen coil to a collapsed dense globule [5.2]. These phenomena, first investigated long ago by *Flory* [5.86], have attracted a great deal of attention in analytical [5.87-100], numerical [5.101-110] and experimental investigations [5.111-114] in connection with protein folding. Our understanding of the universal aspects of the collapse transition within the context of the theory of critical phenomena made some progress recently: *De Gennes* proposed that the collapse transition should exhibit characteristics of a tricritical point [5.92]. Following this suggestion several renormalization group calculations have been devoted to this subject [5.93,94,100]. Evidence for tricritical behavior has been reported recently from finite-size scaling analyis of Monte-Carlo-generated data [5.109,110].

Although the most interesting case is the collapse in $d = 3$ dimensions, the critical properties (scaling, critical exponents, etc.) are most clearly seen in $d = 2$ dimensions. This difficulty is due to the fact that at the marginal dimensionality $d = 3$ the critical properties are governed by logarithmic singularities, which are hard to analyze conclusively from Monte Carlo data [5.110]. Experimentally two-dimensional collapse can be realized if the polymers are strongly adsorbed on a flat surface or if the solution is confined in a slit.

The polymer chain configurations are represented by self-avoiding walks on a square lattice. The long-range attractive van der Waals interaction is modeled by a constant energy $\varepsilon < 0$ for every nearest-neighbor contact of the chain with itself [5.109]. Internal energy U, specific heat C, and intrinsic density ρ are given by

$$U = \varepsilon <n>/N \tag{5.23a}$$

$$C = \varepsilon^2 (<n^2> - <n>^2)/Nk_B T^2 \tag{5.23b}$$

$$\rho = N/6<S_N^2> \quad , \tag{5.23c}$$

where $<S_N^2>$ is the radius of gyration (5.12), and $<n>$ is the average number of nearest-neighbor contacts. The temperature variation of the quantities (5.23) is shown in Fig.5.8.

Identifying $k_B T_t(N)/\varepsilon$ with the positions of the specific heat maxima, it is shown [5.109] that the size dependence of the critical temperature T_t is described by

$$k_B [T_t(\infty) - T_t(N)]/\varepsilon \simeq 3.1 N^{-\phi}t \quad , \tag{5.24}$$

where $k_B T_t(\infty)/\varepsilon = 1.3 \pm 0.06$ and ϕ_t is the crossover exponent. Assuming generalized finite-size scaling [5.115] at a tricritial point [5.116,117] for an isolated polymer [5.91] leads to a density of the form

$$\rho = N^{1-2\nu}t f(N^\phi t_\tau) \quad , \tag{5.25}$$

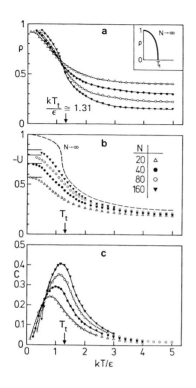

Fig.5.8. Raw data of monomer density ρ (a), internal energy U (b) and specific heat C (c) plotted versus temperature for chain lengths N = 20,40,80, 160. The suggested asymptotic behavior (N → ∞) is indicated qualitativeley for ρ and U

where ν_t is the correlation length exponent at the transition temperature T_t ($\tau \equiv |T - T_t|/T_t$). For the considered case d = 2, approximate values of ν_t and ϕ_t have been calculated using ε-expansion techniques [5.93,118]

$$\nu_t \simeq 0.5055 \quad , \quad \phi_t \simeq 0.6364 \quad . \tag{5.26}$$

The scaling function f(x) exhibits for large x (i.e., $\tau \ll 1$, but N → ∞) the infinite-chain critical behavior [5.91]

$$f(x) \propto x^\mu \quad \begin{cases} \mu = \mu^+ = 2(\nu_t - \nu)/\phi_t & T > T_t & \text{(5.27a)} \\ \mu = 0 & T = T_t & \text{(5.27b)} \\ \mu = \mu^- = 2(\nu_t - \nu_c)/\phi_t & T < T_t \quad , & \text{(5.27c)} \end{cases}$$

where ν and ν_c are the well-known correlation length exponents in the coil and globule states, respectively,

$$\nu = 3/4 \quad , \quad \nu_c = 1/2 \quad . \tag{5.28}$$

Thus one obtains the asymptotic behavior

$$\rho = \begin{cases} B^+ N^{1-2\nu_\tau \mu^+} & T > T_t & \text{(5.29a)} \\ B^0 N^{1-2\nu_t} & T = T_t & \text{(5.29b)} \\ B^- \tau^{\mu^-} & T < T_t \quad , & \text{(5.29c)} \end{cases}$$

where $B^\pm$, B^0 are the critical (nonuniversal) amplitudes for the infinite chain. A similar scaling ansatz for the specific heat is given by

$$C = N^{\alpha\phi}g(N^\phi\tau) \quad , \tag{5.30}$$

where the tricritical exponent $\alpha_t \simeq 1$ has been calculated [5.118] to first order in $\varepsilon = 3 - d$. The scaling function $g(x)$ asymptotically yields the infinite-chain critical behavior $g(x) \to A^\pm x^{-\alpha}$, which gives $C = A^\pm \tau^{-\alpha}$ (A^+ for $T > T_t$, A^- for $T < T_t$).

The raw data of the monomer-density ρ presented in Fig.5.8 are plotted again in Fig.5.9 according to the scaling assumption (5.25). In agreement with (5.25), the scaled data all lie on two smooth curves, for $T > T_t$ and $T < T_t$, respectively. The two

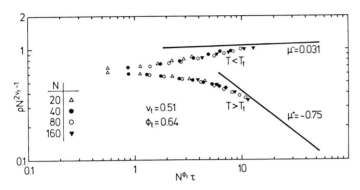

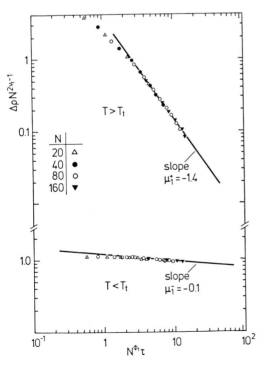

Fig.5.9. Finite-size scaling plot of the monomer density ρ according to (5.25) for various N at temperatures $T > T_t$ (*lower part*) and $T < T_t$ (*upper part*)

Fig.5.10. Finite-size scaling plot for the correction to the asymptotic behavior of the monomer density ρ according to (5.31) for various N at temperatures $T > T_t$ (*upper part*) and $T < T_t$ (*lower part*)

solid lines correspond to the asymptotic forms given by (5.27-29). The slow convergence of the Monte Carlo data to the asymptotic solid lines is due to corrections to the leading asymptotic terms (5.29), which are obviously important for $N^\phi\tau < 10$. In Fig.5.10 it is shown that the data are well described for $x > 1$ by

$$\rho(x)N^{2\nu-1} = B^\pm x^{\mu^\pm} + B_1^\pm x^{\mu_1^\pm} \qquad (x = N^\phi\tau) \quad , \tag{5.31}$$

where $\mu_1^+ \simeq -1.4$ and $\mu_1^- \simeq -0.1$. In Fig.5.10, $\Delta\rho$ is defined as $\Delta\rho N^{2\nu-1} \equiv B^\pm x^{\mu^\pm} - \rho(x)N^{2\nu-1}$.

The finite-size scaling analysis of specific heat including first-order correction terms is discussed in [5.109]. The crossover behavior for free energy and entropy when $d = 2$ has not yet been discussed in the literature. But a successful attempt to examine these quantities has been undertaken recently for $d = 3$ [5.110].

Adsorption of a Single Polymer Chain

A copious amount of theoretical work has been carried out on the adsorption of a single polymer at an interface, i.e., under the influence of a hard wall on the configurations of a chain in the presence of short-range attractive interactions between monomers and the wall. Most of the early works were concerned with an ideal chain without excluded volume effect [5.119,120]. Here, however, I shall discuss the good solvent case, where theoretical predictions are more difficult. Numerical calculations using Monte Carlo [5.120,121] and exact enumeration techniques [5.122-127] have proven to be very useful to examine the critical and asymptotic properties suggested by analytical approaches [5.120,128-134].

The quantities of interest in the adsorption problem are the mean-squared end-to-end distance $\langle R_N^2 \rangle$ and its perpendicular and parallel components with respect to the wall, $\langle R_N^2 \rangle_\perp$ and $\langle R_N^2 \rangle_{||}$, respectively,

$$\langle R_N^2 \rangle = \langle (x_1 - x_N)^2 + (y_1 - y_N)^2 \rangle$$

$$\langle R_N^2 \rangle_\perp = \langle (z_1 - z_N)^2 \rangle$$

$$\langle R_N^2 \rangle = \langle R_N^2 \rangle_{||} + \langle R_N^2 \rangle_\perp \quad , \tag{5.32}$$

the total adsorption energy E_N between the monomers and the wall, which is the number of segments or vertices of a lattice polymer lying inside or adjacent to the wall, and the partition function $Z_N(T)$.

Similar to the collapse transition discussed above, a finite-size scaling analysis of Monte-Carlo-generated data of the above quantities has been carried out recently [5.120]. The assumed scaling behavior for the end-to-end distance is given by:

$$\langle R_N^2 \rangle = N^{2\nu} f^2(\tau N^\phi)$$

$$\langle R_N^2 \rangle_\perp = N^{2\nu} f_\perp^2(\tau N^\phi)$$

$$\langle R_N^2 \rangle_{||} = N^{2\nu} f_{||}^2(\tau N^\phi) \quad , \tag{5.33a}$$

where ϕ is the crossover exponent and $\tau = (T - T_a)/T_a$ is the temperature distance to the critical temperature T_a, below which a finite fraction of monomers is in an adsorbed layer of thickness ξ (independent of N) at the surface (adsorbed phase). The scaling functions have the following asymptotic behavior [5.120]:

$$f_\perp(x) \propto \begin{cases} \text{const} & x \to \infty \\ \text{const} & x \to 0 \\ x^{-\nu/\phi} & x \to \infty \end{cases} \tag{5.33b}$$

$$f(x) \propto f_{||}(x) \propto \begin{cases} \text{const} & x \to \infty \\ \text{const} & x \to 0 \\ x^{(\nu_{d=2}-\nu)/\phi} & x \to \infty \quad , \end{cases} \tag{5.33c}$$

where $\nu_{d=2} = 3/4$ and $\nu \simeq 0.59$ are the correlation length exponents for two and three dimensions respectively. These scaling predictions are confirmed by Monte Carlo studies of self-avoiding walks on the tetrahedral lattice with a free surface [5.120]. In Fig.5.11 the end-to-end distance and its parallel and perpendicular components are plotted versus the scaling variable τN^ϕ for various chain lengths up to $N = 100$. The data collapse on a single line for $\phi \simeq 0.58$ and $T_a \simeq 2.28$. This confirms previous exact enumeration estimates of ϕ and T_a by *Ishinabe* [5.126] on the same lattice model, who disproved *De Gennes'* [5.129] suggestion of $\phi = 1 - \nu \simeq 0.41$; *Ishinabe*'s estimate of ϕ is in fair agreement with recent calculations by ε-expansion methods carried out to order ε^2 which gave $\phi \simeq 0.605$ [5.134].

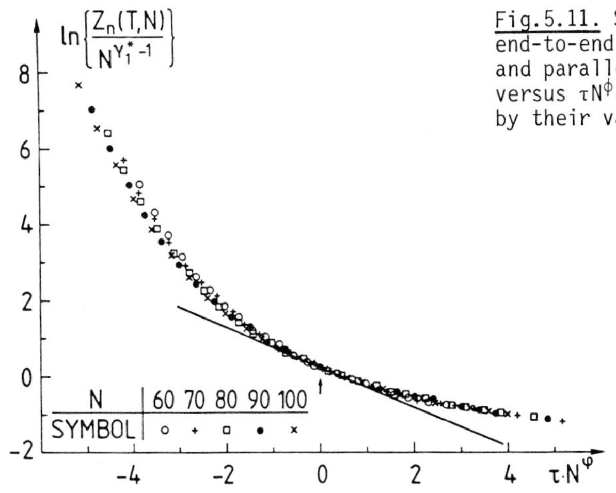

Fig.5.11. Scaling plot for the mean-square end-to-end distance and its perpendicular and parallel components as defined by (5.33) versus τN^ϕ. The quantities are normalized by their value at $T \to \infty$ [5.120]

The crossover scaling for the energy which is given by

$$-E_N/N^\phi = h(\tau N^\phi) \tag{5.34a}$$

with

$$h(x) \propto \begin{cases} x^{-1} & x \to \infty \\ \text{const} & x \to 0 \\ x^{(1-\phi)/\phi} & x \to \infty \end{cases} \tag{5.34b}$$

and the corresponding asymptotic behavior $(N \to \infty)$

$$-E_N/N \propto \begin{cases} 0 & T \geq T_a \\ |\tau|^{(1-\phi)/\phi} & T < T_a \end{cases} \tag{5.34c}$$

have also been confirmed by Monte Carlo calculation [5.120].

The crossover scaling behavior of the partition function for chains with at least the first monomer sitting on the surface is given by [5.120]:

$$Z_N(T) \propto q^N N^{\gamma_1^*-1} g(\tau N^\phi) \tag{5.35a}$$

with

$$g(x) \propto \begin{cases} x^{(\gamma_1-\gamma_1^*)/\phi} & x \to \infty \\ A + Bx & x \to 0 \\ |x|^{(\gamma_{d=2}-\gamma_1^*)/\phi} \exp(|x|^{1/\phi}) & x \to -\infty \, , \end{cases} \tag{5.35b}$$

where q is an effective coordination number for the self-avoiding lattice random walk under consideration. The surface exponent $\gamma_1 \simeq 0.68$ was first estimated by exact enumeration [5.122], and later confirmed as well by analytical [5.130-133], exact enumeration [5.135] and Monte Carlo methods [5.120]. (For definitions of surface exponents and their values and scaling laws in general, see [5.136].) Further $\gamma_{d=2} \simeq 4/3$ is the two-dimensional bulk "susceptibility" exponent [5.3]. In Fig.5.12

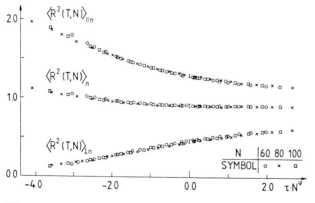

Fig.5.12. Scaling of the normalized partition function $Z_N(T)/N^{\gamma_1^*-1}$ versus τN^ϕ. The best fit is obtained using $\gamma_1^* \simeq 1.44$ [5.120]

the Monte Carlo data from [5.120] is presented for the partition function versus τN^{ϕ} for various chain lengths. Again the agreement between theory and computer experiment is good, assuming a surface exponent $\gamma^{*} \simeq 1.44$ which disagrees with the ε-expansion result [5.134] $\simeq 1.32$ by about 8%.

5.3.3 Chain Morphology in Concentrated Solutions and in the Bulk

The Amorphous State

The arrangement of long polymer chains in condensed phases (such as the liquid or amorphous state, including glasses) still remains a subtle problem. While it has been widely accepted that the configurations of the chains at a macroscopic level (i.e., $q^{-2} \gg \langle S_N^2 \rangle$) are Gaussian and ideal ($\langle S_N^2 \rangle \propto N$), two divergent viewpoints have been propounded with regard to the local structure ($q^{-1} > 100$ A) : (i) In the "bundle models" it is assumed that the chains are organized in small bundles, nodules, meander arrays, or paracrystals [5.137]. In a strict sense, therefore, the system is not amorphous. (ii) The alternative "coil model" holds that amorphous polymers are devoid of all short-range order, even at a level approaching the diameter of the chain [5.138]. Long-ranged Gaussian behavior originally proposed by *Flory* [5.139] has been found in numerous experiments [5.140-142] and in Monte Carlo simulations [5.143-153, 5.23] of lattice and continuum models for high temperatures. Even for polymer films in $d = 2$ dimensions, the ideal behavior $\langle S_N^2 \rangle \propto N$ has been observed by simulations [5.152,153]. But here, in contrast to $d = 3$, the chains are slightly swollen and strongly segregated [5.153], in agreement with *De Gennes'* suggestions [5.3]. A typical snapshot picture of a configuration of 25 disk-necklace chains each 30 disks long is shown in Fig.5.13. The hard disks each of diameter

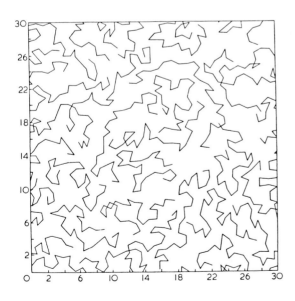

Fig.5.13. Snapshot picture of a typical configuration inside the basic cell for hard-disk rod model with N = 30 disks for each of the 25 chains. Circles for the disks are omitted

h = 0.6ℓ are omitted in the picture. Except for a few cross links, the chains are segregated. Of course, this two-dimensional model of a polymer film neglects the
finite thickness of realistic films. It would be of interest to perform simulations for the crossover between the dimensionalities d = 2 and d = 3 in the melt at
finite temperatures. Possibly thin films with thicknesses comparable with the
bulk-phase end-to-end distances of the polymers become thermodynamically unstable
and break up into islands.

Computer simulations for studying the short-range correlations between chains in
amorphous polymeric systems have not yet been performed as extensively as for the
long-ranged properties discussed above. A recent Monte Carlo investigation of systems of 31 n-alkane chains $C_{30}H_{62}$ with actual structural parameters, including bond
angles, and realistic intermolecular potentials, provided evidence against local
order in agreement with *Flory*'s "coil model" [5.154].

Interfaces

The configurational properties of interfaces such as amorphous layers in semicrystalline polymers [5.155,156] or as "free" interfaces as in adsorbed polymers [5.157]
or in membranes [5.158] have been investigated recently using statistical methods.
Only a small number of Monte Carlo investigations have been devoted to this subject.

Semicrystalline polymers consist of many three-dimensional crystalline lamellae
separated by amorphous layers. From the fact that a sequence of about 200 bonds suffices to span the lamellae from one face to the other, whereas an entire chain contains of the order of 10^4 bonds, one may conclude that a given chain must either
pass repeatedly through the same lamella or must be engaged in a number of them.
One of the basic problems here is related to the small interfacial region between
the crystalline and amorphous structure: what is the probability distribution for
the reentry of a chain to the crystallite from which it emanates, located n lattice
sites from the point of the preceding exit? Obviously this property strongly determines the mechanical and physical properties of semicrystalline polymers. A first
step in investigating this problem using Monte Carlo methods was recently undertaken [5.159]. To interpret small-angle neutron scattering experiments on semicrystalline polyethylene [5.160,161], the scattering function has been calculated by
Monte Carlo simulation for various morphological models allowing adjacent reentry
according to various trial probabilities.

The interphase between lamellar crystallites and the adjoining amorphous region
in semicrystalline polymers resembles the hydrophobic layer in amphiphilic phases,
such as *lipid monolayers or bilayers*. The hydrocarbon tails are highly ordered near
the interfacial plane of the polar hydrophilic head groups [5.162,163]. The order
decreases with distance from the plane. Two possible reasons for the gradient of
disorder have been suggested. (1) Due to the polar heads, the chains are cooperati-
vely tilted with respect to the normal of the interfacial plane. This should in-

crease available volume with distance, and thus increase disorder. (2) Similar to nematic liquid crystals, an orientation-dependent intermolecular attraction could be responsible for increasing anisotropy towards the polar plane. Although such monolayers have been widely studied experimentally [5.164], it is not only recently that studies have attempted to explain the properties of monolayers using statistical physics [5.158] and simulation techniques especially. Recent molecular dynamic calculations [5.165] have been performed on small lipid bilayer models, consisting of two layers of 4 decane polymers each. The head groups are confined in the z direction by harmonic springs, the carbon groups interact via a Lennard-Jones potential, and the intramolecular interaction is taken into account by some kind of dihedral potential function. However, it is not clear whether such small systems with periodic boundary conditions can give some reasonable information on the behavior of systems consisting of many chains. Recently Monte Carlo calculations have been performed on systems of up to 10 chains comprising up to 15 bonds each, but sacrificing some reality in the chain structure [5.166]. All bonds were allowed to assume three states and excluded volume interactions are taken into account. Results are given for the order parameter $S_n = 3(<\cos^2\theta_n> - 1)/2$, where θ_n is the angle between the polar plane and the n-th carbon group. In both investigations the agreement with experiment is fair. But unfortunately many of the simulation results involve a significant amount of model (nonuniversal) information, which masks their real relation to the basic statistics of chain conformations in lipid monolayers and bilayers.

5.3.4 Phase Transitions at High Concentrations

Our present understanding of the critical properties of polymeric systems in the bulk or at higher concentrations is very poor. Examples are the liquid-crystalline transitions [5.167-170], the glass transition, interfacial transitions in adsorption phenomena and in lipid interfaces [5.171], and the melting transition. Except for the melting transition problem, recently investigated by Monte Carlo methods, none of the topics above have been studied by simulation techniques.

Melting Transition

A long-standing problem [5.172-178] in the theory of polymer chains is the phase transition of randomly disordered self-avoiding polymer chains, occupying all sites of a given lattice, to an ordered phase were all chains are oriented parallel. The transition is caused by the different thermodynamic weightings of *gauche* and trans bonds (energy $\varepsilon > 0$ for the *gauche* state and zero for the trans state). On the simple cubic or square lattice these bonds are geometrically defined as the $90°$ and the $180°$ angles, respectively, between two successive segments of one chain.

In an early paper, *Flory* [5.172] suggested that the lattice models should exhibit a complete ordering (crystallization) of the chains primarily as a result of the increasing inflexibility of the chains with lower temperatures. His arguments are based on a mean-field calculation of the probability distribution of *gauche* bonds for the disordered (high-temperature) state, from which he obtained the partition function. Due to the neglect of highly ordered configurations, the total number of configurations became less than unity at a certain temperature T_{FH} (i.e., the free energy became negative for $T < T_{FH}$). This result, he concluded, though approximative, reflects a physical reality: "It rests on the self-evident impossibility of packing long rigid chains, or rods, to high density in disordered array, an assertion that transcends the artificialities of any model" [Ref.5.178, p.4511]. In the spirit of these arguments, he recently combined [5.178] his free-energy expression for the disordered state with an approximate expression of the free energy for the ordered state proposed by *Gujrati* and *Goldstein* [5.176,177]. The two curves intersect discontinuosly at the critical temperature $T_{FG} = 1.525$, indicating a first-order phase transition. The fraction of *gauche* bonds g at the critical point is $g_+ = 0.509$ in the disordered phase and $g_- = 0.072$ in the ordered phase.

Another important attempt to tackle the problem has been done by *Nagle* [5.173]. He reinterpreted some exactly solvable two-dimensional models (KDP, F and dimer models) as polymeric models. These reinterpretations of well-known statistical models revealed some artificial behavior: the existence of short cyclic polymers, at least above the critical temperature, and the variability of chain length with temperature. Except for the KDP model, all other models have higher-order transition. Although none of those models is an ideal polymer model, they provide a reasonable starting point for further analytical calculations.

To clarify this problem at least approximately, Monte Carlo calculations have been performed recently [5.179] on the melting of the square-lattice polymer model. Using the reptation sampling technique, four different systems ($N = 11,21,31,41$) of N chains occupying N-1 lattice sites each on a $N \times N$ lattice have been simulated. The orientational order parameter <s> and the internal energy (= fraction of *gauche* bonds) <g> have been estimated by

$$\langle s \rangle = (\langle n_y \rangle - \langle n_x \rangle)/N(N - 2) \tag{5.36a}$$

$$\langle g \rangle = \langle G \rangle/N(N - 3) \quad , \tag{5.36b}$$

where $\langle n_y \rangle$, $\langle n_x \rangle$ denote the average number of segments in y and x directions respectively, and <G> the average number of *gauche* bonds. In Figs.5.14,15 the estimates of <g> and the average of the absolute value of the order parameter <|s|> are presented. For the larger systems $N = 21,31,41$ one observes first-order phase transitions, whereas for $N = 11$ (which could be considered rather as a polymer solution at concentration 0.9) no discontinuity is observed. To support the first-order character of the transition, hysteresis experiments have been performed which are shown in

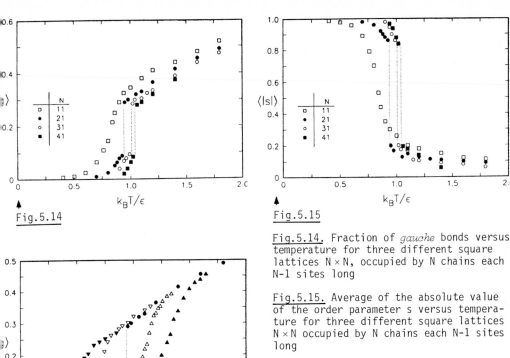

Fig.5.14

Fig.5.15

Fig.5.14. Fraction of *gauche* bonds versus temperature for three different square lattices N × N, occupied by N chains each N-1 sites long

Fig.5.15. Average of the absolute value of the order parameter s versus temperature for three different square lattices N × N occupied by N chains each N-1 sites long

◀ **Fig.5.16.** Hysteresis data for the fraction of *gauche* bonds of the square lattice polymer model N = 21 as described in the text. △ denotes $\Delta t = 4 \times 10^4$; ▲, $\Delta t =$ 8000; •, equilibrium data from Fig.5.14; up-ward (downward) pointing triangle, heating (cooling)

Fig.5.16. Averages are taken at successive temperatures over the same time interval Δt and using the final state at each temperature T as the starting configuration for the next temperature $T \pm \delta T$. Figure 5.16 shows hysteresis data for N = 21 with $\delta T = 0.02$ and $\Delta t = 4 \times 10^4$ (open triangle) and 8000 (full triangle). One Monte Carlo step has been defined as N^2 attempted motions. Triangles pointing upwards or downwards correspond to heating $(T + \delta T)$ and cooling $(T - \delta T)$, respectively. The full circles represent the equilibrium data from Fig.5.14 where $\Delta t = 10^7$ per $\delta T = 0.02$ near the transition point.

It is important to note that according to Figs.5.14,15 the jumps of energy and order parameter at the critical point are practically the same comparing N = 21,31,41. Thus I suggest that the observed jump is approximately that for the infinite system: $g_+ \simeq 0.29$, $g_- \simeq 0.09$, $|s|_- = s_- \simeq 0.85$, $|s|_+ \simeq 0.18$, $s_+ \simeq 0$. The disagreement between the mean-field value of g_+ and the estimate from Monte Carlo data is large. One might suggest that the *Flory-Huggins* approximation is poor for two-dimensional lattices concerning prediction of numerical values. The transition temperature extrapolated to $1/N \to 0$ is about $T_c = 1.16 \pm 0.06$, which is about 20% lower than the values from [5.172,173,178].

5.4 Polymer Dynamics

In this section an overview is given of the present theoretical understanding of irreversible processes in amorphous polymeric systems. Accordingly, the chain dynamics associated with conservative vibrational motions in either crystalline or liquid states will not be discussed.

5.4.1 Brownian Dynamics of a Single Chain

The most successful models of the dynamical properties of polymers represent extensions of the theory of Brownian motion [5.40]. The first successful attempt to incorporate the Brownian diffusion forces in a dynamical theory of the bead-spring model by converting the problem into that of the harmonic oscillator, without destroying the essential topological connectivity of the chain was by *Rouse* [5.30]. Extension to the case of pre-averaged hydrodynamic interactions were performed three years later by *Zimm* [5.33] and later by others [5.5,180-189].

Hydrodynamic Interactions

Monte Carlo simulations including hydrodynamic forces have not yet been performed. The most promising attempt in this direction would be the application of the Brownian-dynamics technique (Sect.5.2) proposed recently by *Ceperley* and co-workers [5.22,37] to the bead-spring model including hydrodynamic interactions. So far, only a few molecular dynamics calculations had been devoted to the hydrodynamical aspect of polymer motions [5.190-192].

The Free-Draining Limit

In concentrated solutions and undiluted polymers of not too great chain length, hydrodynamic interactions are screened out. Neglecting entanglement effects, the polymers can be considered dynamically as independent from each other (free draining). The first attempt to study the dynamics of a free-draining polymer model by means of Monte Carlo simulations was performed by *Verdier* and *Stockmayer* [5.29]. They considered a kink-jump motion (Sect.5.2) of a cubic lattice polymer chain. The surprising result of this "lattice kink-jump" dynamics was that the introduction of the excluded volume condition caused a strong slowing down of the relaxation of the squared end-to-end distances. In their original work the relaxation time τ was defined as the time required for a smooth graph of R_N^2 against the number of Monte Carlo cycles to approach its final average value to within $1/e$ of the difference between the values of the initial stretched-out configuration and the final equilibrium configuration. In subsequent work [5.193-195] the equilibrium autocorrelation function of the squared end-to-end distance

$$\phi_{RR}(t) = [<R_N^2(t_0)R_N^2(t_0 + t) - <R_N^2>^2]/[<R_N^4> - <R_N^2>^2] \qquad (5.37)$$

168

was considered. The simulations lead to $\tau \propto N^3$, in contrast to the conjecture [5.182]

$$\tau_{RR} \propto <R_N^2>/D \propto N^{2\nu+1} \quad , \tag{5.38}$$

where $D \propto 1/N$ is the center of mass diffusion constant. It has been shown recently by *Hilhorst* et al. [5.31,196] that the discrepancy is a consequence of the inability of the "lattice kink-jump" model to permit local extrema in chain conformations to pass each other or to disappear without passage of these extrema all the way to one of the ends of the chain. This type of "defect motion" is closely related to the motion of a flexible chain in a dense, randomly distributed system of fixed obstacles, introduced some years ago by *De Gennes* [5.197]. Subsequent Monte Carlo calculations [5.32,198] with the cubic-lattice model including "crankshaft" kinetics to avoid pure reptation-like motion supported this interpretation. The power law (5.38) has been confirmed by calculating the halftime $\tau_2 \propto \tau$ of the autocorrelation function (5.37), $\phi_{RR}(\tau_2) = 1/2$ assuming an exponential decay of ϕ_{RR} in that time range [5.22]. Another more general definition of the relaxation time τ is given by [5.17]

$$\tau_{RR} = \int_0^\infty dt \ \phi_{RR}(t) \quad . \tag{5.39}$$

This expression has been estimated by Monte Carlo simulations of the bead-rod model self-interacting via a Lennard-Jones potential [5.17]. At high temperatures the results are consistent with (5.38).

Local Modes in Flexible Chains

In contrast to the above discussion on the long-time behavior of free-draining polymers, this paragraph is concerned with local rapid relaxation processes at a molecular level. The very fastest processes even in dilute solution are related to conformational transitions of the chain backbone from one rotational isomeric state to another, and clearly nearly correspond to free-draining conditions as discussed above. As a particular bond rotates, the attached tails cannot rigidly follow without experiencing a huge frictional resistance. Consequently, local motions are correlated only over a short range determined by transition probabilities between the rotational isomeric states of consecutive bonds. One of the most important tasks is to derive from torsional potentials dynamic quantities such as transition probabilities, relaxation times, etc. Many investigations of this problem have been performed [5.199-201]. Standard Monte Carlo simulations have not yet been performed, but molecular dynamics simulations have been shown to be very helpful in understanding the cooperative short-time motion of realistic polymer chains [5.200].

5.4.2 Entanglement Effects

The dynamics of flexible chains in good solvents are now relatively well understood for all cases where the monomer-solvent interaction is dominant compared either with the friction between different chains or with internal barrier effects inside one chain: hydrodynamic interactions play an essential role for long chains and excluded volume effects renormalize the elastic restoring forces. The fact that the chain cannot intersect itself or other chains ("topological interaction") is a more subtle effect, important for dilute polymer solution under theta or poor solvent conditions (self-intersection) in concentrated polymer solution and in polymer melts.

Self-Entanglement Effects

Less attention has been paid to self-entanglement effects on isolated chains. In a good solvent, it is assumed that the chain makes only a few knots on itself. Then the hydrodynamic model is probably nearly correct. In a theta solvent, it is assumed [5.202] that there are many self-knots, and it is not clear how far topological interactions dominate the dynamics compared to hydrodynamic interactions. Analytically, the analysis of highly self-entangled isolated chains, probably relevant to globular proteins and ring polymers, is a most delicate exercise. Elucidating the situation to some extent, Monte Carlo studies have been reported recently on the self-entanglement effects on the dynamics of the bead-rod model [5.203]. Neglecting excluded volume potentials for the sake of clarity, only topological interaction (i.e., segmental motions where chain intersections appear are rejected) were included in the calculations of the center-of-mass diffusion constant D of the chain

$$D = \lim_{t \to \infty} g_{CM}(t)/6t\ell^2 \quad , \tag{5.40}$$

where the time-correlation function $g_{CM}(t)$ of the center of mass $R_{CG}(t) = \sum_{i=1}^{N} r_i(t)/N$ is defined as

$$g_{CM}(t) = \langle [R_{CM}(0) - R_{Cm}(t)]^2 \rangle \quad . \tag{5.41}$$

The other quantity of interest is the reduced time-correlation function of one monomer optionally near the middle of the chain ($n \simeq N/2$) to avoid end effects

$$g_r(t) = \langle [r_n(0) - r_n(t) + R_{CM}(0) - R_{CM}(t)]^2 \rangle \quad . \tag{5.42}$$

In the absence of topological interactions ($g^0(t)$, $g^0_{CM}(t)$) as well as in the presence of entanglements ($g_r(t)$, $g_{CM}(t)$), the dynamics are of Rouse type [5.30,204] (Fig.5.17), where

$$g_r(t) \propto t^{\frac{1}{2}} \tag{5.43a}$$

$$g_{CM}(t) \propto t \qquad \text{for} \quad t \to \infty \tag{5.43b}$$

$$D \propto 1/N \tag{5.43c}$$

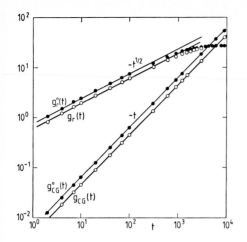

Fig.5.17. Log-log plot of the correlation functions $g_r(t)$, $g_r(t)$, $g_{CG}(t)$, $g_{CG}(t)$ as defined in the text, for $N = 160$

with $g_r(\infty) \propto <S_N^2> \propto N$, and (5,37,38)

$$\phi_{RR}(t) \propto \exp(-t/\tau_{RR}) \qquad \text{for } t \to \infty \qquad\qquad (5.43d)$$

$$\tau_{RR} \propto N^2 \quad . \qquad\qquad (5.43e)$$

But there is a strong effect on the value of the diffusion constant of the entangled chain: it is reduced by about 25% compared to the nonentangled case [5.203]. In poor solvents, where the isolated chains are expected to exist in a compact, globular state, the self-entanglement effect might be more important. This is indicated by Monte Carlo calculations on $g_r(t)$ for a bead-rod chain with self-interaction via Lennard-Jones potential at very low temperatures, where the chain is collapsed to a very dense globule [5.18]. There self-intersections are approximately excluded by the "effective" hard core of the Lennard-Jones potential $\sigma = 0.89$ [compare (5.3)]. [It can be shown that for bead-rod models with hard spheres of diameter $h/\ell \geq (3/4)^{\frac{1}{2}}$, topological intersections are excluded in kink-jump processes.] Here, the dynamics are characterized by reptation laws, at intermediate times $(t < \tau_{RR})$

$$g_r(t) \propto t^{\frac{1}{4}} \qquad\qquad (5.44a)$$

$$g_{CM}(t) \propto t^{\frac{1}{2}} \qquad\qquad (5.44b)$$

and for long times $(t > \tau_{RR})$

$$g_r(t) \to g_r(\infty) \propto N \qquad\qquad (5.45a)$$

$$g_{CM}(t) \propto t \qquad\qquad (5.45b)$$

$$D \propto 1/N^2 \qquad\qquad (5.45c)$$

$$\phi_{RR}(t) \propto \exp(-t/\tau_{RR}) \tag{5.45d}$$

$$\tau_{RR} \propto N^3 \ , \tag{5.45e}$$

originally found by *De Gennes* [5.205] for a Rouse chain moving between fixed obstacles. Although the reptation model is important for a few experimental realizations (e.g., gel chromatography), the reptation concept has become even more important in recent discussions on the dynamics of polymer melts. This is discussed in the following paragraph.

Entanglement Effects in Polymer Melts

The mechanical behavior of highly entangled polymeric liquids has been examined experimentally for many years [5.4]. The viscosity exhibits a crossover from $\eta \propto N$ to $\eta \propto N^{3.4}$ with increasing chain length at a critical value N_e. Early attempts to measure the diffusion constant by NMR methods led to $D \propto N^{-1.7}$ [5.206,207] for $N \gg N_e$, while more recently it has been found that $D \propto N^{-2.0\pm0.1}$ [5.208] for polyethylene. Attempts to measure the relaxation time have not yet led to clear answers [5.209], values between N^3 and $N^{3.5}$ so far having been produced. Various concepts to explain the dynamics of such systems have been proposed [5.210]. But the reptation concept was the first successful atttempt to include explicitly the topological effect of entanglement constraint in a melt by the use of the "tube model", where the tube is the topological skeleton of the chain under consideration, diffusing through slip links or entanglement points with other chains. First propounded by *De Gennes* [5.205], the concept was later developed by *Doi* and *Edwards* [5.211], and by *Curtiss* and *Bird* [5.212]. However, since the tube model leads to somewhat weaker exponent in viscosity and relaxation time $\eta \propto \tau \propto N^3$, the situation is still controversial and unsatisfactory.

In this situation, it was hoped that Monte Carlo simulations could shed some light on these questions. Encouraging are recent studies of an ensemble of $n = 10$ bead-rod chains (Fig.5.2b) consisting of $N = 16$ rods each [5.19,213]. As an arbitrary interaction between the beads a Lennard-Jones potential (5.3) with $\sigma/\ell = 0.4$ was assumed. Simulations have been performed for two temperatures ($k_B T/\varepsilon = 3,0.4$) and two concentrations [$c = 2.5$ and $c = 10$, where $c = nN/(L/\ell)^3$, and L is the linear size of the periodic cell]. Dynamics were simulated by the kink-jump dynamics. In the transition probability, the Lennard-Jones potential via appropriate Boltzmann factors has been included, while all rotations during which any link in the system would be intersected have been excluded. The topological interaction is quite effective: the relaxation became slower by an order of magnitude. A log-log plot of $g_r(t)$ and $g_{CM}(t)$, shown in Fig.5.18, reveals the following results: for $k_B T/\varepsilon = 3$ and $c = 2.5$ one observes a Rouse behavior similar to isolated chains (5.43a,b); only for high concentration ($c = 10$) or low temperature ($k_B T/\varepsilon = 0.4$) is there some evi-

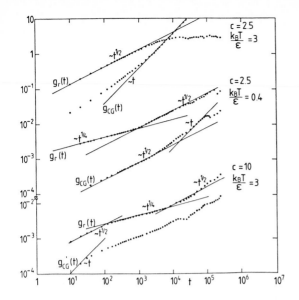

Fig.5.18. Log-log plot of the correlation functions $g_r(t)$ (5.42) and $g_{CM}(t)$ (5.41) versus t for N = 16 and several temperatures and concentrations

dence for reptation behavior (5.44a,b) at intermediate times. However, this reptation phenomenon occurs in a regime where the absolute magnitude of the displacements is very small, $g_r(t)/\ell^2 \approx 10^{-2}$. Hence it seems reasonable to interpret these simulated states as essentially glassy states, where bead positions are practically frozen in over a long time. However, a final answer to this question can be achieved only by calculating the glass transition temperature of the present model.

To compare the Monte Carlo results with recent experiments on polydimethylsiloxane (PDMS) melt using quasielastic neutron small-angle scattering [5.213], the incoherent and coherent dynamic structure factors of a single chain

$$S_{inc}(\mathbf{q},t) = <\sum_{i=1}^{N} \exp\{i\mathbf{q}.[\mathbf{r}_i(0) - \mathbf{r}_i(t)]\}>/N \qquad (5.46a)$$

$$S_{coh}(\mathbf{q},t) = <\sum_{i=1}^{N} \sum_{j=1}^{N} \exp\{i\mathbf{q}.[\mathbf{r}_i(0) - \mathbf{r}_j(t)]\}>/N \qquad (5.46b)$$

have been calculated. It has been shown by *De Gennes* [5.204] that for the Rouse chain the dynamic coherent structure factor is given by

$$S_{coh}(\mathbf{q},t)/S_{coh}(\mathbf{q},0) = f[q^2\ell^2(Wt)^{\frac{1}{2}}/6] \quad , \qquad (5.47)$$

where W is an appropriate transition probability of a bead, and f(u) is a function with $f(u) \propto \exp(-u)$ for $u \gg 1$. In Fig.5.19 the normalized dynamic coherent structure factor of experimental results (open symbols) and Monte Carlo results (full symbols) are plotted versus the Rouse variable $u = q^2\ell^2(Wt)^{\frac{1}{2}}/6$. Both experimental and simulation data collapse on a single curve, consistent with the Rouse model (5.47). The

173

full line represents the Rouse curve for an infinite chain [5.213]. The broken curves represent a recent theory of *De Gennes* [5.214,215] for the dynamic structure factor of a single reptating chain in polymer melts

$$-\ln[S_{coh}(\mathbf{q},t)/S_{coh}(\mathbf{q},0)] \propto \begin{cases} q^2 \ell^2 u \;\; ; \quad u \ll 1 \\ -\ln[1 - (qD/6)^2] - \{[(qD/6)^2 - 1]u\}^{-1} \;\; ; \quad u \gg 1 \;\; , \end{cases} \quad (5.48)$$

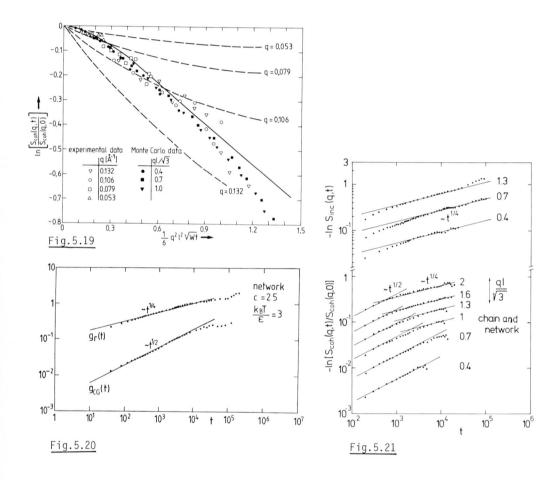

Fig.5.19

Fig.5.20

Fig.5.21

Fig.5.19. Structure factor plotted versus the Rouse variable $q^2\ell^2(Wt)^{\frac{1}{2}}/6$. Full curve: Rouse model [5.204]; broken curves: reptation model [5.125] (for details see [5.213])

Fig.5.20. Correlation functions of monomer displacement $g_r(t)$ and center of mass $g_{CM}(t)$ versus time for a chain at $k_BT/\varepsilon = 3$ in an environment of other frozen-in chains at $c = 2.5$

Fig.5.21. Log-log plot of $S_{inc}(q,t)$ (upper part) and $S_{coh}(q,t)$ (lower part) versus time for a chain at $k_BT/\varepsilon = 3$ in a frozen-in network by other chains of $c = 2.5$, for several values of q

where u is the Rouse variable, and D is the "effective tube diameter" in units of ℓ. Both experimental and Monte Carlo data disagree with *De Gennes'* predictions.

To show that the simulated chains are not too short to detect reptation at all, the results of Monte Carlo calculations of a single chain (of same chain length $N = 16$) moving in an environment of fixed obstacles have been compared to the predictions of De Gennes' reptation model (5.44). The fixed obstacles have been represented by a frozen configuration of the melt with initial parameters $c = 2.5$ and $k_B T/\epsilon = 3$. Figure 5.20 demonstrates the reptation laws (5.44a,b) in agreement with theory. Even the dynamic structure factors exhibit reptation effects (Fig.5.21): a pronounced $t^{\frac{1}{4}}$ law is observed.

Similar Monte Carlo calculations of the diffusion of one chain moving through a permanent network have been reported recently [5.216]. The stochastic motion of one cubic lattice polymer has been achieved by the kink-jump procedure, similar to the original method of *Verdier* and *Stockmayer* [5.29]. To simulate the motion of one chain under "melt conditions," where the excluded volume effect is screened out [5.139,217], the self-avoiding walk condition was neglected. They found Rouse behavior for the free chain (no obstacles), and reptation behavior in the presence of randomly occupied sites (= fixed obstacles).

More recent simulations of the dynamics of polymer melts with chains moving on a tetrahedral lattice [5.218] at a concentration of 34% and with chains comprising 200 lattice sites each and including self-avoiding walk constraints did not support the reptation law $g(t) \propto t^{\frac{1}{4}}$. A more detailed review on the dynamics of polymeric liquids is given in [5.219].

5.5 Conclusions and Outlook

Due to the lack of exact results for self-avoiding polymer chain models, Monte Carlo studies have been extremely helpful in investigating the athermal and thermodynamic properties of long flexible macromolecules. Especially the application of finite-size scaling analysis to Monte-Carlo-generated data has proven to be very powerful in clarifying the critical behavior of several phase-transition phenomena of single chains (collapse, adsorption). Very recent simulations have provided promising starting points for investigations of many-chain systems. Our understanding of some long-standing equilibrium problems (melting transition) as well as dynamical problems (dynamics of melts) has been considerably advanced by Monte Carlo simulations. Many interesting and outstanding problems such as interfacial transitions, liquid-crystalline transitions, the configurational statistics of many-chain interfaces, and the dynamics of segregation phenomena should be attacked in the future.

Acknowledgements. The author would like to thank D.Y. Yoon for his hospitality at IBM San Jose Research Laboratory, where this review was written.

References

5.1 P.J. Flory: *Statistical Mechanics of Chain Molecules* (Interscience, New York 1969)

5.2 P.J. Flory: *Principles of Polymer Chemistry* (Cornell University Press, Ithaca, New York 1971)

5.3 P.G. de Gennes: *Scaling Concepts in Polymer Physics* (Cornell University Press, Ithaca, New York 1979)

5.4 J.D. Ferry: *Viscoelastic Properties of Polymers* (Wiley, New York 1980)

5.5 W.H. Stockmayer: In *Fluides Moleculeires*, ed. by R. Balian and G. Weill (Gordon and Breach, New York 1976), p.107

5.6 See, e.g., C. Domb: Adv. Chem. Phys. **15**, 229 (1969); D.S. McKenzie: Phys. Rep. **27**, 35 (1976) and references therein

5.7 F.T. Wall, J.J. Erpenbeck: J. Chem. Phys. **30**, 634 (1985); S. Windwer: In *Markov Chains and Monte Carlo Calculations in Polymer Science*, ed. by G.G. Lowry (Marcel Dekker, New York 1970)

5.8 M.N. Rosenbluth, A.W. Rosenbluth: J. Chem. Phys. **23**, 356 (1955)

5.9 J. Mazur, F.L. McCrackin: Chem. Phys. **49**, 648 (1986)

5.10 S.J. Fraser, M.A. Winnik: J. Chem. Phys. **70**, 575 (1979)

5.11 W. Kuhn: Kolloid Z. **76**, 258 (1936; **87**, 3 (1939)

5.12 A. Baumgärtner, K. Binder: J. Chem. Phys. **71**, 2541 (1979)

5.13 A. Baumgärtner: Z. Physik **42**, 265 (1981)

5.14 K. Kremer, A. Baumgärtner, K. Binder: Z. Phys. **40**, 331 (1981)

5.15 A. Baumgärtner: J. Chem. Phys. **76**, 4275 (1982)

5.16 I. Webman, J.L. Lebowitz, M.H. Kalos: Phys. Rev. **21**, 5540 (1980)

5.17 A. Baumgärtner: J. Chem. Phys. **72**, 871 (1980)

5.18 A. Baumgärtner: J. Chem. Phys. **73**, 2489 (1980)

5.19 A. Baumgärtner, K. Binder: J. Chem. Phys. **75**, 2994 (1981)

5.20 W. Bruns, R. Bansal: J. Chem. Phys. **74**, 2064 (1981)

5.21 M. Bishop, M.H. Kalos, H.L. Frisch: J. Chem. Phys. **70**, 1299 (1979)

5.22 D. Ceperley, M.H. Kalos, J.L. Lebowitz: Macromolecules **14**, 1472 (1981)

5.23 M. Bishop, D. Ceperley, H.L. Frisch, M.H. Kalos: J. Chem. Phys. **72**, 3228 (1980)

5.24 W. Bruns, R. Bansal: J. Chem. Phys. **75**, 5149 (1981)

5.25 M. Bishop, D. Ceperley, H.L. Frisch, M.H. Kalos: J. Chem. Phys. **76**, 1557 (1982)

5.26 J. des Cloizeaux: J. Physique **37**, 431 (1976)

5.27 F.T. Wall, F. Mandel: J. Chem. Phys. **63**, 4592 (1975)

5.28 I. Webman, J.L. Lebowitz, M.H. Kalos: J. Physique **41**, 579 (1980)

5.29 P.H. Verdier, W.H. Stockmayer: J. Chem. Phys. **36**, 227 (1962)

5.30 P.E. Rouse: J. Chem. Phys. **21**, 1272 (1953)

5.31 H.J. Hilhorst, J.M. Deutch: J. Chem. Phys. **63**, 5153 (1975)

5.32 M. Lax, C. Brender: J. Chem. Phys. **67**, 1785 (1977)

5.33 B.H. Zimm: J. Chem. Phys. **24**, 269 (1956)

5.34 K. Binder (ed.): In *Monte Carlo Methods in Statistical Physics*, Topics Curr. Phys., Vol.7 (Springer, Berlin, Heidelberg, New York 1983)

5.35 F. Geny, L. Monnerie: J. Poly. Sci. Phys. Ed. **17**, 131 (1979)

5.36 K. Kremer, A. Baumgärtner, K. Binder: J. Phys. A**15**, 2879 (1982)

5.37 D. Ceperley, M.H. Kalos, J.L. Lebowitz: Phys. Rev. Lett. **41**, 313 (1978)

5.38 J.G. Kirkwood: *Macromolecules* (Gordon and Breach, New York 1976)

5.39 R. Zwanzig: Adv. Chem. Phys. **15**, 325 (1969)

5.40 S. Chandrasekhar: Rev. Mod. Phys. **5**, 1 (1943)

5.41 S.F. Edwards: Proc. Roy. Soc. London **85**, 613 (1965)

5.42 P.G. de Gennes: Phys. Lett. A**38**, 339 (1972)

5.43 J. des Cloiseaux: Phys. Rev. A**10**, 1665 (1974)

5.44 J.C. Le Guillou, J. Zinn-Justin: Phys. Rev. Lett. **39**, 95 (1977)

5.45 M.E. Fisher: Revs. Mod. Phys. **46**, 597 (1974)

5.46 E. Brezin, J.C. Le Guillou, J. Zinn-Justin: Phys. Rev. D**8**, 434, 2418 (1973)

5.47 C. Domb, A.J. Barrett: Polymer **17**, 179 (1976)

5.48 S.-K. Ma: Phys. Rev. Lett. **37**, 461 (1976)

5.49 Z. Friedman, J. Felsteiner: Phys. Rev. B**15**, 5317 (1977)

5.50 R.H. Swendsen: Phys. Rev. Lett. **42**, 859 (1979); Phys. Rev. B**20**, 2080 (1979)

5.51 K. Binder: Phys. Rev. Lett. **47**, 693 (1981)

5.52 A. Baumgärtner: J. Phys. A**13**, L39 (1980)

5.53 P.G. de Gennes: Riv. Nuovo Cimento **7**, 363 (1977)
5.54 M. Gabay, T. Garel: J. Physique Lett. **39**, 123 (1978)
5.55 Y Oono: J. Phys. Soc. Japan **47**, 683 (1979)
5.56 S.-K. Ma: *Modern Theory of Critical Phenomena*(Benjamin Redding, 1976)
5.57 T.W. Burkhardt, J.M.J. Leeuwen (eds.): *Real-Space Renormalization* (Springer, Berlin, Heidelberg, New York 1982)
5.58 H.E. Stanley, P.J. Reynolds, S. Redner, F. Family: In [Ref. 5.57, Chap.7]
5.59 B. Farnoux, F. Boue, J.P. Cotton, M. Daoud, G. Jannink, M. Nierlich, P.G. de Gennes: J. Physique **39**, 77 (1978)
5.60 P. Debye: J. Phys. Colloid Chem. **51**, 18 (1947)
5.61 M.E. Fisher, B.J. Hiley: J. Chem. Phys. **34**, 1253 (1961)
5.62 M.F. Sykes: J. Chem. Phys. **39**, 410 (1963)
5.63 C. Domb, J. Gillis, G. Wilmers: Proc. Phys. Soc. **85**, 625 (1965); **86**, 426 (1965)
5.64 J.L. Martin, M.F. Sykes, F.T. Hioe: J. Chem. Phys. **46**, 3478 (1966)
5.65 A.J. Guttman, M.F. Sykes: J. Phys. C6, 945 (1973)
5.66 S.G. Whittington, R.E. Trueman, J.B. Wilker: J. Phys. A**8**, 56 (1975)
5.67 S. Redner: J. Phys. A**13**, 3525 (1980)
5.68 A. Baumgärtner: J. Chem. Phys. **76**, 4275 (1982)
5.69 M.E. Fisher: J. Chem. Phys. **44**, 616 (1966)
5.70 J. des Cloiseaux: J. Physique **41**, 223 (1980)
5.71 J. Mazur: J. Chem. Phys. **43**, 4354 (1965)
5.72 I. Webman, J.L. Lebowitz: J. Physique **41**, 579 (1980)
5.73 M. Daoud, P.G. de Gennes: J. Physique **38**, 85 (1977)
5.74 C. Brender, M. Lax, S. Windwer: J. Chem. Phys. **74**, 2576 (1981)
5.75 Yi-der Chen: J. Chem. Phys. **74**, 2034; **75**, 2447, 5160 (1981)
5.76 W.R. Bauer, F.H.C. Crick, J.H. White: Sci. Am. **243**, 110 (1980)
5.77 S. Redner: J. Phys. A**12**, L239 (1979)
5.78 F.L. McCrackin, J. Mazur: Macromolecules **14**, 1214 (1981)
5.79 D. Stauffer, A. Coniglio, M. Adam: Adv. Polymer Sci. **44**, 103 (1982)
5.80 T.M. Birshtein, A.M. Skvortsov, A.A. Sariban: Macromolecules **9**, 888 (1976)
5.81 A.A. Sariban, T.M. Birshtein, A.M. Skvortsov: Polymer Sci. USSR **19**, 1976, 2977 (1977)
5.82 M. Levitt, A. Warshel: Nature **253**, 694 (1975)
5.83 S.H. Northrup, J.A. McCammon: Biopolymers **19**, 1001 (1980)
5.84 D.C. Rapaport, H.A. Scheraga: Macromolecules **14**, 1238 (1981)
5.85 W.R. Krigbaum, S.F. Liu: Macromolecules **15**, 1135 (1982)
5.86 P.J. Flory: J. Chem. Phys. **17**, 303 (1949)
5.87 W.A. Stockmayer: Macromol. Chemie **35**, 54 (1960)
5.88 S.F. Edwards: In *Critical Phenomena*, ed. by M.S. Green, J.V. Sengers (Nat. Bur. Stand.Miscell. Pub.) **273**, 225 (1966)
5.89 I.M. Lifshitz: Sov. Phys. JETP **28**, 1280 (1969)
5.90 C. Domb: Polymer **15**, 259 (1974)
5.91 P.G. de Gennes: J. Physique Lett. **36**, L55 (1975)
5.92 P.G. de Gennes: J. Physique Lett. **39**, L299 (1978)
5.93 J.M. Stephen: Phys. Lett. A**53**, 363 (1975)
5.94 D.J. Burch, M.A. Moore: J. Phys. A**9**, 435 (1976)
5.95 M.A. Moore: J. Phys. A**10**, 305 (1977)
5.96 I.M. Lifshitz, A.Yu. Grosberg, A.R. Khokhlov: Rev. Mod. Phys. **50**, 683 (1978)
5.97 A. Malakis: J. Phys. A**12**, 305 (1979)
5.98 C.B. Post, B.H. Zimm: Biopolymers **18**, 281 (1979)
5.99 I.C. Sanchez: Macromolecules **12**, 980 (1979)
5.100 B. Duplantier: J. Physique Lett. **41**, L409 (1980)
5.101 M.E. Fisher, B.J. Hiley: J. Chem. Phys. **34**, 1253 (1967)
5.102 F.L. McCrackin, J. Mazur, C.L. Guttman: Macromolecules **6**, 8591 (1973)
5.103 R. Finsy, M. Janssen, A. Belleman: J. Phys. A**8**, L106 (1975)
5.104 A.M. Skvortsov, T.M. Birshtein, A.A. Sariban: Polymer Sci. USSR **18**, 3124 (1976)
5.105 D.C. Rapaport: J. Phys. A**10**, 637 (1977)
5.106 J.G. Curro, D.W. Schaefer: Macromolecules **13**, 1199 (1980)
5.107 I. Webman, J.L. Lebowitz, M.H. Kalos: Macromolecules **14**, 1495 (1981)
5.108 J. Tobochnik, I. Webman, J.L. Lebowitz, M.H. Kalos: Macromolecules **15**, 549 (1982)

5.109 A. Baumgärtner: J. Physique **43**, 1407 (1982)
5.110 K. Kremer, A. Baumgärtner, K. Binder: J. Phys. A**15**, 2879 (1982)
5.111 C. Cuniberti, U. Bianchi: Polymer **15**, 346 (1974)
5.112 E.L. Slagowski, B. Tsai, D. McIntyre: Macromolecules **9**, 687 (1976)
5.113 M. Nierlich, J.P. Cotton, B. Farnoux: J. Chem. Phys. **69**, 1379 (1978)
5.114 G. Swislow, S. Sun, I. Nishio, T. Tanaka: Phys. Rev. Lett. **44**, 796 (1980)
5.115 M.E. Fisher: *Proceedings of the International Summer School Enrico Fermi.* Course LI (Academic, New York 1971)
5.116 E.K. Riedel: Phys. Rev. Lett. **28**, 675 (1972)
5.117 R.B. Griffiths: Phys. Rev. B**7**, 545 (1973)
5.118 M.J. Stephen, J.L. McCauley: Phys. Lett. A**44**, 89 (1973)
5.119 P.G. de Gennes: Rep. Prog. Phys. **32**, 187 (1969)
5.120 E. Eisenriegler, K. Kremer, K. Binder: J. Chem. Phys. **77**, 6296 (1982)
5.121 F.L. McCrackin: J. Chem. Phys. **47**, 1980 (1967)
5.122 M. Lax: Macromolecules **7**, 660 (1974)
5.123 S.G. Whittington: J. Chem. Phys. **63**, 779 (1975)
5.124 G.M. Torrie, K.M. Middlemiss, S.H.P. Bly, S.G. Whittington: J. Chem. Phys. **65**, 1867 (1976)
5.125 A. Bellemans, J. Orban: J. Chem. Phys. **75**, 2454 (1981)
5.126 T. Ishinabe: J. Chem. Phys. **76**, 5589 (1982)
5.127 T. Ishinabe: J. Chem. Phys. **77**, 3171 (1982)
5.128 M.N. Barber: Phys. Rev. B**8**, 407 (1973)
5.129 P.G. de Gennes: J. Physique **37**, 1445 (1976)
5.130 A.J. Bray, M.A. Moore: J. Phys. A**10**, 1927 (1977)
5.131 M.N. Barber, A.J. Guttman, K.M. Middlemiss, G.M. Torrie, S.G. Whittington: J. Phys. A**11**, 1833 (1978)
5.132 K. de Bell, J.W. Essam: J. Phys. C**13**, 4811 (1980)
5.133 J.S. Reeve, A.J. Guttman: Phys. Rev. Lett. **45**, 1581 (1980)
5.134 H.W. Diehl, S. Dietrich: Phys. Rev. B**24**, 2878 (1981)
5.135 L. Ma, K.M. Middlemiss, S.G. Whittington: Macromolecules **10**, 1415 (1977)
5.136 K. Binder: In *Phase Transitions and Critical Phenomena*, Vol.8, ed. by C. Domb, J.L. Lebowitz (Academic, New York 1983) p.2
5.137 P.H. Geil: J. Macromol. Sci.-Phys. Ed. **12**, 173 (1976)
5.138 P.J. Flory: Faraday Disc. Chem. Soc. **68**, 14 (1979)
5.139 P.J. Flory: J. Chem. Phys. **17**, 303 (1949)
5.140 R.G. Kirste, W.A. Kruse, J. Schelten: Makromol. Chemie **162**, 299 (1973)
5.141 D.G. Ballard, J. Schelten, G.D. Wignall: Eur. Polym. J. **9**, 965 (1973)
5.142 J.P. Cotton, D. Decker, H. Benoit, B. Farnoux, J. Higgins, G. Jannink. R. Ober, C. Pigot, J. des Cloiseaux: Macromolecules **7**, 863 (1974)
5.143 A. Bellemans, M. Janssens: Macromolecules **7**, 809 (1974)
5.144 E. de Vos, A. Bellemans: Macromolecules **7**, 812 (1974); **8**, 651 (1975)
5.145 J.G. Curro: J. Chem. Phys. **61**, 1203 (1974); **64**, 2496 (1976); Macromolecules **12**, 463 (1979)
5.146 F.T. Wall, J.C. Chin, F. Mandel: J. Chem. Phys. **65**, 2231 (1976); **66**, 3143 (1977)
5.147 F.T. Wall, W.A. Seitz: J. Chem. Phys. **67**, 3722 (1977); Proc. Natl. Acad. Sci. USA **76**, 8 (1979)
5.148 A.M. Skvortsov, A.A. Sariban, T.M. Birshtein: Polymer Sci. USSR **19**, 1169 (1977)
5.149 T.A. Weber, E. Helfand: J. Chem. Phys. **71**, 4760 (1979)
5.150 T.M. Birshtein, A.A. Sariban, A.M. Skvortsov: Polymer **23**, 1481 (1982)
5.151 H. Okamoto: J. Chem. Phys. **64**, 2686 (1976)
5.152 M. Bishop, D. Ceperley, H.L. Frisch, M.H. Kalos: J. Chem. Phys. **75**, 5538 (1982)
5.153 A. Baumgärtner: Polymer **23**, 334 (1982)
5.154 M. Vacatello, G. Avitabile, P. Corradini, A. Tuzi: J. Chem. Phys. **73**, 548 (1980)
5.155 P.J. Flory, D.Y. Yoon: Nature **272**, 226 (1978)
5.156 L. Mandelkern: Faraday Disc. Chem. Soc **68**, 310 (1979)
5.157 P.G. de Gennes: Macromolecules **14**, 509 (1981); **15**, 492 (1982)
5.158 K.A. Dill, P.J. Flory: Proc. Natl. Acad. Sci. USA **77**, 3115 (1980)
5.159 D.Y. Yoon, P.J. Flory: Polymer **18**, 509 (1977); Faraday Disc. Chem. Soc. **68**, 288 (1979)

5.160 J. Schelten, D.G. Ballard, G.D. Wignall, G. Longman, W. Schmatz: Polymer **17**, 751 (1976)
5.161 M. Stamm, E.W. Fischer, M. Dettenmaier, P. Convert: Faraday Disc. Chem. Soc. **68**, 263 (1979)
5.162 S.J. Singer, G.L. Nicolson: Science **175**, 720 (1972)
5.163 C. Tanford: *The Hydrophobic Effect* (Wiley, New York 1980)
5.164 J. Seelig: Quart. Rev. Biophys. **10**, 353 (1977)
5.165 P. van der Ploeg, H.J.C. Berendsen: J. Chem. Phys. **76**, 3271 (1982)
5.166 H.L. Scott: Biochim. Biophys. Acta **469**, 264 (1977)
5.167 P.J. Flory: Proc. Roy. Soc. London A**234**, 73 (1956)
5.168 P.J. Flory, G. Ronca: Mol. Cryst. Liq. Cryst. **54**, 289 (1979); **54**, 311 (1979)
5.169 G. Ronca, D.Y. Yoon: J. Chem. Phys. **76**, 3295 (1982)
5.170 A.Yu. Grosberg, A.R. Khokhlov: Adv. Poly. Sci. **41**, 53 (1981)
5.171 J.F. Nagle: Ann. Rev. Phys. Chem. **31**, 157 (1980)
5.172 P.J. Flory: Proc. Roy. Soc. London A**234**, 60 (1956)
5.173 J.F. Nagle: Proc. Roy. Soc. London A**337**, 569 (1974)
5.174 M. Gordon, P. Kapadia, A. Malakis: J. Phys. A**9**, 751 (1976)
5.175 A. Malakis: J. Phys. A**13**, 651 (1980)
5.176 P.D. Gujrati, M. Goldstein: J. Chem. Phys. **74**, 2596 (1981)
5.177 P.D. Gujrati: J. Stat. Phys. **28**, 441 (1982)
5.178 P.J. Flory: Proc. Natl. Acad. Sci. USA **79**, 4510 (1982)
5.179 A. Baumgärtner, D.Y. Yoon: J. Chem. Phys. **79**, 521 (1983)
5.180 M. Bixon: Ann. Rev. Phys. Chem. **27**, 65 (1976)
5.181 R. Zwanzig: J. Chem. Phys. **60**, 2717 (1974)
5.182 S.F. Edwards: In *Fluides Moleculares*, ed. by R. Balian and G. Weill (Gordon & Breach, New York 1976) p.151
5.183 P.G. de Gennes: Macromolecules **9**, 587 (1976)
5.184 D. Jasnow, M.A. Moore: J. Physique Lett. **38**, 467 (1977)
5.185 J. des Cloiseaux: J. Physique Lett. **39**, 151 (1978)
5.186 M. Daoud, G. Jannink: J. Physique **39**, 331 (1978)
5.187 M.G. Brereton, S. Shah: J. Phys. A**11**, L111 (1978)
5.188 G. Weill, J. des Cloiseaux: J. Physique **40**, 99 (1979)
5.189 R.S. Adler, K.F. Freed: J. Chem. Phys. **70**, 3119 (1979)
5.190 J.A. McCammon, J.M. Deutch: Biopolymers **15**, 1397 (1976)
5.191 M. Fixman: J. Chem. Phys. **69**, 1527, 1538 (1978)
5.192 B.H. Zimm: Macromolecules **13**, 592 (1980)
5.193 P.H. Verdier: J. Chem. Phys. **45**, 2122 (1966)
5.194 D.E. Kranbuehl, P.H. Verdier: J. Chem. Phys. **56**, 3145 (1972)
5.195 P.H. Verdier: J. Chem. Phys. **59**, 6119 (1973)
5.196 H. Boots, J.M. Deutch: J. Chem. Phys. **67**, 4608 (1977)
5.197 P.G. de Gennes: J. Chem. Phys. **55**, 572 (1971)
5.198 D.E. Kranbuehl, P.H. Verdier: J. Chem. Phys. **71**, 2662 (1979)
5.199 E. Helfand: J. Chem. Phys. **71**, 5000 (1979)
5.200 E. Helfand, Z.R. Wasserman, T.A. Weber: Macromolecules **13**, 526 (1980)
5.201 J. Skolnick, H. Helfand: J. Chem. Phys. **72**, 5489 (1980)
5.202 F. Brochard, P.G. de Gennes: Macromolecules **10**, 1157 (1977)
5.203 A. Baumgärtner: Polymer **22**, 1308 (1981)
5.204 P.G. de Gennes: Physics **3**, 37 (1967)
5.205 P.G. de Gennes: J. Chem. Phys. **55**, 572 (1971)
5.206 D.W. McCall, C.M. Huggins: Appl. Phys. Lett. **7**, 153 (1965)
5.207 J.E. Tanner, K.-J. Liu, J.E. Anderson: Macromolecules **4**, 586 (1971)
5.208 J. Klein: Nature **271**, 143 (1978)
5.209 G. Kraus, K.W. Rollmann: J. Poly. Sci. Poly. Symp. **48**, 87 (1974)
5.210 W. Graessley: Adv. Poly. Sci. (1982)
5.211 M. Doi, S.F. Edwards: J. Chem. Soc. Faraday Trans. **274**, 1789, 1802, 1818 (1978); **275**, 38 (1979)
5.212 C.F. Curtiss, R.B. Bird: J. Chem. Phys. **74**, 2016, 2026 (1981)
5.213 D. Richter, A. Baumgärtner, K. Binder, B. Ewen, J.B. Hayter: Phys. Rev. Lett. **47**, 109 (1981); **48**, 1695 (1982)
5.214 P.G. de Gennes: J. Chem. Phys. **72**, 4756 (1980)

5.215 P.G. de Gennes: J. Physique **42**, 735 (1981)
5.216 S.F. Edwards, K.E. Evans: J. Chem. Soc. Faraday Trans. **77**, 1891, 1913, 1929 (1981)
5.217 S.F. Edwards: J. Phys. A**8**, 1670 (1975)
5.218 K. Kremer: Macromolecules **16**, 1632 (1983)
5.219 A. Baumgärtner: Ann. Rev. Phys. Chem. **35**, 419 (1984)

6. Simulation of Diffusion in Lattice Gases and Related Kinetic Phenomena

K. W. Kehr and K. Binder

With 26 Figures

This chapter reviews various Monte Carlo studies of dynamical properties of lattice gas models, which serve to simulate self-diffusion of tagged particles in interstitial and substitutional alloys, surface diffusion of adsorbate atoms in adsorbed monolayers, etc. These systems serve as archetypical models of order-disorder and unmixing phase transitions, and are well suited to study the basic aspects of associated kinetic phenomena near equilibrium as well as far from equilibrium, such as nucleation of ordered domains from a disordered phase, their diffusion-controlled growth, coarsening of domain structures by diffusion of domain walls, etc. Earlier Monte Carlo work on related problems, such as the kinetics of nucleation and phase separation [6.1], or the kinetics of crystal growth [6.2], has been a unique tool for checking analytical theories on this subject, and has stimulated beautiful new experiments. We feel that the more recent work reviewed here will again be very stimulating for a variety of fields, from the statistical thermodynamics of irreversible processes to materials science. The main emphasis of this article is on kinetic phenomena near equilibrium, i.e., on simulations of diffusion in lattice gases, which have not yet been reviewed, except for [6.3].

6.1 General Aspects of Monte Carlo Approaches to Dynamic Phenomena

In the importance-sampling Monte Carlo method, a Markov chain of system configurations (or phase space points R_ν, respectively) is generated by a stochastic process, using a transition probability $W(R \to R')$ which satisfies a detailed balance condition to ensure approach towards thermal equilibrium

$$W(R \to R')P_{eq}(R) = W(R' \to R)P_{eq}(R') \quad , \tag{6.1}$$

with $P_{eq}(R) = \exp(-\mathcal{H}/k_B T)/\text{Tr}\{\exp(-\mathcal{H}/k_B T)\}$, $\mathcal{H}$ being the Hamiltonian of the system considered. This sequence of states $R \to R' \to R''$ ..., generated one after the other from the respective previous state can simply be interpreted as a dynamic evolution of the system, if a time t is associated with the sequential label ν of these states. Then the probability $P(R,t)$ that a state R occurs at time t satisfies a master equation [6.1], where W is now interpreted as a transition probability per unit time

$$\frac{d}{dt} P(\mathbf{R},t) = - \sum_{\mathbf{R}'} W(\mathbf{R} \rightarrow \mathbf{R}')P(\mathbf{R},t) + \sum_{\mathbf{R}'} W(\mathbf{R}' \rightarrow \mathbf{R})P(\mathbf{R}',t) \quad . \tag{6.2}$$

It is important to note that the detailed balance condition (6.1) does not specify $W(\mathbf{R} \rightarrow \mathbf{R}')$ uniquely: the precise choice of $W(\mathbf{R} \rightarrow \mathbf{R}')$ is hence determined from other physical considerations about the process one wishes to simulate. For instance, suppose we have in mind an interstitial alloy (such as H in Pd) at high enough temperatures, where all interactions are neglected except that double occupancy of the sites of the interstitial lattice is forbidden. Then the transition $\mathbf{R} \rightarrow \mathbf{R}'$ may consist of jumps of an interstitial atom from one lattice site to another, and we may consider jumps with a rate Γ_{nn} towards nearest-neighbor sites, and with a rate Γ_{nnn} towards next-nearest-neighbor sites, etc. In this case, all these rates are put to zero if a site to which a particles wishes to jump is already occupied, and (6.1) is trivially fulfilled. Of course, the (trivial) equilibrium properties of this model are completely independent of the particular assumption about Γ_{nn}, Γ_{nnn}, etc., although they do influence dynamic properties as the self-diffusion constant. The same is also true for more complicated models with nontrivial static properties: many different dynamic models belong to the same static model, and the choice of a particular dynamical model is to a large extent arbitrary. It will in practice be dictated by physical considerations from other sources (e.g., experimental evidence that Γ_{nnn} is negligibly small in comparison with Γ_{nn} or something like that).

At this point it is quite obvious that the description of dynamic processes via Monte Carlo methods is a phenomenological modeling of what one thinks are the essential steps of a coarse-grained description, rather than a first-principles approach. In the latter, one would have to deal with the Schrödinger equation of all the particles forming the host lattice and the interstitials. Even the first-principle description of one diffusing interstitial along such lines would be a formidable problem, and since the time scale at which jumps typically occur is orders of magnitude larger than the phonon times, the phenomenological hopping-model description is sufficient for many purposes. In other systems such as superionic conductors, these time scales are not really so different: nevertheless the hopping model is frequently used as a crude approximation or at least as an interesting limiting case.

Concentrating on an important subset of the degrees of freedom while all the other subsets act only as a heat bath, that induces random changes $\mathbf{R} \rightarrow \mathbf{R}'$, is a simplification enabling so many rather different physical problems to be modeled to some extent by a Monte Carlo process.

6.2 Diffusion in Lattice-Gas Systems in Equilibrium

6.2.1 Self-Diffusion in Noninteracting Two- and Three-Dimensional Lattice Gases

In this subsection we consider particles which occupy the sites of two- and three-dimensional regular lattices with finite concentration $c(0 < c < 1)$. Double occupancy is excluded, otherwise the particles are noninteracting. The equilibrium state of these lattice gases is characterized by random occupation of the sites. The dynamics of this class of models consists of transitions of the particles with rate Γ to empty nearest-neighbor sites. In the noninteracting case the collective diffusion, i.e., the decay of a density gradient, can be reduced to a single-particle problem [6.4], and the coefficient of collective diffusion is identical to the diffusion coefficient $D_{s.p.}$ of a single particle in the empty lattice.

The main point of interest of the noninteracting lattice gas is the correlated random walk of individual, tagged particles. While the total state change of the lattice gas is a Markov process, the motion of a tagged particle is no longer Markovian, and its random walk is correlated. This fact was first recognized by *Bardeen* and *Herring* [6.5] in the context of metal physics, corresponding to the limit $c \to 1$. In metals, diffusion is caused by thermal vacancies with concentration $c_v \ll 1$ ($c_v \hat{=} 1 - c$). When a tagged atom has exchanged with a vacancy there is a large probability that it exchanges again with this vacancy. Hence a strong backward correlation exists for the motions of the tracer. Without these correlations the diffusion coefficient of the tracer would be $D_t = c_v D_{s.p.}$. The backward correlations are taken into account by defining a correlation factor f, and the tracer diffusion coefficient is $D_t = f c_v D_{s.p.}$. The correlation factor f can be derived in the metal physics limit from the random walk of the vacancy which exchanges with the tracer [6.6,7]. There is no point in simulating diffusion in regular crystals in the limit $c \to 1$, except to derive some auxiliary quantities of theoretical models, such as parameters of the "encounter model" [6.8]. Hence we completely ignore this important field here.

Tracer diffusion becomes less well understood when one abandons the limitation $c_v \ll 1$ and considers general concentrations of the lattice gas. The physical origin of the correlation effects in the motion of tagged particles is essentially the same as in the limit $c \to 1$. If the particle has made a jump, there is a vacancy behind it immediately after the jump with probability 1, whereas in the other directions vacancies are present with probability 1-c. Hence there is again a backward correlation and a corresponding reduction of the tracer diffusion coefficient. The obvious generalization of the expression for the tracer diffusion coefficient of the noninteracting lattice gas is

$$D_t = (1 - c)D_{s.p.} f(c) \quad , \tag{6.3}$$

where 1-c is the availability factor of vacancies of the noninteracting lattice gas and f(c) the correlation factor. Furthermore $(1-c)D_{s.p.}$ also represents the "mean-field" tracer diffusion coefficient in the absence of correlations (Sect.6.2.3).

Simulation of the tracer diffusion coefficient in a lattice gas is straightforward. In the noninteracting case it suffices to occupy randomly the sites of the lattice with a given concentration of particles to provide an equilibrium configuration. Each particle of the lattice gas is considered as tagged and its initial as well as actual position is registered. At each elementary step, a particle is selected at random, then a random neighboring site is chosen. If the neighboring site is empty, the particle is moved to this site. In this way the attempted jumps are simulated as a Poisson process. The introduction of a time unit is arbitrary; however, a convenient definition is the Monte Carlo step (MCS) per particle. During one MCS each particle attempts one jump, *on the average*. This definition corresponds to choosing $\Gamma = 1/z$ for the noninteracting case, where z is the number of nearest-neighbor sites.

The diffusion coefficient is obtained from

$$D_t(n) = \frac{1}{2dn} \langle R_n^2 \rangle \quad , \tag{6.4}$$

where $\langle R_n^2 \rangle$ is the average mean-square displacement of all tagged particles after n MCS, d the dimensionality. D_t is defined theoretically as the limit $n \to \infty$ of (6.4), although in practice $D_t(n)$ fluctuates with n. No improvement of accuracy is achieved by increasing the number of steps n, since the root mean-square deviations of $\langle R_n^2 \rangle$ also grow with n. On the other hand, the deviations of $D_t(n)$ from its correct value are proportional to the inverse square root of the particle number, hence large particle numbers of the lattice gases should be used. An additional possibility to increase the estimated accuracy of D_t is to average (6.4) over different initial starting times, or, equivalently, over a range of steps,

$$\bar{D}_t = \frac{1}{n_2 - n_1} \sum_{n'=n_1}^{n_2} D_t(n') \quad . \tag{6.5}$$

This average has been performed in practically all investigations of tracer diffusion. It is useful to divide by the mean-field expression for the tracer diffusion coefficient, $(1-c)D_{s.p.}$, and to present data for the correlation factor only.

We now survey the numerical simulations of the correlation factor f(c) of the noninteracting two- and three-dimensional lattice gas. To our knowledge the first simulation was that by *De Bruin* and *Murch* [6.9] who determined f(c) for the simple cubic lattice, Fig.6.1. Both exactly known limits $f(c \to 0) = 1$ and $f(c \to 1) = 0.653 \ldots$ are nicely reproduced. This and the following papers of Murch and co-workers also contain simulations of the interacting case, Sect.6.2.4. Further data on f(c) for the sc lattice were given by *Murch* and *Thorn* [6.10]. The fcc lattice was studied by *Murch* [6.11]; additional data for f(c) of the fcc lattice were given in [6.12], see Fig.6.2 for the results of this paper. *Murch* and *Thorn* [6.14] simulated diffusion on the honeycomb lattice in two dimensions, where $f(c \to 1) = 1/3$. It was found that f(c) varied approximately linearly with c, in contrast to the three-dimensional case.

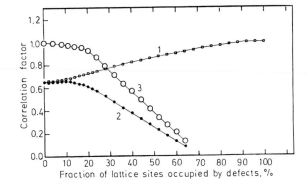

Fig.6.1. Correlation factor as a function of concentration for tracer diffusion in a simple cubic lattice gas. (1) represents f(c) for nearest-neighbor jumps of the particles on the sc lattice, (2) and (3) refer to more complicated interstitial processes. From [6.9] after omission of an insert

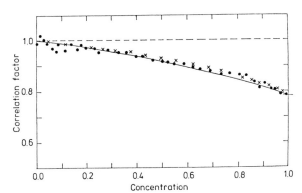

Fig.6.2. Correlation factor as a function of concentration of an fcc lattice gas. The points represent the results of the simulation, the crosses the results according to (6.8), the full line the theory of *Sankey* and *Fedders* [6.13], and the dashed line the mean-field value. The exactly known value f(1) = 0.78146... is indicated by a bar [6.12]

Unpublished data [6.15] exist for the simple square lattice, which are shown in Fig.6.3. Some of the lattice structures have not yet been simulated; however, one expects similar behavior to that of the structures already studied.

The first successful theory for f(c) in two- and three-dimensional lattice gases over the whole concentration range was given by *Sankey* and *Fedders* [6.13]. They introduced a complicated diagrammatic formulation of the problem of tracer diffusion in a lattice gas. However, they did not evaluate their final expression for f(c) in a fully self-consistent way, resulting in a small deviation of their f(c → 1) from the exact values. A different derivation of f(c) was given by *Naka-zato* and *Kitahara* [6.16], which corresponds to a self-consistent treatment. Their result is

$$f(c) = \frac{1 + <\cos\theta>}{1 + [(2 - 3c)/(2 - c)]<\cos\theta>} \ , \tag{6.6a}$$

where $<\cos\theta>$ is the average angle of successive jumps of the tracer in the limit c → 1. Equation (6.6a) contains the usual expression for the correlation factor in the limit c → 1

$$f = \frac{1 + <\cos\theta>}{1 - <\cos\theta>} \ . \tag{6.6b}$$

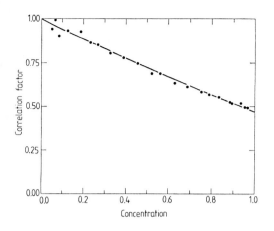

Fig.6.3. Correlation factor as a function of concentration of a simple square lattice gas. 2500 sites were introduced and 1400 MCS/particle taken in the simulation. The line gives the theory of *Nakazato* and *Kitahara* [6.16]

Expansion of (6.6a) up to second order yields Fedder's result. The result (6.6a) has been rederived recently by truncation of the hierarchy of master equations for the motion of a tracer atom by *Tahir-Kheli* and *Elliott* [6.17], and *Schroeder* [6.18]. Previous work includes treatment of the case $c_v \ll 1$ with matrix methods by *Koiwa* [6.19], the work of *Benoist* et al. [6.20] and *Wolf*'s analysis in terms of a modified encounter model [6.21]. However, the approximate analytical expression proposed by Wolf disagrees with (6.6a) and the numerical simulations. Figures 6.1-3 include the theory of Sankey and Fedders, and Nakazato and Kitahara (which cannot be distinguished on the scales of the figures). The (self-consistent) theory is exact near c = 0 and c = 1. Comparison with the simulation shows that it is a very good approximation at all concentrations.

We turn to the studies of self-diffusion as a dynamical process. The general aim is the determination of the time-dependent conditional probability $P(\boldsymbol{\ell},t)$ of finding a tagged particle at site $\boldsymbol{\ell}$ at time t, when it started at the origin at t = 0. The Fourier transform of this quantity is the incoherent dynamical structure function which is experimentally accessible, e.g., by inelastic incoherent neutron scattering. A general formulation for determining $P(\boldsymbol{\ell},t)$ has again been provided by *Sankey* and *Fedders* [6.13], however, no explicit results have been given. Also the procedure of *Nakazato* and *Kitahara* [6.16] yields in principle an expression for the Fourier transform of $P(\boldsymbol{\ell},t)$. Two more phenomenological approaches will be described now. *Ross* and *Wilson* [6.22] expressed $P(\boldsymbol{\ell},t)$ as a sum over the product of the probability of finding n attempted jumps by time t, and the spatial distribution over the sites of the lattice after n attempted jumps. The spatial distribution of the tracer particle after n attempted jumps has been determined by Monte Carlo simulation on an fcc lattice. Only preliminary results have been published in [6.22]. Those authors found that the incoherent scattering law of a lattice gas on an fcc lattice is approximately one Lorentzian at low frequencies, with an effective width determined by the jump rate, blocking factor 1-c, and the correlation factor f(c). They also showed that the incoherent scattering function approaches the mean-field form at high frequencies.

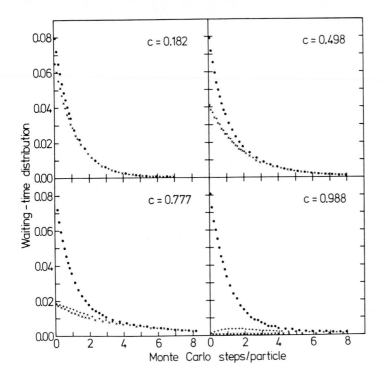

Fig.6.4. Waiting-time distribution for backward jumps (dots) and forward jumps to nearest-neighbor sites (plus signs) and fourth-neighbor sites (crosses). The jump rates are normalized to $\Gamma = 1/12$ [6.12]

In [6.12] the time dependence of the random-walk process of the tracer particles was determined in detail in the form of "waiting-time distributions." Waiting-time distributions (WTD) are basic quantities of random-walk theory [6.23]; the work described below seems to be the first simulation of such quantities in a nontrivial physical context. The waiting-time distribution $\psi(t)dt$ of a particle which has performed its last jump at $t = 0$ is defined as the probability that it performs its next jump between t and $t + dt$ after having waited until time t. The WTD are easily numerically estimated by determining the time intervals between successive jumps separately for each tagged particle. A distinction was also made between backward jumps of a tracer particle, and forward jumps to neighbor sites of the order $(1) - (4)$, relative to the previous jump at $t = 0$. The WTD for a forward jump is expected to be nearly exponential, corresponding approximately to a Poisson process with mean rate $(1-c)\Gamma$. This has been confirmed by simulations, Fig.6.4. The WTD for a backward jump should reflect the behavior of the vacancy which is left behind the tagged particle immediately after its jump. With increasing time this vacancy can also be filled by other particles, resulting in a strongly time-dependent WTD for backward jumps, as is seen in Fig.6.4. From the WTD a time-dependent conditional jump rate can be extracted which is displayed in Fig.6.5. One sees that

187

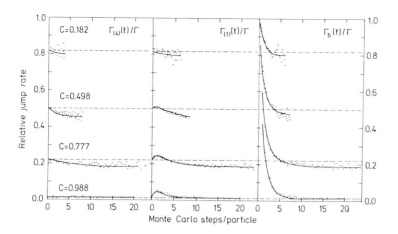

<u>Fig.6.5.</u> Time-dependent jump rates $\Gamma(t)/\Gamma$ of tracer particles, with the condition that no jump has occurred by time t. (4) corresponds to jumps to fourth-nearest neighbors, (1) to jumps to first neighbors, and (b) to backward jumps. The points give the results of the simulations and the curves represent the theoretical expressions of [6.12]

the jump rate for backward jumps approaches its unblocked value Γ for $t \to 0$, whereas the jump rates for other jumps approach $(1-c)\Gamma$ in this limit. The time dependence of the backward jump rate can be deduced from the stochastic process of filling the vacancy [6.12]. As is seen in Fig.6.5 the rate $\Gamma(t)$ for jumps to nearest-neighbor sites of the previous site of the particle at $t = 0^-$, where the vacancy exists at $t = 0^+$, is initially enhanced. This is due to the possibility, peculiar to the fcc structure, that the vacancy hops to this site (1) and enables the jump of type (1) of the tagged particle. This process has also been quantitatively deduced in [6.12]. The asymptotic behavior of the jump rate $\Gamma(t)$ for $t \to \infty$ has been only partially clarified [6.12].

The Fourier transform of the self-correlation function $P(\ell, t)$ can be expressed in terms of the Fourier transforms of the WTD, however, the expressions are rather lengthy for the fcc lattice, even in the main symmetry directions. Much simpler is the resulting expression for the frequency-dependent diffusion coefficient

$$D_t(\omega) = (1 - c)D_{s.p.} f(\omega) \quad , \tag{6.7}$$

where

$$f(\omega) = \mathrm{Re}\left\{\frac{1 + \Delta(\omega)}{1 - \Delta(\omega)}\right\} \tag{6.8}$$

and the combination of the Fourier transforms of the WTD gives

$$\Delta(\omega) = \int_0^\infty dt\ \exp(-i\omega t)[\psi_4(t) + 2\psi_3(t) - 2\psi_1(t) - \psi_b(t)] \quad . \tag{6.9}$$

In the limit $\omega = 0$, $\Delta(0)$ is identical to the average of the angle between successive jumps of the tagged particle, $<\cos\theta>(c)$. Hence we recover in this limit the result for the correlation factor in the form of (6.6b) but now with the average $<\cos\theta>$ at finite concentration.

In [6.12] $\Delta(0)$ was determined from the integrals over the WTD; $f(c)$ can be estimated from $\Delta(0)$. Figure 6.2 shows that there is good agreement between $f(c)$ estimated in this indirect way, and $f(c)$ estimated directly from the simulation of self-diffusion. It should be pointed out that the use of the WTD between consecutive jumps of a tagged particle implies that only correlations between such successive jumps are included; also the determination of the conditional probability and $D_t(\omega)$ from the WTD employs this approximation. The good agreement between the directly simulated $f(c)$ and $f(c)$ determined via (6.7-9) shows that the approximation of correlated consecutive jumps is good in the fcc lattice at all concentrations, not only near $c = 0$ and $c = 1$ where it becomes exact.

Up to now no direct simulations of the frequency-dependent diffusion coefficient or of the full conditional probability have been performed. It should be emphasized that the static diffusion coefficient $D_t(0)$ is most sensitive to the proper inclusion of all correlations. Nevertheless, it is desirable that the frequency dependence of (6.7-9) be tested by a simulation, so that the transition to the high-frequency behavior with $f(\omega \to \infty) = 1$ can be verified. Particularly interesting seems to be the behavior of the Fourier transform of $P(\ell,t)$ near the edge of the Brillouin zone, where *Tahir-Kheli* and *Elliott* [6.17] predict rapid variations as a function of concentration.

6.2.2 Anomalous Diffusion in One-Dimensional Lattices

In this subsection we consider linear chains occupied by a finite concentration $c(0 < c < 1)$ of particles. The dynamics is the same as for two- and three-dimensional lattices, each particle can hop to an empty neighbor site with rate Γ. The single-filing constraint, i.e., the condition that the particles cannot pass each other, leads to a vanishing tracer diffusion coefficient in the infinite chains. A formal expression of the same fact is $f(c) = 0$ for the linear chain.

In view of the vanishing tracer diffusion coefficient one must study the mean-square displacement of tagged particles directly. It was pointed out by mathematicians [6.24,25] that the mean-square displacement of a tagged particle on the concentrated linear chain should be proportional to $t^{\frac{1}{2}}$, instead of the usual t, for long times. The first numerical investigation of tracer diffusion in the linear chain was made by *Richards* [6.26], who indeed found proportionality of $<X_n^2>$ with $n^{\frac{1}{2}}$ at a concentration of $c = 1/2$. Richards also gave qualitative arguments for the reduced mean-square displacement in terms of the density fluctuations against which the test particle has to move. Shortly afterwards *Fedders* [6.27] was able to derive the asymptotic law for $\Gamma t \gg 1$

$$<X^2>(t) = \frac{2(1 - c)}{c}\left(\frac{\Gamma t}{\pi}\right)^{\frac{1}{2}}$$

(6.10)

from the diagrammatic approach of [6.13]. The lattice spacing has been put equal one, a = 1. Fedders recognized a 30% difference between his asymptotic law and Richards' numerical results. A physical justification of (6.10) was given by *Alexander* and *Pincus* [6.28] who related the diffusion of a tagged particle to the collective diffusion of the other particles, and obtained (6.10) up to an erroneous numerical factor π. For short times, the mean-square displacement of a tagged particle should behave as

$$<X^2>(t) = 2(1 - c)\Gamma t \quad .$$

(6.11)

Meanwhile the behavior of the mean-square displacement at arbitrary times was derived by van Beijeren. He deduced the velocity correlations of the motion of a tracer particle by considering the temporal history of the vacancy that is present immediately after the jump of the tracer; this vacancy is indeed responsible for all backward correlations. He found an explicit, although still approximate expression for the velocity correlation function of the tracer in the Laplace domain and thus also the mean-square displacement in the Laplace domain. The ensuing asymptotic laws for $<X^2>(t)$ agree with (6.10,11); at arbitrary times $<X^2>(t)$ can be obtained by numerical inverse Laplace transformation. The theory has been described in [6.29] and compared with numerical simulations. Figure 6.6 shows that there is very satisfactory agreement between the theory and simulation at different concentrations and all times. The crossover between (6.10,11) is seen nicely. Matching of both equations yields the estimate $\Gamma t^* = (1/\pi c^2)$ for the crossover time.

In [6.29] the influence of the boundary conditions has also been studied. The case of reflecting boundary conditions is rather simple: the mean-square displacement of a tagged particle reaches a finite value after an initial rise according to (6.11) and some transient behavior. The final mean-square displacement can be found

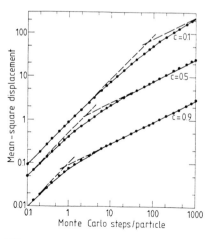

Fig.6.6. Mean-square displacement of tracer particles on linear chains of 64 000 sites as a function of time. The points have been obtained by simulations, the full lines represent the theory of van Beijeren and the dashed lines the asymptotic behavior of (6.10,11) [6.29]

from combinatorial considerations. One obtains asymptotically for finite chains with N sites after averaging over all particles

$$<\Delta X^2>_{as} \approx \frac{(1 - c)}{3c} N \quad . \tag{6.12}$$

Figure 6.7 shows that this behavior is well obeyed for different concentrations and chain lengths. The more interesting case is that of periodic boundary conditions. Now a particle can also encircle the one-dimensional chain. If displacements are measured along the circle, without reduction to the original chain, the displacement of a tagged particle may grow with time indefinitely. It turns out that the mean-square displacement increases linearly with time for large times,

$$<X^2>(t) \xrightarrow[t \to \infty]{} 2D_t t \tag{6.13}$$

with a tracer diffusion coefficient

$$D_t = \frac{(1 - c)}{c} \frac{\Gamma}{(N - 1)} \quad . \tag{6.14}$$

Figure 6.8 demonstrates that the mean-square displacement asymptotically reaches a linear behavior. The full lines represent the extension of van Beijeren's theory to the case of periodic boundary conditions, the dashed lines indicate the asymptotic law (6.13,14). Figure 6.9 shows the N dependence of the diffusion coefficient D_t deduced from numerical simulations. We mention that the problem of tracer diffusion in a one-dimensional lattice gas with periodic boundary conditions is related to the problem of tracer diffusion through pores in membranes, which can be filled by other particles. In fact, the tracer diffusion coefficient of single-file diffusion through membrane pores is identical with that given in (6.14) and was first derived in this context, see [6.30] and the references cited therein.

While tracer diffusion is anomalous in the one-dimensional lattice gas, collective diffusion is as simple as in higher dimensions. Also in this case it is equivalent to a single-particle problem and the coefficient of collective diffusion is simply $D_{coll} = D_{s.p.}$. The velocity autocorrelation function of all particles consists of a δ-like spike at the time origin, as has been verified by the numerical simulations of *Richards* [6.26]. The collective diffusion becomes more complicated when the linear chain consists of two different types of sites A, B which are periodically arranged, have different equilibrium energies and transition rates to neighboring sites. In this case the velocity autocorrelation function exhibits a negative contribution at general times, and the diffusion coefficient is reduced accordingly [6.26].

An interesting question is that of the crossover between one-dimensional and higher-dimensional lattice gases. Consider a d-dimensional lattice gas with one

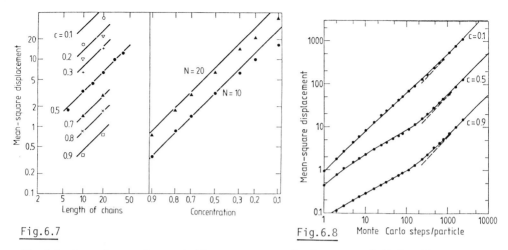

Fig.6.7

Fig.6.8   Monte Carlo steps/particle

Fig.6.7. Dependence of asymptotic mean-square displacement of finite linear chains on (a) number of sites N and (b) $(1-c)/c$. The lines correspond to (6.12)

Fig.6.8. Mean-square displacement of tracer particles on linear chains of 20 sites with periodic boundary conditions. For further remarks, see Fig.6.6. The dashed lines correspond to diffusion with D_t given in (6.14)

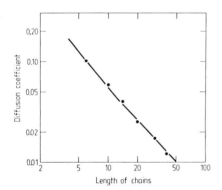

Fig.6.9. Dependence of the diffusion coefficient of tracer particles around chains with periodic boundary conditions on chain length for $c = 0.5$. The line corresponds to (6.14)

direction with large jump rates $\Gamma_{\parallel}$, and d-1 directions with small jump rates $\Gamma_{\perp}$. It is to be expected that for intermediate times the mean-square displacement shows a power-law behavior $\propto t^{\alpha}$ where α is near 1/2, and that for large times diffusional behavior $\propto t$ appears. A particularly simple model system has been investigated by *Kutner* et al. [6.31], where they considered two coupled chains with rates $\Gamma_{\parallel}$ along the chains and $\Gamma_{\perp}$ between the chains. The mean-square displacement shows the features sketched above. Further studies of this crossover behavior are suggested.

6.2.3 Tracer Particles with Different Jump Rates and the Percolation Conduction Problem

It is interesting to consider the diffusion of tracer atoms in a noninteracting lattice gas where the tracer particles have a different jump rate (Γ') compared to that of the background particles (Γ). Physical realizations of this model are given, for instance, by hydrogen in metals with some isotopes added (D,T, or μ^+). The primary interest in this model is, however, theoretical: very different physical situations can be realized by varying the ratio of the jump rates and the concentration c of the background particles. It is evident that these models are easily simulated on the computer, except for the need of larger systems and large computing times to produce good statistics. For the same reason it is necessary to introduce a small but finite concentration c'(c' $\ll$ c) of tracer particles.

We begin with the situation of immobile background particles ($\Gamma = 0$). *De Gennes* [6.32] compared the motion of a tracer particle in a lattice where a portion of the sites are blocked by other particles with the random walk of an ant in a labyrinth. If the concentration of the background particle is larger than $1-c_{pv}$, where c_{pv} is the percolation concentration of the vacancies, a test particle will find itself in finite vacancy clusters only, and no long-range diffusion is possible. At the percolation threshold when $c = 1 - c_{pv}$ diffusion will be anomalous while for $c < 1-c_{pv}$ diffusion of a tracer particle will be possible, if it is in the infinite vacancy cluster. These statements hold for infinite systems; in finite systems with periodic boundary conditions a particle may show diffusion already at $c \gtrsim 1-c_{pv}$ in a vacancy cluster connected by the boundary conditions. Figure 6.10 shows the behavior of the mean-square displacement below and above the percolation threshold of the vacancies.

We shall outline here the main points of the expected critical behavior of the diffusional properties near the percolation threshold of the vacancies $c_v = c_{pv}$ and

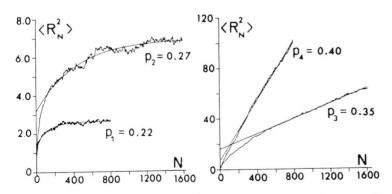

Fig.6.10. Mean-square displacement of tracer particles as a function of the number of steps where the sites of an sc lattice are partially blocked from occupation. The vacancy concentrations have been indicated, their percolation concentration is $p_c \hat{=} c_{pv} = 0.312...$, [6.33]

refer to a forthcoming survey for more details [6.34]. Phenomenologically, the main quantities of interest have the following leading power-law behavior

i)   For $c_v < c_{pv}$ $<R_\infty^2> \propto (c_{pv} - c_v)^{-m}$ (6.15a)

ii)   For $c_v > c_{pv}$ $<R^2>(t) \propto t(c_v - c_{pv})^x$ (6.15b)

iii)   For $c_v = c_{pv}$ $<R^2>(t) \propto t^k$.   (6.15c)

Apart from the early work of *Brandt* [6.35], the first systematic simulation of diffusion of test particles in a partially blocked lattice near the percolation threshold was made by *Mitescu* and *Roussenq* [6.36], see also [6.33,37]. They obtained the following critical indices:

$x = 0.99 \pm 0.2$ $(d = 2)$

$x = 1.72 \pm 0.03$ $(d = 3)$

$m = 2.5 \;\; \pm 0.2$ $(d = 2)$

$m = 1.1 \;\; \pm 0.4$ $(d = 3)$.

Vicsek [6.38] simulated the ac conductivity for $c_v < c_{pv}$ in simple-square and simple cubic lattices and obtained $m = 2.4 \pm 0.3$ $(d = 2)$ and $m = 1.2 \pm 0.2$ $(d = 3)$. However, his value of x, deduced indirectly from the approach to the asymptotic mean-square displacement would be only $x = 0.6$ in $d = 2$ *and* $d = 3$. Recently *Mitescu* and *Roussenq* reanalyzed their data and obtained $m = 1.65 \pm 0.05$ in three dimensions [6.34]. Unfortunately Mitescu and co-workers did not present data on the time dependence of $<R^2>(t)$ at the percolation threshold for $1 \ll \Gamma t \ll |c_v - c_{pv}|^{x-m}$. [They analyzed mainly the short-time behavior and obtained $k = 0.78 \pm 0.02$ $(d = 2)$ and $k = 0.63 \pm 0.02$ $(d = 3)$; a value of $k = 0.49 \pm 0.01$ was found in $d = 3$ from intermediate times.] The exponent k was determined recently in [6.39] from simulations on an fcc lattice in the time range $600 \leq \Gamma t \leq 6000$ and $k = 0.51 \pm 0.02$ was found (Fig.6.11).

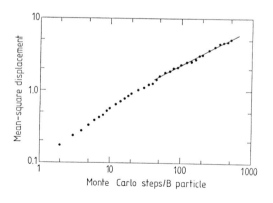

Fig.6.11. Mean-square displacement of tracer particles as a function of time where the sites are partially blocked with concentration $c = 0.807$, in double logarithmic representation. The straight line has a slope of 0.51 [6.39]

To interpret the data, we note the close similarity between the diffusion of the tracer particles near the vacancy percolation threshold and the percolation conductivity problem suggested by *de Gennes* [6.32]. If we consider the *vacant* sites of the lattice to consist of conducting material and if current can flow between neighboring conducting sites, the system exhibits a conductivity σ for $c_v > c_{pv}$ which behaves as [6.40]

$$\sigma \propto (c_v - c_{pv})^\mu \quad . \tag{6.16}$$

The precise relation between the conductivity exponent μ and the exponents introduced above was the subject of some debate [6.34], finally it was clarified by *Gefen* et al. [6.41]. These authors established a scaling form of the mean-square displacement of a tracer particle which can start at an arbitrary vacant site (i.e., which may belong to a larger or smaller vacancy cluster). The results of their scaling analysis are

$$x = \mu \tag{6.17a}$$

$$m = 2\nu - \beta \tag{6.17b}$$

$$k = \frac{2\nu - \beta}{2\nu - \beta + \mu} \quad , \tag{6.17c}$$

β and ν being the usual critical indices of percolation [6.42]; in $d = 2$ they are presumably given exactly by $\beta = 5/36$ and $\nu = 4/3$ [6.43]. The most recent estimate of the critical indices β and ν in three dimensions is the one obtained by *Gaunt* and *Sykes* [6.44] from series expansions, $\beta = 0.454 \pm 0.08$ and $\nu = 0.88 \pm 0.02$. The resulting values of $m = 2.53$ ($d = 2$) and $m = 1.31 \pm 0.03$ ($d = 3$) are in good agreement with the original analyses by Mitescu and co-workers, and with Vicsek. The conductivity exponent was directly determined by *Kirkpatrick* [6.45] with the result $\mu = 1.10 \pm 0.05$ ($d = 2$) and $\mu = 1.62 \pm 0.05$ ($d = 3$). From these values one predicts an exponent $k = 0.70 \pm 0.01$ in $d = 2$ and $k = 0.45 \pm 0.02$ in $d = 3$. The three-dimensional value is somewhat less than the above-mentioned value of $k = 0.52 \pm 0.02$. However, the numerical results of [6.39] must be regarded as provisional since the samples were not sufficiently large and the time intervals should also be taken larger. If one took the numerical value $k = 0.51$, one would predict a conductivity exponent of $\mu = 1.26$ in $d = 3$. This value is definitively below the value found by Kirkpatrick. At this point we refer to another determination of k (or μ) in $d = 3$ from the investigation of the case of a moving background, see below.

The other region of interest is $c_v > c_{pv}$ where long-range diffusion of tracer particles is possible if they are in an infinite cluster. This region has also been treated by *Brandt* [6.34] who produced some preliminary data on the tracer diffusion constant as a function of the concentration of the background particles. The problem of diffusion of a particle in a lattice where some sites are inaccessible plays an important role in exciton transport in binary crystals ([6.46] and references therein). The simulations pertaining to exciton transport concentrated mainly on the determination of the number of distinct sites visited, and on the effect of correlated random walk on this quantity. The diffusion of tracer particles in a lattice with fixed inaccessible sites was studied in [6.39] for an fcc lattice. The data were presented in form of a correlation factor $f_R(c)$ by writing the diffusion coefficient as

$$D_t = (1 - c_{tot})\Gamma' f_R(c) \quad . \tag{6.18}$$

Here $c_{tot} = c + c'$ and the lattice constant has been taken as $a = 1$. The theory of *Nakazato* and *Kitahara* [6.16] mentioned above also includes the dependence on the ratio of the jump rates, $\gamma = \Gamma'/\Gamma$. If this ratio is infinite, the following expression for $f_R(c)$ results:

$$f_R(c) = \left(1 + \frac{1 - f}{f} \frac{c}{1 - c}\right)^{-1} \quad , \tag{6.19}$$

where f is the correlation factor of tracer particles with equal jump rates in the limit $c \to 1$ (6.6b). Figure 6.12 shows the results of [6.39] together with (6.19). Agreement is satisfactory between theory and simulation below $c \lesssim 0.5$. It can be concluded that an adequate theoretical description of tracer diffusion for this case exists, as long as one is far away from the percolation threshold.

We now turn to the general case of moving background and tracer particles with different jump rates. Essentially two groups of simulations have been made. The first group was directed towards an explanation of NMR data in mixtures on H with

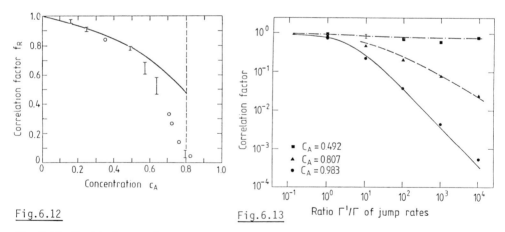

Fig.6.12   Fig.6.13   Ratio Γ'/Γ of jump rates

Fig.6.12. Residual correlation factor $f_R(c)$ according to (6.19). The circles and error bars give the results of the simulations, the full line represents (6.19) and the dashed line indicates the percolation threshold [6.39]

Fig.6.13. Correlation factor for diffusion of tracer atoms with jump rate Γ' in a lattice gas with particles at concentration c and jump rates Γ, below, near and above the percolation threshold of the vacancies. The curves represent theoretical expressions explained in the text

D in niobium. *Fukai* and co-workers [6.47,48] assumed specific forms of the hydrogen/ deuterium interactions and obtained results in agreement with their measurements. In the second simulation [6.39], diffusion in a noninteracting lattice gas with a wide range of ratios γ and concentrations $(0 < c < 1)$ was explored and c' was always small compared to c. The results were expressed in terms of a generalized correlation factor, defined in analogy with (6.18)

$$D_t = (1 - c_{tot})\Gamma'f(c,\gamma) \quad . \tag{6.20}$$

It was checked that the dependence of $f(c,\gamma)$ on c' was of minor importance. Figure 6.13 shows the results from [6.39] regarding $f(c,\gamma)$ as a function of γ in three different regimes. There is common behavior in the limit $\gamma \to 0$. Here the tracer experiences a rapidly fluctuating background, hence all correlations in the random walk of the tracer decay quickly, and $f = 1$, i.e., mean-field behavior occurs. At high particle concentrations the correlation factor can be well represented by a formula which follows from the work of *Manning* [6.7]:

$$f(c \to 1,\gamma) = \frac{f}{f + (1 - f)\gamma} \quad . \tag{6.21}$$

In the limit $c \to 1$ diffusion of the tracer is effected by isolated single vacancies. In this case the correlations between different jumps of the tracer can be reduced to the correlations of consecutive jumps, and the usual derivations of the correlation factor extended to the case $\gamma \neq 1$. It should be mentioned that (6.21) can also be obtained from [6.16] in the limit $c \to 1$. However, this theory fails for general $c > 1-c_{pv}$ and $\gamma \gg 1$. On the other hand, it provides a reasonable description for $c \ll 1-c_{pv}$ as is seen from the curve for the lowest concentration in Fig.6.13. Also the results given in Fig.6.12 support the validity of this theory in this region.

Finally, near the percolation threshold of vacancies an asymptotic power-law behavior of $f(1-c_{pv},\gamma)$ appears. The curve given in Fig.6.13 was obtained by a fit of the data with a modified form of (6.21); its asymptotic behavior is

$$f(1 - c_{pv}, \gamma \to \infty) \propto \gamma^{-w} \tag{6.22}$$

and $w = 0.58 \pm 0.01$ for $c = 0.807$. The scaling form for $\langle R^2 \rangle(t)$ developed by *Gefen* et al. [6.41] can be extended to the case of a moving background and the following prediction for the critical behavior can be made [6.39]:

$$\langle R^2 \rangle(t) \propto t\,\gamma^{-\frac{\mu}{2\nu-\beta+\mu}} \quad . \tag{6.23}$$

Taking the critical indices for percolation given before, one obtains $w = 0.55 \pm 0.02$ while the extrapolation of the simulation to $c = 1-c_{pv}$ gives $w = 0.56 \pm 0.02$. Hence there is good agreement between prediction and simulation. It appears that the study of this general model of a moving background allows another independent verification of the critical conductivity index of percolation.

6.2.4 Self-Diffusion and Collective Diffusion in Interacting Lattice Gases, Including Systems with Order-Disorder Phase Transitions

It is of interest to consider also interaction of the lattice gas particles, both for theoretical reasons and for practical purposes, since most real systems show interaction effects. A great variety of interactions of the lattice gas particles is possible. Even if the interactions are known, they do not yet fully determine the jump rates. However, the jump rates in the interacting systems must satisfy the condition of detailed balance, cf. (6.1). Some remarks pertaining to their choice have been made in Sect.6.1.

In the interacting case, the preparation of the samples must be made in a different way from the noninteracting case. Now a distinct preparation step is necessary to achieve thermodynamic equilibrium of the lattice gas. The problem is that particle conservation delays the approach to equilibrium, since it can occur through diffusion only. A more rapid approach to equilibrium is achieved when the condition of fixed particle number is replaced by the condition of fixed chemical potential during the preparatory step. The procedure is more easily visualized in magnetic language. In the preparatory step, flips of nonconserved, interacting spins are executed with transition rates in accordance with the interactions, and the magnetic field is kept fixed. At the end of this step the magnetization, corresponding to the particle number, is determined, and kept fixed in the diffusion step to follow [6.12,49]. This preparation technique is still insufficient when phase separation occurs or ordered phases appear, see below.

One of the earliest simulations of diffusion in interacting lattice gases was motivated by experiments of diffusion of oxygen in nonstoichiometric UO_{2+x}, and the interactions were chosen to model the properties of this system [6.50]. *De Bruin* and *Murch* [6.9] considered the effect of vacancy interactions (exclusion of double occupancy of nearest-neighbor sites by vacancies), the influence of motions of Willis-type defects, and of an interstitialcy mechanism on the tracer diffusion of oxygen in UO_{2+x}. A similar investigation was applied to $Fe_{1-x}S$ [6.51].

Later simulation work concentrated mainly on the Ising model of nearest-neighbor interactions of the particles of the lattice gas. In the terminology of spin models, we have an Ising model with exchange of interacting spins which are conserved; this is the *Kawasaki* model [6.52]. As mentioned above, there is still freedom in the choice of the transition rates of the particles in the interacting system. *Murch* and *Thorn* and some other groups, e.g. [6.53], chose as transition probability of a particle to jump from site i to a vacant site ℓ_i

$$P(i \rightarrow \ell_i) = \exp(-\Delta E/k_B T) \quad , \tag{6.24}$$

where

$$\Delta E = \begin{cases} \varepsilon(\ell - z + 1) & \text{for repulsion} \quad (\varepsilon < 0) \\ \varepsilon \ell & \text{for attraction} \quad (\varepsilon > 0) \end{cases} \quad .$$

Here z is the number of nearest-neighbor *sites* of the lattice and ℓ the number of nearest-neighbor *particles* in the initial situation. Our group chose

$$P(i \rightarrow \ell_i) = \frac{1}{2} \left(1 - \tanh \frac{\delta \mathcal{H}}{2k_B T} \right) \quad , \tag{6.25}$$

where $\delta \mathcal{H}$ is the energy difference between the final and initial situations. In the simulation, a random number z is compared with P. If $z < P$ the jump is performed, if $z > P$ no jump occurs. While both choices fulfil the condition of detailed balance (6.1) and thus lead to the same equilibrium configurations, the first choice is asymmetric with respect to interchange of particles and vacancies, e.g., in the case $\varepsilon > 0$ and the limit $c \rightarrow 0$ a step of a single particle is much less probable than a step of a vacancy in the limit $c \rightarrow 1$. We reiterate that the choice of the transition probabilities is partly arbitrary and that additional criteria must be used to fix it.

To obtain a diffusion constant with dimensions we must convert transition probabilities into transition rates. We have already defined a Monte Carlo step (MCS) per particle and use the same time unit here also. The choice of transition probabilities of (6.24) corresponds to assigning a rate $\Gamma = 1/z$ in units of MCS/particle to the transition between two specific sites in the case of no interactions. It is convenient to normalize the jump rates in the single-particle limit $c \rightarrow 0$, i.e., to introduce an extra factor $\exp[|\varepsilon|(z-1)]$ in the repulsive case for the choice (6.24). This has been done in the work of Murch et al. The second choice (6.25) requires the assignment $\Gamma = 1/2z$ for the transition rate between two sites in the non-interacting case. No further normalization is needed here.

Murch and Thorn have decomposed the tracer diffusion coefficient in the interacting case in the following way (given in a slightly modified form and valid for cubic crystals with lattice spacing a):

$$D_t = VWfa^2 \quad . \tag{6.26}$$

Here V is the vacancy availability factor, W the effective rate of performed jumps and f the correlation factor. The vacancy availability factor reflects an equilibrium property of the lattice gas; it is related to the nearest-neighbor short-range order parameter α_1 through

$$V = (1 - c)(1 - \alpha_1) \quad . \tag{6.27}$$

Estimates of V or α_1 can be made in Monte Carlo simulations without paying attention to dynamic quantities. The effective rate at which possible jumps are performed on the average is given by

$$W = \tau_s^{-1} <P(i \rightarrow \ell_i)>_T / V \quad . \tag{6.28}$$

If the MCS/particle is employed as the time unit, the time scale factor is $\tau_s = z$ with z the number of nearest-neighbor sites. In simulations $<P(i \rightarrow \ell_i)>$ is estimated from the quotient of the number of performed jumps to the number of all at-

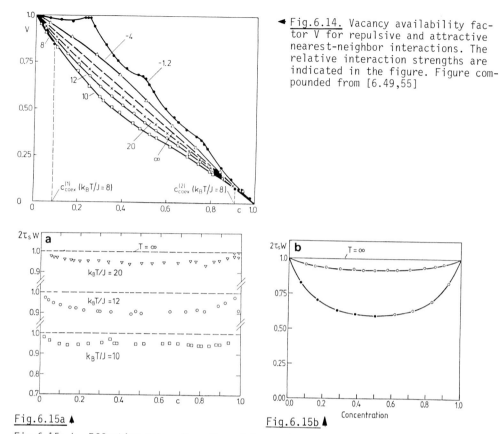

◄ Fig.6.14. Vacancy availability factor V for repulsive and attractive
nearest-neighbor interactions. The
relative interaction strengths are
indicated in the figure. Figure compounded from [6.49,55]

Fig.6.15a ▲   Fig.6.15b ▲

Fig.6.15a,b. Effective rate of performed jumps W as a function of concentration.
(a) Attractive nearest-neighbor interactions. Figure compounded from [6.49,55];
(b) repulsive nearest-neighbor interactions; the circles corresponds to kT/J = -20
and the full points to kT/J = -4

tempted jumps. This procedure amounts to determining the time average over all
successful hopping events, As can be verified in examples by application of de-
tailed-balance arguments, the average is identical to an ensemble average over the
different possible jump rates.

Since the vacancy availability factor V is an equilibrium quantity, its value
should be independent of the choice of the transition rates. Data on V in a lattice gas in an sc lattice for attractive and repulsive interactions were given by
Murch and *Thorn* [6.10] and for an attractive lattice gas in an fcc lattice in
[6.54]. Data on V of the fcc lattice gas have also been obtained in [6.49] for attractive and in [6.55] for repulsive interactions, with good agreement with previous work. Some typical data on V are shown in Fig.6.14. One sees that repulsive
interaction increases the vacancy availability factor while attraction decreases
it below the noninteracting value 1-c. Also the two-phase region in the case of
attraction has been omitted for the reasons given above. The effective rate of per-

formed jumps W is influenced by the choice of the transition probability. The data presented by *Murch* and *Thorn* [6.10] show an asymmetry between $c \to 0$ and $c \to 1$; their W increases with c for repulsion and decreases with attraction. We prefer to work with the symmetrical definition (6.25). In Fig.6.15 we present typical data on W both for attraction and repulsion, at higher temperatures where no ordering effects have occurred. In both cases one observes a symmetrical reduction of W below its value for the noninteraction system. It is not completely clear whether different choices of the transition probability can influence the correlations which appear in the random walk of a tracer atom as discussed above. It is plausible that no or only little influence exists. *Murch* and *Thorn* found a slight decrease of the correlation factor f(c) on the sc lattice gas for attractive interaction, and a more pronounced decrease in the case of repulsive interaction [6.10]. The strongest decrease was found at a temperature below the order-disorder transition, and at a concentration of $c = 1/2$. See below for further comments on diffusion in ordered structures.

In the fcc lattice gas, *Murch* and *Thorn* found a slight reduction of f(c) in the attractive case [6.54], compared to the noninteracting value, with considerable scatter of the individual data points, especially around $c = 0.7$. *Kutner* et al. [6.49] found within the accuracy of the data no influence of the attractive interaction on the correlation factor, Fig.6.16. In the repulsive case of an fcc lattice gas, there is also only a minor influence of the interactions of f(c) as long as one is in the regions of disorder of the phase diagram, Fig.6.17. Specific application of tracer diffusion in an fcc lattice in the interacting case has been made in [6.56] to carbon diffusion in austenite.

Tracer diffusion in two-dimensional lattice gases with interactions has been investigated by *Murch* and *Thorn* to model Na diffusion in the superionic conductors β" and β alumina [6.57,58]. The pertinent lattice is honeycomb structured in both cases with equivalent sites for β" alumina and two inequivalent sites for β alumina The vacancy availability factor V shows the same qualitative behavior as in three-dimensional lattice gases. The effective rate W of performed jumps also exhibits similar behavior to the data of Murch and Thorn on three-dimensional lattices. The correlation factor f(c) now shows a light *increase* with increasing interaction strength, and a decrease with increasing repulsion, Fig.6.18. Again the lowest curve with a pronounced minimum was taken with an interaction strength which leads to an ordered structure at $c = 1/2$. We also refer to the discussion of diffusion in ordered structures below. One aim of the simulations was to assess the validity of the path probability method of *Sato* and *Kikuchi* [6.59] for diffusional problems. While there is very good agreement between the results for V and W and the numerical data, there is a clear discrepancy between the predictions for f and the simulations, attributable to the nature of the approximations employed in that method. The case of two energetically inequivalent sites which are periodically arranged on the honey-

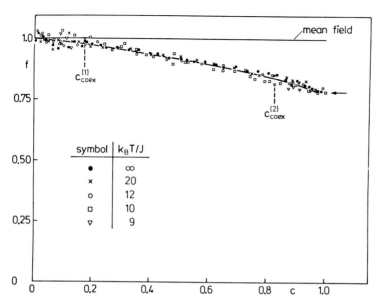

Fig.6.16. Correlation factor as a function of concentration for fcc lattice gases with attractive interactions of various strenghts, from [6.49]. The theory of *Sankey* and *Fedders* [6.13] is indicated by a full line

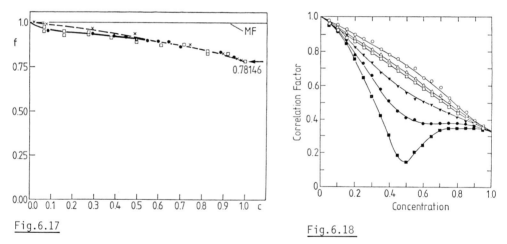

Fig.6.17

Fig.6.18

Fig.6.17. Correlation factor as a function of concentration for fcc lattice gases with repulsive interaction of strengths kT/J = -20 (crosses), -6 (dots), and -4 (squares). The dashed line indicates the theory of *Sankey* and *Fedders* [6.13]. Figure taken from [6.55]

Fig.6.18. Correlation factor for tracer diffusion in a honeycomb-lattice gas, redrawn from Figs.3,4 of [6.57] but with omission of some theoretical results. Relative interaction strengths $k_B T/J$: ■ -0.3, ● -0.5, ▼ -1,0, □ ∞, △ 1.0, ○ 0.5

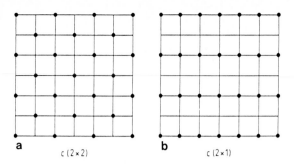

a c(2×2) b c(2×1)

Fig.6.19a,b. c(2 × 2) and c(2 × 1) structure of square lattice gas at c = 1/2 under the interaction conditions described in the text. The c(2 × 1) structure shown is degenerate with one vertical orientation of the lines

comb lattice can be related to the case of a lattice of equivalent sites by simple thermodynamic considerations. This was done in [6.58] and the relevant quantities V,M,f were determined by the simulations. Again, very good agreement of the data for V and W, but not for f, with the results of the path probability method was found.

We now turn to the diffusion of particles in ordered structures which can appear in the lattice gases when there are repulsive nearest-neighbor interactions, corresponding to Ising models with antiferromagnetic ordering. The most simple such models are the lattice gases on square lattices and on a simple cubic lattice. Both models have an ordered structure for c = 1/2, repulsive nearest-neighbor interactions, and low temperature. For the square lattice the ordered structure is a square lattice again with lattice constant $a' = a\sqrt{2}$; for the sc lattice the ordered structure is an fcc lattice. More interesting and complicated structures appear when there are competing interactions. For instance, when there are nearest- and next-nearest neighbor interactions J_{nn} and J_{nnn} in a square-lattice gas, the structure depends on the ratio $R = J_{nnn}/J_{nn}$. Diffusion studies have been made for R = 1 where the c(2 × 1) structure comprises a stable structure at low temperatures for c = 1/2 (Fig.6.19b) and for R = -0.465 where c(2 × 2) is the stable structure for c = 1/2 (Fig.6.19a). More information on the possible structures and phase diagrams of these systems can be found in [6.60]. In the case of an fcc lattice gas interesting structures appear already for repulsive nearest-neighbor interactions only, since competing effects appear as a consequence of possible "triangular" connections of nearest-neighbor sites in this lattice structure. The structures and the phase diagram of an fcc lattice gas with repulsive nearest-neighbor interactions have been elucidated in [6.61,62]. The completely ordered structures at low temperatures are an sc lattice of occupied sites at c = 1/4, alternating occupied and empty planes perpendicular to a (100) direction at c = 1/2, and an sc lattice of vacant sites at c = 3/4. Clearly all these structures will be reflected in the behavior of the diffusion coefficients.

To simulate diffusion in ordered structures, special care is necessary for the preparatory step. As pointed out above, it is advantageous in the interacting case to prepare an interacting lattice gas at equilibrium by fixing the chemical potential instead of the particle number. If a lattice gas were prepared in this way in a region of the phase diagram with an ordered structure where the particles are initially added at random positions, very likely domain walls, grain boundaries, etc., would appear which would heal out at low temperatures only after very long times, beyond any reasonable computing times. Hence the lattice gas must be prepared in the ordered regions of the phase diagrams by starting from an ordered structure which can be created by providing a modulated chemical potential for the initial occupation process. The correct equilibrium distribution of vacancies is then obtained by thermalization steps. It should be pointed out that a large system at finite temperature will have the type of defects mentioned above, but with low concentrations only at low temperatures. Whereas a neglectful preparation step generates such defects at high concentrations, not corresponding to equilibrium.

The previous investigations of tracer diffusion in interacting lattice gases include cases where the ordered phases are stable. However, since the preparation technique discussed above was not implemented, the results must be considered as qualitative only. This remark refers to the lowest temperature investigated in the sc lattice in [6.10] where a strong dip in f(c) at c = 1/2 was found, and in the honeycomb lattice where an analogous effect was seen [6.57]. In [6.63] where diffusion in the c(2 × 2) structure of a square lattice gas with competing interactions was studied, neither data on self-diffusion nor on collective diffusion were obtained directly. It is also not established that the Darken equation and analogous relations can be used to predict the diffusivities in these ordered structures correctly.

A recent investigation by *Sadiq* and *Binder* [6.64] of the square lattice gas with competing interactions used the structure-adapted preparation procedure outlined above. They investigated mainly the case R = 1, where at low temperature and c around 1/2 alternating filled and empty lines appear, with some vacancies, and interstitials between the lines. (We refrain from reviewing the results for the disordered case at higher temperature, and refer to the original paper [6.64] for much of the details.) Now clear one-dimensional effects appear in the tracer diffusion, Fig.6.20. At the concentration of c = 0.47 and very low temperatures the filled lines have vacancies (i.e., the filled lines are occupied with c' = 0.94) and one-dimensional diffusion occurs. The mean-square displacement shows an aproximate $t^{\frac{1}{2}}$ behavior, according to the discussion of Sect.6.2.2. At concentration c = 0.53 there are a few extra particles (effective concentration c' = 0.06) between the filled rows, which then perform one-dimensional diffusion within the empty lines, although it is not as evident as before. At c' = 0.06 and the times covered by the simulation one is still below the crossover time t^* from t^1 to a $t^{\frac{1}{2}}$ behavior. The full lines in Fig.6.20 were obtained by applying the theory of van Beijeren and readjusting the prefactor. This theory cannot be applied directly since the particles have additional interaction, not included in the theory. For example, at c' = 0.06 two mobile particles will experience an additional repulsive interaction

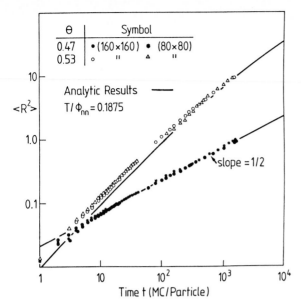

θ	Symbol	
0.47	• (160×160)	• (80×80)
0.53	○ "	△ "

Analytic Results ———

$T/\phi_{nn} = 0.1875$

$\langle R^2 \rangle$

slope = 1/2

Time t (MC/Particle)

Fig.6.20. Mean-square displacement of tracer particles as a function of time in a simple-square lattice gas with a c(2 × 1) ordered structure. The points are the results of the simulation and the curves are explained in the text [6.64]

J_{nn} when they are at neighboring sites. In the other case the vacancies have a tendency to stay apart.

Also the recent work of *Kutner* et al. [6.55] on diffusion in an fcc lattice gas with repulsive nearest-neighbor interactions showed clearly the influence of the structure at low temperature. Figure 6.21 (taken from [6.55]) shows the behavior of the effective rate of performed jumps at temperatures in the ordered regions of the phase diagram near $c = 1/4$, $1/2$, and $3/4$. One recognizes pronounced minima of W due to the ordered structures. The detailed behavior of V and W especially near $c = 1/4$ and $1/2$ is discussed in [6.55]. Figure 6.22 reproduces the results for the correlation factor $f(c)$ at the lowest temperature investigated. It exhibits strongly nonmonotonic behavior as a function of concentration with minima near $c = 1/4$, $1/2$, followed by a sharp rise. The behavior of the correlation factor near $c = 1/4$ and $1/2$ can be understood by analysis of the effective diffusion processes in the ordered structures. At concentrations slightly above $c = 1/2$ ($1/2^+$) and at low temperatures there are some extra particles inbetween the planes mentioned above. These particles perform two-dimensional diffusion between the planes, as has been verified directly. In the low-concentration limit of the mobile particles one expects $f(c' \to 0) = 1$, as has been found. At concentration $c = 1/2^-$, at low temperatures one has some vacancies in the full planes which then effect the diffusion process. In the limit $c' \to 1$ the correlation factor for a square lattice is $f = 0.467...$, in agreement with the data. At $c = 1/4^+$ there is an sc superlattice of particles with some extra particle inbetween. Again a correlation factor of $f = 1$ is expected and observed from the motion of these extra particles. The three-dimensional nature of their motion has also been checked. At $c = 1/4^-$ the sc superlattice

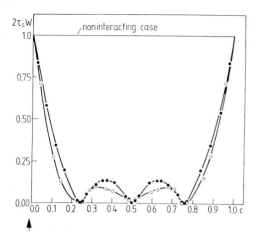
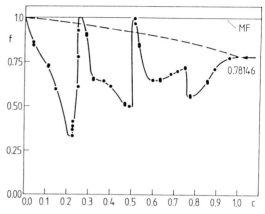

Fig.6.21. Effective rate of performed jumps W in an fcc lattice gas with repulsive nearest-neighbor interactions of strengths kT/J = -1.2 (full circles) and kT/J = -0.8 (open circles) [6.55]

Fig.6.22. Correlation factor as a function of concentration for fcc lattice gases with repulsive nearest-neighbor interaction of strength kT/J = -0.8. The full line is a guide to the eye, the dashed line represents the theory of *Sankey* and *Fedders* [6.13]. Figure taken from [6.55]

has some vacancies. It is easy to see that these vacancies can move with equal rates to nearest and next-nearest sites of the sc lattice by an "interstitial" process of the particles in this superlattice. The correlation factor for this process is $f(c' \rightarrow 1) = 0.809...$. However, there is an additional factor 2 to consider between the determination of W in the real lattice, and the effective diffusion process in the sc lattice, since half of the jumps of the particles in the real lattice to not lead to a jump in the sc lattice. A detailed analysis is given in [6.55]. The value $f/2 = 0.404...$ compares favorably with the simulations, Fig.6.22. In conclusion we want to emphasize the model character of these investigations. We expect similar effects to appear and analogous considerations to apply whenever diffusion occurs in lattice gases with ordered structures.

We finally report on simulations of collective diffusion in interacting lattice gases. While the coefficient of collective diffusion D_{coll} is identical to the single-particle diffusion coefficient in the noninteracting lattice gas, it is influenced by the interactions. Estimating D_{coll} in the course of Monte Carlo simulations of lattice gases is not as simple as estimating a tracer diffusion coefficient. There are two major methods of determining D_{coll}. The first introduces a gradient of the chemical potential by providing two planes with different chemical potentials and thus different average particle numbers of these planes. Particles then flow from the one plane to the other and D_{coll} is obtained from the particle current divided by the negative gradient of the chemical potential. Boundary effects are reduced by using periodic boundary conditions. *Murch* [6.65] has checked the accuracy of this method by relating D_{coll} to the tracer diffusion coefficient

D_t determined independently, using the Darken equation whereby the other parameters were also obtained independently. The second method monitors the decay of nonequilibrium density profiles. It was applied to surface diffusion by *Bowker* and *King* [6.53], who set up a sharp density profile at t = 0 and observed its decay. As a function of the concentration $D_{coll}(c)$ was obtained from the density profiles at different times by a Boltzmann-Matano analysis. However, the nonrandom method of particle selection for the diffusion step might introduce spurious correlations. Also no periodic boundary conditions were used. The statistical scatter of the data was relatively high. A more refined version of monitoring the decay of density disturbances was developed by *Kutner* [6.4] and *Kutner* et al. [6.49] for collective diffusion in fcc lattice gases. In this version, only a small disturbance of the lattice gas density is set up by applying a cos-like additional chemical potential during the preparation step and switching it off at t = 0. This method yields D_{coll} fairly directly at one concentration, and no Boltzmann-Matano analysis is necessary. The accuracy was first checked for the noninteracting lattice gas [6.4] and found very satisfactory; the method was then applied to the interacting case [6.49].

There are other methods to deduce D_{coll} from numerical simulations which will be briefly mentioned. *Sadiq* [6.66] obtained D_{coll} from time-displaced current correlation functions in equilibrium by using Kubo formulas; this method apparently requires large computing times. In addition, D_{coll} can also be inferred indirectly from the drift resulting from an external force on the particles, either by using a generalized Einstein relation [6.67] or by relating it to tracer diffusion via the generalized Darken equation [6.68]. *Reed* and *Ehrlich* [6.69] deduced D_{coll} from a Monte Carlo simulation of a two-dimensional lattice gas by an analysis of the time correlations of concentration fluctuations. Finally, *Zwerger* [6.67] suggested that D_{coll} can be obtained by determining the diffusion constant of the center-of-mass of all particles and multiplying it with simple thermodynamic equilibrium susceptibilities. While simple in principle this method also seems to require large computing times.

We now describe the results of simulations of the coefficient of collective diffusion. *Bowker* and *King* [6.53] determined D_{coll} by employing the choice of transition probabilities of (6.24). They confirm that D_{coll} is independent of concentration for the case of no interactions, strong increase of D_{coll} with concentration in the case of repulsion, and a strong decrease of D_{coll} with concentration in the case of attraction. These features could be expected from the corresponding behavior of the effective jump rate W. *Murch* [6.65] also used (6.24) and deduced D_{coll} as a function of concentration and interaction strength in a three-dimensional sc lattice gas. His data show the same qualitative behavior as discussed above, with very good statistical accuracy. *Kutner* et al. [6.49] employed definition (6.25) for the transition rates, hence their results are symmetric around c = 1/2, i.e., against interchange of particles and vacancies. Kutner et al. were interested in the behavior of

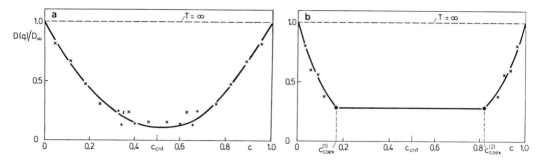

Fig.6.23. Coefficient of collective diffusion as a function of concentration. (a) Temperature above the critical temperature T_C for phase separation, $kT/J = 9.9$. (b) Temperature below T_C and $kT/J = 9.0$. The crosses represent the results of the simulations, the curves are guides to the eye, except for the tieline across the two-phase region, which is explained in the text [6.49]

$D_{coll}(c)$ in a range of attractive interaction strengths which include the critical temperature for phase separation in the lattice gas. Figure 6.23 shows the behavior of $D_{coll}(c)$ slightly above and below the critical ratio $k_BT_c/J = 9.79$ for phase separation. Below the critical temperature, data points were taken outside of the two-phase region only, i.e., outside the coexistence curve. This is required since inside the two-phase region the detailed kinetics of phase separation are so complicated that one is not assured that a typical equilibrium state has been reached even after long preparation times. This fact excluded the possibility of obtaining meaningful results for tracer diffusion, short-range order parameter, etc., within the two-phase region. However, a simple consequence of the symmetry of $D_{coll}(c)$ against interchange of c with 1-c and its equality at both sides of the coexistence region is that it must be concentration independent within the two-phase region. This assertion has been used in drawing the horizontal line in Fig.6.23b. Figure 6.23a shows qualitatively the effect at one temperature of "critical slowing down" of the collective diffusion when the critical point is approached from above. In [6.49] a mean-field treatment of $D_{coll}(c)$ and also a scaling analysis of its behavior have been given. The results of the mean-field theory which gives critical slowing down with a classical power-law behavior $(T-T_c)^{-1}$ have not been included in Fig.6.23a since the mean-field and the actual critical temperature differ appreciably ($k_BT_c^{MF}/J = 12$ compared to 9.79). The statistical scatter of the data, especially on approaching T_c, did not allow the scaling behavior D_{coll} (c = 1/2) $\propto (T-T_c)^{-\gamma}$ to be verified but the data are consistent with it. Reference [6.49] also discusses the wave-number dependent corrections to $D_{coll}(c)$ which can appear when applying a disturbance of finite wavelength. Finally, in [6.55] collective diffusion was investigated for repulsive interaction, with choice (6.25) of the transition probabilities. Only one temperature was covered where ordering of the particles with different sublattice structures already occurs (see below). Figure 6.24 shows that diffusion is enhanced compared to the noninteracting case in the concentration

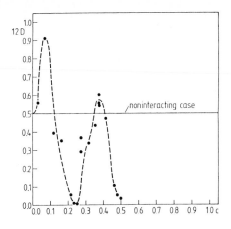

Fig.6.24. Coefficient of collective diffusion as a function of concentration for repulsive nearest-neighbor interaction of strength $kT/J = -1.2$. The dashed line is a guide to the eye. Data were taken up to $c = 1/2$ only because of the symmetry of $D_{coll}(c)$ against the interchange of c with 1-c [6.55]

regions between the ordered phases, while it is strongly suppressed for the ordered structures. This behavior is in accordance with the qualitative considerations of the influence of the structures.

In this review we have omitted tracer and collective diffusion of particles in binary alloys, where the diffusion process is promoted by vacancies and the A and B atoms have different exchange rates with the vacancies. This is an important problem in view of its applications to metal physics. We refer to [6.70-73] and the references contained therein for a description of the simulation work done on these systems, both in ordered and disordered phases.

6.3 Diffusion and Domain Growth in Systems far from Equilibrium

In this section we are concerned with the simulation of processes which occur as a consequence of abrupt changes of external parameters in an interacting many-body system of the type considered in the previous section, e.g., if the temperature in a lattice gas model is abruptly changed, one may simulate "quenching experiments" where one brings a system from a state in the disordered phase to a region of the phase diagram where thermal equilibrium would require either a two-phase coexistence of several phases, or the appearance of an ordered phase. Computer experiments which model the corresponding "nucleation" or "spinodal decomposition" processes have been a valuable tool for assessing the validity of various theoretical concepts [6.1,74]. We shall briefly review the most recent developments in this field (Sect.6.3.1), with particular emphasis on the experimentally relevant "late-stage scaling behavior" (Sect.6.3.2). We also include related work on simple relaxational models in our discussion. Then we proceed to the related problems of the kinetics of the formation of ordered phases (Sect.6.3.3) and random structures formed by aggregation or gelation processes (Sect.6.3.4). Although work on the latter problems has only just begun, it seems clear that a lot of interesting questions can be asked, and Monte Carlo simulation is the adequate tool for handling such problems.

6.3.1 Nucleation, Spinodal Decomposition, and Lifshitz-Slyozov Growth

Let us consider a lattice gas with a nearest-neighbor attractive interaction. For temperatures T less than a critical temperature T_c a two-phase coexistence region occurs bounded by the two branches $c_{coex}^{(1)}(T)$, $c_{coex}^{(2)}(T)$ (cf. also Figs.6.14-17 for a study of diffusion *outside* this region) of the coexistence curve. If the system for time $t < 0$ is held in equilibrium in the one-phase region at $T_0 > T_c$ (e.g., at infinite temperature) and at time $t = 0$ the temperature is switched to $T < T_c$ such that $c_{coex}^{(1)}(T) < c < c_{coex}^{(2)}(T)$, the initially homogeneous system has to unmix. One supposes that in the "metastable regime" the unmixing starts by strong and localized fluctuations ("heterophase fluctuations," "droplets"), while in the "unstable regime" weak and delocalized fluctuations ("homophase fluctuations") grow instead of decaying. This latter mechanism is called "spinodal decomposition," the former "nucleation," and the boundary between the metastable and unstable regime is called the "spinodal curve" (for more details about all these concepts, see [6.74-76]).

One basic question, however, already concerns the very existence of a well-defined spinodal curve. In the mean-field limit it is defined as the locus of inflection points of the free-energy of one-phase states within the two-phase region. In a system with short-range forces, however, one must make more precise what one means by such a free energy. One possibility is to define a coarse-grained free energy f_L by considering the probability $P_L\{\psi_i\} \propto \exp(-L^d f_L/k_B T)$ that an order parameter ψ_i occurs in the i-th d-dimensional cell of linear dimension L. Recent Monte Carlo studies of f_L in the regime where L distinctly exceeds the correlation length ξ of concentration fluctuations shows, however, that the spinodal curve distinctly depends on L and tends towards the coexistence curve as $L \to \infty$ [6.77]. This result is clear evidence that a unique (i.e., L-independent) spinodal curve does not exist for short-range systems, and hence also the transition from nucleation to spinodal decomposition must be rather gradual. But the situation is different in systems with long-range forces. Studying the equivalent-neighbor Ising model where q neighbors interact with the same strength, where q ranged from 6 (nearest-neighbor case on the simple cubic lattice) up to 3374 (interaction range 7 lattice constants), it was found [6.78] that with increasing interaction range metastable states could be observed nearly up to the mean-field spinodal. The inverse susceptibility of these states strongly decreased as the spinodal was approached, as predicted from mean-field theory. This validity of the mean-field prediction is understood from the fact that with increasing range of interaction the correlation length (and hence also the width of the droplet interface) is very large, and nucleation is then suppressed. In close vicinity of the mean-field spinodal where nucleation is still visible, one finds that the nucleation events start with rather "ramified" droplets, which must first become compact before they can grow and form domains of the new phase [6.79]. The size distribution of droplets near the spinodal curve follows a "lattice animal" distribution [6.80], while closer to the coexistence curve the size distribution follows the prediction of the classical "capillarity approximation" [6.1,74]

$$n_s \propto \exp(hs - \Gamma s^{2/3}) \quad s \to \infty \quad , \tag{6.29}$$

where n_s is the number of droplets of size s, h is the magnetic field in suitable units, and Γ is related to the surface tension associated with a flat planar interface. The droplet size distribution has also been studied for the nearest-neighbor Ising model at $T/T_c = 0.59$ [6.81]. These authors were able to simulate rather large systems (168^3 sites) and measured both the size of the critical cluster and the nucleation rate itself as a function of h in the single spin-flip kinetic Ising model, applying the multispin-coding technique (App. 1.A). While in related previous work (Fig.6.3 of [6.1]) it was possible to estimate n_s in the range $n_s \gtrsim 10^{-7}$, in [6.81] meaningful data could be recorded for three more orders of magnitude $\{n_s \gtrsim 10^{-10}\}$. While the data of [6.81] agree with (6.29) for $10^{-5} \lesssim n_s \lesssim 10^{-8}$ (assuming a spherical shape of the droplet in relating Γ to the surface tension), deviations become apparent for larger droplet sizes. A tentative interpretation of these deviations might be that for large droplet sizes the typical droplet shape at low T resembles a cube rather than a sphere in a lattice model, so enhancing Γ. Then the free-energy barrier for nucleation is larger than the prediction of the classical capillarity approximation for spherical droplets. In the critical region, however, where lattice anisotropy should be unimportant, *Furukawa* and *Binder* [6.82] obtained an energy barrier $\Delta F^*/k_B T_c$ which approaches the classical behavior only for such large clusters that $\Delta F^*/k_B T_c \gtrsim 50$ (or larger), while at $\Delta F^*/k_B T_c \approx 25$ the energy barrier is reduced to about one-half its classical value. This work is based on measuring the enhancement of the chemical potential over its value at the coexistence curve due to phase coexistence in a finite volume at constant density of the lattice gas, applying a method proposed in [6.83] (see also Chap.1). The contribution of the droplet surface is inferred from a thermodynamic analysis of this two-phase coexistence without a microscopic characterization of the droplet [6.82]. In contrast, [6.81] is based on "counting droplets" which are just defined by contours around neighboring reversed spins. Such a droplet definition would not work near the critical point of three-dimensional Ising systems, due to difficulties with the "percolation" of droplets [6.84,85] (for a discussion of percolation, see Chap.8). Work using a modified droplet definition [6.86] for which the critical point and percolation transition coincide is clearly desirable. In addition, it seems necessary to calculate the surface tension in the critical region more precisely than done in [6.87] before a final statement regarding the validity of classical nucleation theory in the critical region can be made.

In the two-dimensional case *Binder* and *Kalos* [6.88] studied two-phase coexistence at low temperatures and found that the surface free energy of the clusters was distinctly enhanced in comparison to classical theory, but could also be represented by the same form ($\Gamma s^{\frac{1}{2}}$ in this case). Later on exact calculations of the equilibrium shape of droplets in the Ising model [6.89] showed that the enhancement of the prefactor Γ is due to noncircular droplet shape resulting from the anisotropy of surface tension (it depends on the orientation of the interface, because an interface in the diagonal direction of a square lattice involves twice as many broken bonds as one in x or y direction). More recent work [6.90] confirmed the results of [6.88] and showed that $\ell n\, n_{s,k} \propto sf(k/s)$, where k is the number of bonds in the cluster. The "scaling function" $f(k/s)$ was estimated and shown to be consistent with a "droplet behavior" for large s, i.e., $n_s \propto \exp(hs - \Gamma s^{\frac{1}{2}})$.

Binder and *Kalos* [6.88] also studied the diffusion constant of droplets as a function of droplet size and interpreted their results in terms of crossover between several competing mechanisms: due to evaporation-condensation events of single atoms at the droplet surface the droplet moves randomly, and this effect may lead to droplet coagulation events which contribute to the phase-separation process [6.91]. In related work [6.92] the diffusion constant of two- and three-dimensional percolation clusters was also studied.

While the work described so far is addressed only to particular aspects of nucleation (the droplet free energy, droplet diffusivity, etc.), one can also study the kinetics of nucleation processes as a whole, and the associated dynamics of the relaxation into as well as out of the metastable state [6.93]. For the single-spin flip Ising model, corresponding Monte Carlo simulations have been reviewed in [6.1]. For the spin-exchange Ising model (magnetization being conserved), which is isomorphic to the lattice gas model in Sects.6.2.1,4, careful simulations are now available for a lattice of size 50^3, $T/T_c = 0.59$, and several concentrations [6.94, 95]. The results are in qualitative, though not completely quantitative, agreement with extensions of nucleation theory, where the time-dependent equations for the cluster concentrations are numerically solved [6.91,96,97]. These treatments are consistent with droplet growth during the later stages of phase separation according to the theory of *Lifshitz* and *Slyozov* [6.98]. While near the coexistence curve the simulations show that the mean droplet linear dimension L(t) grows with time as $L(t) \propto t^{1/3}$ as predicted [6.98], in the regime where a transition from nucleation to spinodal decomposition occurs, as well as in the unstable regime itself, the behavior may be interpreted in terms of the Lifshitz-Slyozov theory [6.99] but is also consistent with an asymptotic exponent of about 0.2-0.25 instead of 1/3 [6.100]. More effort is needed to establish conclusively the power laws describing the late stages of droplet growth.

Apart from these problems relating to late times, there is also some discussion concerning transient effects of droplet formation near T_c [6.101] since there the collective diffusion constant is very small. On the basis of phenomenological considerations consistent with simulations, *Furukawa* [6.101] suggests that nucleation starts to occur at a relative distance $\delta c/[c_{coex}^{(2)} - c_{coex}^{(1)}] \propto |1 - T/T_c|^{-\phi} t^{-3/8}$ from the coexistence curve, where t is the observation time after the quench and the exponent $\phi \approx 0.47$ for the three-dimensional lattice gas system.

Another interesting approach to nucleation based on purely kinetic consider-ations suggests considering the average rate of growth (or shrinking) of a droplet in a surrounding supersaturated gas, and thereby identifying the critical droplet size (where this rate vanishes) [6.102]. This concept is also tested by preliminary simulations [6.102].

With respect to spinodal decomposition, most effort has been devoted to study the "scaling behavior" in the late stages, Sect. 6.3.2. Another problem which aroused recent attention are quenches to zero temperature [6.103,104]. At $c = 0.5$ one finds that the system quickly settles down in a metastable pattern, where the ratio between the number of bonds and the coordination number seems to be (at least nearly) independent of lattice type and dimensionality, at least for $d \geq 3$.

Finally we wish to mention that spinodal decomposition has been studied in mo-dels which are more complicated than just the nearest-neighbor lattice gas model. *Kawasaki* [6.105,106] performed a computer simulation of a two-dimensional model of a magnetic binary alloy and studied the time-evolution changes caused by the in-fluence of the magnetic interactions. Also the dynamics of phase separation in two-dimensional tricritical systems was studied [6.107-109]. As for the ordinary short-range models [6.1], no time domain was observed where the linearized theory of spinodal decomposition would be valid. Again the characteristic lengths in the problem increase with time according to power laws, at least approximately. In con-trast to the simple lattice gas model, where the phase diagram and the dynamics of concentration fluctuations are symmetric with respect to the critical concentra-tion ($c = 1/2$), there is no such symmetry with respect to a tricritical point, and indeed interesting asymmetry effects were observed [6.108,109].

With the molecular dynamics technique the phase separation in Lennard-Jones fluids has also been studied [6.110-113]. While the results in the two-dimensional case [6.112,113] are qualitatively similar to the Ising model studies, in the three-dimensional case [6.110,111] it was claimed that the results agree well with the linearized theory of spinodal decomposition [6.114]. However, since the system stu-died was rather small (1372 atoms, while the Ising studies typically use lattices of 124 000 sites) a careful study of finite-size effects is needed to substantiate this conclusion.

6.3.2 Late-Stage Scaling Behavior

Many theoretical concepts about the late stages of phase separation imply that the equal-time structure factor $S(k,t)$ at time t after the quench should not depend on the wave vector k and time t separately, but rather should have a scaling property [6.91,93,115-118]

$$S(k,t) = [L(t)]^d \tilde{S}(kL(t)), \ d = \text{dimensionality}, \ L(t) \propto t^x \ , \tag{6.30}$$

where x is an exponent (which is 1/3 according to the Lifshitz-Slyozov theory
[6.98]) and $\tilde{S}(z)$ is a scaling function, which recently has been calculated approximately [6.119]. Recent simulations have shown that (6.30) gives a very good account
of the data, particularly if the exponent x is treated as an adjustable parameter
[6.99,100]. Subsequent simulations showed the validity of (6.30) also for phase separation in tricritical systems [6.107-109]. These simulations in turn now stimulated
a large amount of experimental work, and indeed scaling behavior has been established for the phase separation of both fluid [6.120-121] and solid [6.122-125]
mixtures. Since the original phenomenological theories [6.115,116,96] were in fact
motivated by the simulations described in [6.1], prediction (6.30) is a nice example
of how simulations may be fruitful for further development of both theory and experiment. Although (6.30) looks simple, a first-principle derivation of it as yet
is still completely lacking.

6.3.3 Diffusion of Domain Walls and Ordering Kinetics

Next we consider the time evolution of systems which are quenched from the disordered phase to a temperature below the transition temperature where an ordered
phase appears. Since the order parameter of the system is degenerate (twofold degeneracy in an Ising magnet, where the magnetization in the absence of fields may
point either up or down; p-fold degeneracy in the p-state Potts model [6.126] in
which each lattice site may be in one of the p states and an energy is won if two
neighboring sites are in the same state, etc.), the system forms ordered domains which
are separated by domain walls. As time goes on, these domains coarsen and thus the
(unfavorable) excess free energy due to the walls is reduced. The kinetics of domain
growth is relevant in surface sciences [6.127] and metallurgy [6.128], for instance,
and various analytical theories have lead to interesting predictions [6.93,129-133].
For a scalar (one-component) order parameter, which hence is twofold degenerate,
rather detailed theories exist [6.93,129,131,133], predicting that the scaling relation (6.30) also holds in this case, by measuring the wave vector k from the
point in the Brillouin zone describing the ordering (where Bragg scattering appears
below the critical temperature). The exponent $x = 1/2$ since the order parameter of
the system is not a conserved quantity. The behavior in the case of p-fold degenerate ordering with $p > 2$ is rather uncertain, however. There it has been suggested
that for $p \geq d + 1$ one rather has a very slow growth law [6.130,132] $L(t) \propto \ln t$.

Simulations have been carried out for nearest-neighbor Ising system on square
[6.134,135] and simple cubic [6.136] lattices. An order-disorder transition is obtained by a repulsive interaction between the particles (Sect.6.2.4). Fixed concentration [6.135] and fixed chemical potential [6.134] have been studied for a square
lattice, and a behavior $L(t) \propto t^{\frac{1}{2}}$ was found in both cases, as well as for $d = 3$
[6.136], and the scaling behavior (6.30) was confirmed [6.135,136], see Fig.6.25

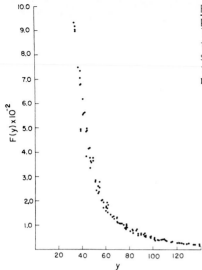

Fig.6.25. Scaled structure factor $t^{-1}S(k,t) \equiv F(y)$ plotted vs the variable $y = kt^{\frac{1}{2}}$ for a 60×60 square lattice quenched at $c = 1/2$ from infinite temperature to $T/T_c = 0.6$, and averaged over 10 runs. The structure factor $S(k,t) = N^{-1}|\Sigma(-1)^r \exp(ik \cdot r)\sigma(r)|^2$ is circularly averaged [$\sigma(r)$ being the spin at site r] [6.135]

for an example [6.135]. A further particularly interesting prediction [6.131] is the result that the prefactor in $L(t) \propto t^{\frac{1}{2}}$ does not involve surface tension, in contrast to the $L(t) \propto t^{1/3}$ law in the conserved case [6.137], and hence there should not be a critical slowing down in this prefactor. This prediction has been checked by simulations where one prepares an initial state containing a roughly spherical domain surrounded by a state of opposite order parameter orientation and watches its decay with time, or related work where one studies the diffusion of planar interfaces in systems with a finite cross section [6.138-140]. The conclusion is that although the prefactor in $L(t) \propto t^{\frac{1}{2}}$ seems to remain nonzero at T_c, there is a strong temperature dependence which so far is not described by the analytic theories [6.138]. These studies [6.138-140] allow detailed checks on the equation describing random local motions of interfaces.

In one dimension an interface simply is a kink separating opposite local orders from each other. For such a situation a coarsening $L(t) \propto \ln t$ [$L(t)$ now is the mean distance between kinks at time t after the quench] was predicted and confirmed by simulations [6.141].

A number of recent studies consider the case $p > 2$ [6.142-144]. *Sahni* and *Gunton* [6.142] considered a two-dimensional lattice gas model on the centered rectangular lattice (100×100), with interaction parameters up to fifth-nearest neighbors, appropriate to simulate the ordering of chemisorbed O on W(110) surfaces. In this case a (2×1) structure forms with a 4-fold degenerate ground state. Rather than the predicted $L(t) \propto \ln t$ law of domain growth [6.132] a power law is observed, but the exponent seems to be rather uncertain. Nevertheless, the scaling (6.30) is again observed. *Sahni* et al. [6.143] study quenches in the p-state Potts model from infinite temperature to (nearly) $T = 0$. They find that $L(t) \propto t^x$ with $x \approx 1/2$

for $p \lesssim 6$ and $x \approx 0.38$ for large p. The prefactor in this relation decreases as $p^{-\frac{1}{2}}$ for large p. *Sadiq* and *Binder* [6.144] studied the square Ising lattice with nearest- and next-nearest neighbor repulsion at $c = 1/2$, which also has a (2×1) structure with 4-fold degenerate ground state (the diffusion near equilibrium in this model has also been studied [6.64], Sect.6.2.4). While at constant chemical potential $x = 1/2$ in agreement with [6.143], at constant concentration the exponent x is in the range $0.3 \lesssim x \lesssim 0.4$. A quantitative determination is difficult, since near T_c there is a crossover to critical relaxation (which is also observed in the studies at constant chemical potential), and near $T = 0$ there is another crossover to a state of frozen-in metastable finite domain sizes. Thus there is hardly any evidence for the theories predicting $L(t) \propto \ln t$ [6.130,132], and as yet one does not understand theoretically the exponents $x < 1/2$ found recently [6.143,144].

6.3.4 Kinetics of Aggregation, Gelation and Related Phenomena

We first consider "diffusion-limited aggregation" [6.145-151], whereby one idealizes the process by which matter irreversibly combines to form dust, soot, dendrites and other random objects, where the rate-limiting step is diffusion of matter towards the aggregate. Using a lattice model, one starts with one site occupied and puts another one at the boundary of the lattice, and lets it now diffuse until it gets to a nearest-neighbor site of an already occupied site, where it then sticks, etc. In this way one forms a random self-similar object (for an example, see Fig.6.26). Defining Hausdorff dimensions [6.152] D by $N \propto R^D$, where N is the number of particles in the object and R its radius of gyration, one finds $D/d \approx 5/6$ for $d = 2-5$. In a "compact" object one has $D = d$, of course, and hence D is a measure of the "ramification" of the object. The theoretical explanation of this Hausdorff dimension is currently of great interest [6.147,148,150].

A somewhat similar problem of creating a random object by aggregation occurs when the additional particles which may be added need not diffuse in the surround-

20 Lattice Constants

Fig.6.26. An aggregate of 3000 particles on a square lattice [6.145]

216

ings of the object before they are added, but they are just put at random to any one of the "perimeter sites" (empty nearest-neighbor sites of the object). This model was introduced by *Eden* [6.153] in the context of the formation of biological structures. Monte Carlo studies [6.154] show that the "Eden clusters" are essentially compact, i.e., $D = d$.

Finally we mention the problem of the "gelation transition" of branched polymers [6.155]. A simple lattice model of irreversible gelation has been studied by Monte Carlo methods recently [6.156]. In this work, one models the free-radical copolymerization process, where the sol consists of small monomers, and the gelation is initiated by radicals. The radicals saturate, opening up a double bond of a monomer and leaving one bond in the monomer unsaturated. This creates a new radical that continues the growth process. In the simulation, $L \times L \times L$ simple cubic lattices (where L ranged from $L = 15$ to $L = 60$, and a finite-size scaling analysis, Chap.1, was performed) with periodic boundary conditions were used, containing a fraction c_2 of bifunctional sites and a fraction $1 - c_2$ of tetrafunctional sites. (A bifunctional or tetrafunctional site can have at most two or four occupied bonds incident, respectively.) The initialization is performed by randomly occupying a fraction c_I of bonds. No adjacent bonds are allowed to be occupied. (Chemically an occupied bond means a broken double bond between carbon atoms. The two free ends of an occupied bond are the radicals or "active centers".) To simulate the growth process then, an active center and adjacent bond are randomly chosen. If the other end of this bond is not forbidden, it is occupied and the active center is shifted to the other end of the bond. In this way more and more bonds are occupied, until at a concentration p_c of bonds an infinite cluster (the gel) forms. The molecular weight distribution $n_s(p)$ of macromolecules containing s occupied sites is sampled. This formation of an infinite cluster at p_c is reminiscent of percolation (Chaps. 1,8), and in fact a recent proposal has been made that gelation and percolation critical phenomena belong to the same universality class [6.157]. The results of the simulations rather suggest that kinetic gelation and percolation belong to different universality classes, however. More work is needed to clarify whether there is just one or perhaps several distinct classes of kinetic gelation.

6.4 Conclusion

In this review, we have described several simple models of diffusion processes in lattices, and used characteristic examples to show the type of information obtainable from simulations. Of course, our description is not exhaustive, and there are conceivably many other related models worthy of study. For instance, even in the case of diffusion in a one-dimensional chain one can consider various other models, e.g., a tagged particle in a chain which is randomly occupied by A and B particles may exchange sites with A, B with frequencies $\nu_{\pm A}$, $\nu_{\pm B}$ [± signs refer to an asymmetry between forward (+) and backward (−) jumps]. This model has rather interesting properties and was also studied by simulations [6.158]. It must be kept in mind, of course, that all the models considered here are approximate, as one restricts the particles to lattice sites, and does not consider the detailed physical properties of the single-jump process itself. If one wishes to be more realistic in this aspect, one has to apply the molecular dynamics technique — which has indeed been used to study self-diffusion in the two-dimensional classical electron gas [6.159] or surface diffusion on tungsten (110) surfaces [6.160], for instance. Such studies have not been considered here. We have restricted attention to lattice models because their study is relatively simple and their applications are still far from fully exploited.

With respect to nonlinear phenomena such as nucleation, kinetics of ordering, etc., we have been relatively brief and restricted attention to very recent work. Although considerable progress has been made, many questions still remain, even with respect to clusters and nucleation in the simple nearest-neighbor Ising model, which has been the subject of so many investigations. A still more systematic investigation of both low and high temperatures has to be done. With respect to ordering kinetics, one probably must study several related models in detail, before one can say what "classes" characterized by different growth exponents exist, etc.

The work on kinetics of aggregation, gelation, etc., mentioned in the last section obviously is closely related to questions regarding chemical kinetics. Although outside the scope of this chapter it is clear that for studying chemical reactions, polymerization kinetics [6.161], etc., Monte Carlo methods are a valuable tool, too.

Acknowledgments. We are particularly indebted to R. Kutner for his continued collaboration on diffusion in lattice gases. We are grateful to A. Sadiq, D. Stauffer, and H. van Beijeren for their cooperation.

References

6.1   K. Binder, M.H. Kalos: In *Monte Carlo Methods in Statistical Physics*, ed. by K. Binder, Topics Current Phys., Vol.7 (Springer, Berlin Heidelberg New York 1979) Chap.6
6.2   H. Müller-Krumbhaar: In *Monte Carlo Methods in Statistical Physics*, ed. by K. Binder, Topics Current Phys., Vol.7 (Springer, Berlin Heidelberg New York 1979) Chap.7
6.3   G.E. Murch: *Atomic Diffusion Theory in Highly Defective Solids* (Trans Tech House, Adermannsdorf 1980)
6.4   R. Kutner: Phys. Lett. **81A**, 239 (1981)
6.5   J. Bardeen, C. Herring: In *Imperfections in Nearly Perfect Crystals*, ed. by W. Shockley (Wiley, New York 1952) p.261
6.6   A.D. Le Claire: Physical Chemistry **10**, 261 (Academic, New York 1970)
6.7   J.R. Manning: *Diffusion Kinetics for Atoms in Solids* (Van Nostrand, New York 1968)
6.8   A.G. Redfield, M. Eisenstadt: Phys. Rev. **132**, 635 (1963)

6.9   H.J. De Bruin, G.E. Murch: Phil. Mag. **27**, 1475 (1973)
6.10 G.E. Murch, R.J. Thorn: J. Phys. Chem. Solids **38**, 789 (1977)
6.11 G.E. Murch: J. Nucl. Mat. **57**, 239 (1975)
6.12 K.W. Kehr, R. Kutner, K. Binder: Phys. Rev. B**23**, 4931 (1981)
6.13 O.F. Sankey, P.A. Fedders: Phys. Rev. B**15**, 3586 (1977)
6.14 G.E. Murch, R.J. Thorn: Phil. Mag. **35**, 493 (1977)
6.15 R. Kutner: Unpublished
6.16 K. Nakazato, K. Kitahara: Progr. Theor. Phys. **64**, 2261 (1980)
6.17 R.A. Tahir-Kheli, R.J. Elliott: Phys. Rev. B**27**, 844 (1983)
6.18 K. Schroeder: Unpublished
6.19 M. Koiwa: J. Phys. Soc. Jpn. **45**, 1327 (1978)
6.20 P. Benoist, J.L. Bocquet, P. La Fore: Acta Met. **25**, 265 (1977)
6.21 D. Wolf: J. Phys. Chem. Solids **41**, 1053 (1980)
6.22 D.K. Ross, D.L.T. Wilson: In *Neutron Inelastic Scattering 1977*, Proc. Intern. Atomic Energy Agency, Vienna 1977 (IAEA, Vienna 1978) Vol.II, p.383
6.23 E.W. Montroll, G.H. Weiss: J. Math. Phys. **6**, 167 (1965)
6.24 T.E. Harris: J. Appl. Prob. **2**, 323 (1965)
6.25 F. Spitzer: Adv. Math. **5**, 246 (1970)
6.26 P.M. Richards: Phys. Rev. B**16**, 1363 (1977)
6.27 P.A. Fedders: Phys. Rev. B**17**, 40 (1978)
6.28 S. Alexander, P. Pincus: Phys. Rev. B**18**, 2011 (1978)
6.29 H. van Beijeren, K.W. Kehr, R. Kutner: Phys. Rev. B**28**, 5711 (1983)
6.30 D.G. Levitt: Bioch. Bioph. Acta **373**, 115 (1974)
6.31 R. Kutner, H. van Beijeren, K.W. Kehr: Phys. Rev. B**30**, 4382 (1984)
6.32 P.G. De Gennes: La Recherche **7**, 919 (1976)
6.33 C.D. Mitescu, H. Ottavi, J. Roussenq: AIP Conf. Proc. **40**, 377 (1979)
6.34 C.D. Mitescu, J. Roussenq: Ann. Isr. Phys. Soc. **5**, 81 (1983)
6.35 W.W. Brandt: J. Chem. Phys. **63**, 5162 (1977)
6.36 C.D. Mitescu, J. Roussenq: C.R. Acad. Sci. Paris **283A**, 999 (1976)
6.37 J. Roussenq: Thèse Université de Provence (1980)
6.38 T. Vicsek: Z. Physik B**45**, 153 (1981)
6.39 K.W. Kehr, R. Kutner, K. Binder: In *Point Defects and Defect Interactions in Metals*, ed. by J.-I. Takamura, M. Doyama, M. Kiritani (University of Tokyo Press, Tokyo 1982) p.582;
      R. Kutner, K.W. Kehr: Phil. Mag. **48**, 199 (1983)
6.40 S. Kirkpatrick: Rev. Mod. Phys. **45**, 574 (1973)
6.41 Y. Gefen, A. Aharony, S. Alexander: Phys. Rev. Lett. **50**, 77 (1983)
6.42 D. Stauffer: Phys. Rept. **54**, 1 (1979)
6.43 B. Nienhuis: J. Phys. A**15**, 199 (1982)
6.44 M.F. Gaunt, D.S. Sykes: J. Phys. A**16**, 783 (1983)
6.45 S. Kirkpatrick: In *Ill-Condensed Matter*, ed. by R. Balian, R. Maynard, G. Toulouse (North Holland, Amsterdam 1979) p.321
6.46 P. Argyrakis, R. Kopelman: Chem. Phys. **57**, 29 (1981)
6.47 S. Kazama, Y. Fukai: Suppl. Trans. Japan Inst. Met. **21**, 173 (1980)
6.48 S. Sugimoto, Y. Fukai: Suppl. Trans. Japan Inst. Met. **21**, 177 (1980)
6.49 R. Kutner, K. Binder, K.W. Kehr: Phys. Rev. B**26**, 2967 (1982)
6.50 G.E. Murch: Phil. Mag. **32**, 1129 (1975)
6.51 G.E. Murch, J.M. Rolls, H.J. De Bruin: Phil. Mag. **29**, 337 (1974)
6.52 K. Kawasaki: Phys. Rev. **145**, 224 (1966); **148**, 375 (1966); **150**, 285 (1966)
6.53 M. Bowker, D.A. King: Surf. Sci. **71**, 583 (1978)
6.54 G.E. Murch, R.J. Thorn: Phil. Mag. **35**, 1441 (1977)
6.55 R. Kutner, K. Binder, K.W. Kehr: Phys. Rev. B**28**, 1846 (1983)
6.56 G.E. Murch, R.J. Thorn: J. Phys. Chem. Solids **40**, 389 (1979)
6.57 G.E. Murch, R.J. Thorn: Phil. Mag. **35**, 493 (1977)
6.58 G.E. Murch, R.J. Thorn: Phil. Mag. **36**, 517 (1977)
6.59 H. Sato, R. Kikuchi: J. Chem. Phys. **55**, 677, 702 (1971)
6.60 K. Binder, D.P. Landau: Surf. Sci. **108**, 503 (1981)
6.61 K. Binder: Phys. Rev. Lett. **45**, 811 (1980)
6.62 K. Binder, J.L. Lebowitz, M.K. Phani, M.H. Kalos: Acta Met. **29**, 1655 (1981)
6.63 G.E. Murch: Phil. Mag. A**43**, 871 (1981)
6.64 A. Sadiq, K. Binder: Surf. Sci. **128**, 350 (1983)
6.65 G.E. Murch: Phil. Mag. A**41**, 157 (1980)

6.66 A. Sadiq: Phys. Rev. B9, 2299 (1974)
6.67 W. Zwerger: Z. Phys. B42, 333 (1981)
6.68 G.E. Murch, R.J. Thorn: Phil. Mag. A40, 477 (1981)
6.69 D.A. Reed, G. Ehrlich: Surf. Sci. 105, 603 (1981)
6.70 H.J. De Bruin, G.E. Murch, H. Bakker, L.P. van der Mey: Thin Solid Films 25, 47 (1975)
6.71 H.J. De Bruin, H. Bakker, L.P. van der Mey: Phys. Stat. Sol. (b) 82, 581 (1977)
6.72 H. Bakker: Phil. Mag. A40, 525 (1979)
6.73 G.E. Murch, S.J. Rothman: Phil. Mag. A43, 229 (1981)
6.74 J.D. Gunton, M. San Miguel, P.S. Sahni: In *Phase Transitions and Critical Phenomena*, Vol.XIII, ed. by C. Comb, J.L. Lebowitz (Academic, New York 1983) p.269
6.75 K. Binder: In *Stochastic Nonlinear Systems in Physics, Chemistry and Biology*, ed. by L. Arnold, R. Lefever, Springer Ser. Synergetics, Vol.8 (Springer, Berlin Heidelberg New York 1981) p.62
6.76 K. Binder: In *Systems Far from Equilibrium*, ed. by L. Garrido, Lecture Notes Phys., Vol.132 (Springer, Berlin, Heidelberg, New York 1980) p.76
6.77 K. Kaski, K. Binder, J.D. Gunton: J. Phys. A16, L623 (1983)
6.78 D.W. Heermann, W. Klein, D. Stauffer: Phys. Rev. Lett. 49, 1262 (1982)
6.79 D.W. Heermann, W. Klein: Phys. Rev. Lett. 50, 1062 (1983)
6.80 D.W. Heermann, W. Klein: Phys. Rev. B27, 1732 (1983)
6.81 D. Stauffer, A. Coniglio, D.W. Heermann: Phys. Rev. Lett. 49, 1299 (1982)
6.82 H. Furukawa, K. Binder: Phys. Rev. A26, 556 (1982)
6.83 H. Meirovitch, Z. Alexandrowicz: Mol. Phys. 34, 1027 (1977)
6.84 H. Müller-Krumbhaar: Phys. Lett. A50, 27 (1974)
6.85 H. Müller-Krumbhaar: In *Monte Carlo Methods in Statistical Physics*, ed. by K. Binder, Topics Current Phys., Vol.7 (Springer, Berlin Heidelberg New York 1979) Chap.5
6.86 A. Coniglio, W. Klein: J. Phys. A13, 2775 (1980)
6.87 K. Binder: Phys. Rev. A25, 1699 (1982)
6.88 K. Binder, M.H. Kalos: J. Statist. Phys. 22, 363 (1980)
6.89 R.K.P. Zia, J.E. Avron: Phys. Rev. B25, 2042 (1982)
6.90 R. Dickman, W.C. Schieve: Physica 112A, 51 (1982)
6.91 K. Binder: Phys. Rev. B15, 4425 (1977)
6.92 H. Gould, K. Holl: J. Phys. A14, L443 (1981)
6.93 C. Billotet, K. Binder: Z. Phys. B32, 195 (1979)
6.94 M. Kalos, J.L. Lebowitz, O. Penrose, A. Sur: J. Statist. Phys. 18, 39 (1978); A. Sur, J.L. Lebowitz, M.H. Kalos, J. Marro: Phys. Rev. B15, 3014 (1977)
6.95 O. Penrose, J.L. Lebowitz, J. Marro, M.H. Kalos, A. Sur: J. Statist. Phys. 19, 243 (1978)
6.96 P. Mirold, K. Binder: Acta Met. 25, 1435 (1977)
6.97 O. Penrose, A. Buhagiar: J. Statist. Phys. 30, 219 (1983)
6.98 I.M. Lifshitz, V.V. Slyozov: J. Phys. Chem. Solids 19, 35 (1961)
6.99 J.L. Lebowitz, J. Marro, M.H. Kalos: Acta Met. 30, 297 (1982)
6.100 J. Marro, J.L. Lebowitz, M.H. Kalos: Phys. Rev. Lett. 43, 282 (1979)
6.101 H. Furukawa: Phys. Rev. A28, 1729 (1983)
6.102 A.M.J. Huiser, J.-P. Marchand, Ph.A. Martin: Helv. Phys. Acta 55, 259 (1982)
6.103 A. Levy, S. Reich, P. Meakin: Phys. Lett. 87A, 248 (1982)
6.104 P. Meakin, S. Reich: Phys. Lett. 92A, 247 (1982)
6.105 T. Kawasaki: Progr. Theor. Phys. 59, 1812 (1978)
6.106 T. Kawasaki: Progr. Theor. Phys. 61, 384 (1979)
6.107 P.S. Sahni, J.D. Gunton: Phys. Rev. Lett. 45, 369 (1980)
6.108 P.S. Sahni, J.D. Gunton, S.L. Katz, R.H. Timpe: Phys. Rev. B25, 389 (1982)
6.109 P.S. Sahni, S.L. Katz, J.D. Gunton: To be published
6.110 F.F. Abraham, D.R. Schreiber, M.R. Mruzik, G.M. Pound: Phys. Rev. Lett. 36, 261 (1976)
6.111 M.R. Mruzik, F.F. Abraham, G.M. Pound: J. Chem. Phys. 69, 3462 (1978)
6.112 F.F. Abraham, S.W. Koch, R.C. Desai: Phys. Rev. Lett. 49, 923 (1982)
6.113 S.W. Koch, R.C. Desai, F.F. Abraham: Phys. Rev. A27, 2153 (1983)
6.114 J.W. Cahn: Trans. Metall Soc. AIME 242, 166 (1968)
6.115 K. Binder, D. Stauffer: Phys. Rev. Lett. 33, 1006 (1974)

6.116 K. Binder, C. Billotet, P. Mirold: Z. Phys. B**30**, 183 (1978)
6.117 H. Furukawa: Phys. Rev. A**28**, 1717 (1983)
6.118 H. Furukawa: Phys. Rev. Lett. **43**, 136 (1979); Phys. Rev. A**23**, 1535 (1981)
6.119 P.A. Rikvold, J.D. Gunton: Phys. Rev. Lett. **49**, 286 (1982)
6.120 C.M. Knobler, N.C. Wong: J. Phys. Chem. **85**, 1972 (1981)
6.121 Y.C. Chou, W.I. Goldburg: Phys. Rev. A**23**, 858 (1981)
6.122 M. Hennion, D. Ronzand, P. Guyot: Acta Met. **30**, 599 (1982)
6.123 A. Craievich, J.M. Sanchez: Phys. Rev. Lett. **47**, 1308 (1981)
6.124 S. Komura, K. Osamura, H. Fujii, T. Takeda: Physica **120**B, 397 (1983)
6.125 P. Guyot, J.P. Simon: In *Solid State Phase Transformations*, ed. by H.I. Aaronson (AIME, New York 1982) p.325
6.126 R.B. Potts: Proc. Camb. Phil. Soc. 48, 106 (1952)
6.127 M.G. Lagally, G.C. Wong, T.M. Lu: Crit. Rev. Solid State Mater. Sci. **7**, 233 (1978)
6.128 J.E. Burke, D. Turnbull: Progr. Met. Phys. **3**, 220 (1952)
6.129 K. Kawasaki, C. Yalabik, J.D. Gunton: Phys. Rev. B**17**, 455 (1978); S.K. Chan: J. Chem. Phys. **65**, 5755 (1977)
6.130 I.M. Lifshitz: Sov. Phys. JETP **15**, 939 (1962)
6.131 S.M. Allen, J.W. Cahn: Acta Met. **27**, 1085 (1979)
6.132 S.A. Safran: Phys. Rev. Lett. **46**, 1581 (1981)
6.133 T. Ohta, D. Jasnow, K. Kawasaki: Phys. Rev. Lett. **49**, 1223 (1982)
6.134 C. Kawabata, K. Kawasaki: Phys. Lett. **65**A, 137 (1978)
6.135 P.S. Sahni, G. Dee, J.D. Gunton, M. Phani, J.L. Lebowitz, M.H. Kalos: Phys. Rev. B**24**, 410 (1981)
6.136 M.K. Phani, J.L. Lebowitz, M.H. Kalos, O. Penrose: Phys. Rev. Lett. **45**, 366 (1980)
6.137 M. San Miguel, J.D. Gunton, G. Dee, P.S. Sahni: Phys. Rev. B**23**, 2334 (1981)
6.138 P.S. Sahni, G.S. Grest, S.A. Safran: Phys. Rev. Lett. **50**, 60 (1983)
6.139 P.S. Sahni, G.S. Grest: J. Appl. Phys. **53**, 8002 (1982)
6.140 S.A. Safran, P.S. Sahni, G.S. Grest: Phys. Rev. B**26**, 466 (1982)
6.141 K. Kawasaki, T. Ohta, T. Nagai: J. Phys. Soc. Jpn. **52**, Suppl. 131 (1983)
6.142 P.S. Sahni, J.D. Gunton: Phys. Rev. Lett. **47**, 1754 (1981)
6.143 P.S. Sahni, G.S. Grest, M.P. Anderson, D.J. Strolovitz: Phys. Rev. Lett. **50**, 263 (1983)
6.144 A. Sadiq, K. Binder: Phys. Rev. Lett. **51**, 674 (1983), and to be published
6.145 T.A. Witten, Jr., L.M. Sander: Phys. Rev. Lett. **47**, 1400 (1981)
6.146 P.A. Meakin: Phys. Rev. A**27**, 604 (1983)
6.147 T.A. Witten, L.M. Sander: Phys. Rev. B**27**, 5686 (1983)
6.148 M. Muthukumar: Phys. Rev. Lett. **50**, 839 (1983)
6.149 H.E. Stanley: J. Phys. Soc. Japan **52**, Suppl. 151 (1983)
6.150 H. Gould, F. Family, H.E. Stanley: Phys. Rev. Lett. **50**, 686 (1983)
6.151 C. Allain, B. Jouhier: J. Physique (Paris) **44**, L421 (1983)
6.152 B. Mandelbrot: *The Fractal Geometry of Nature* (Freeman, San Francisco 1982), and references therein
6.153 M. Eden: Proc. 4th Berkeley Symp. on Math. Statistics and Probability, Vol. IV, ed. by J. Neyman (Univ. Calif. Press, Berkeley 1961), p.229
6.154 H.P. Peters, D. Stauffer, H.P. Hölters, K. Loevenich: Z. Phys. B**34**, 399 (1979)
6.155 P.J. Flory: *Principles of Polymer Chemistry* (Cornell University Press, Ithaca NY 1953)
6.156 H.J. Herrmann, D.P. Landau, D. Stauffer: Phys. Rev. Lett. **49**, 412 (1982); H.J. Herrmann, D. Stauffer, D.P. Landau: J. Phys. A**16**, 1221 (1983)
6.157 D. Stauffer, A. Coniglio, M. Adam: Adv. Polymer Sci. **44**, 103 (1982)
6.158 D.E. Temkin: Soviet Math. Dokl. **13**, 1172 (1972)
6.159 J.P. Hansen, D. Levesque, J.J. Weis: Phys. Rev. Lett. **43**, 979 (1979)
6.160 H.K. McDowell, J.D. Doll: Surf. Sci. **121**, L537 (1982)
6.161 W. Bruns, I. Motoc, K.F. O'Driscoll: *Monte Carlo Applications in Polymer Science*, Lecture Notes in Chemistry, Vol.27 (Springer, Berlin Heidelberg New York 1981)

7. Roughening and Melting in Two Dimensions

Y. Saito and H. Müller-Krumbhaar

With 11 Figures

In two-dimensional systems with a continuous symmetry group long-wavelength fluctuations are dominant. Thus Kosterlitz and Thouless (KT) predicted a unique phase
transition caused by topological defects. We review Monte Carlo (and also molecular
dynamics) studies of two typical phase transitions which are assumed to belong to
this category — the roughening transition of a crystal-vapor interface and the melting transition of a two-dimensional crystal. The former agrees with KT theory,
whereas the latter does not in general. The cause of this different behavior becomes evident in the dual representation as Coulomb gas systems with scalar and
vector charges.

7.1 Introductory Remarks

The thermodynamic behavior of systems in low dimensions is substantially affected
by long-wavelength fluctuations. In particular, dimension two has a unique character in this respect. In the formulation of a locally defined order-parameter field
$\emptyset(r)$ the leading fluctuation term in the energy functional is written as $(\nabla\emptyset(r))^2$
for systems with a continuous symmetry group. This term leads to the divergence of
order-parameter fluctuations and consequently to the destruction of the long-range
order (LRO) [7.1,2]. The absence of LRO is exactly proven for various 2D systems
[7.3-6]. Minimization of the energy functional yields the Laplace equation $\Delta\emptyset(\mathbf{r}) = 0$,
whose eigenfunction solution is logarithmic $\emptyset(r) \sim \ln(r)$ in two dimensions. It is
this logarithmic behavior, unbounded both in the long- and short-range limit, which
appears common in various two-dimensional systems and causes the specific problems.

Some systems such as the 2D-Heisenberg magnet remain in a disordered state with
only a short-range order (SRO) at any finite temperatures [7.7-9]. However, for
some other systems a well-developed *quasi* LRO is possible at low temperatures,
characterized by the slow algebraic decay of the order-parameter correlation function [7.10-12]. The slow decay is also reflected in the divergence of susceptibility at low temperatures [7.13,14]. The quasi-ordered phase is closely similar to
the behavior at a critical point in higher dimensions because of the algebraic decay of correlations at long distances. Since this behavior is not confined to a

single critical temperature but is observed in a whole range of temperatures, one expects these properties to be describable in terms of a line of fixed points in a suitable renormalization scheme.

The quasi-ordered state is distinct from the high-temperature disordered phase with only a SRO, characterized by the rapid exponential decay of the correlation function. Thus one expects a phase transition inbetween. According to *Kosterlitz* and *Thouless* (KT) [7.15,16], the quasi LRO is destroyed by topological defects such as vortices, dislocations, disclinations and so on. Since the systems have only a *quasi* order at low temperatures, their phase transitions have weaker singularities than those of systems with a *true* LRO. Free energy, internal energy and specific heat have only an essential singularity at the transition point [7.15-18], which is practically impossible to detect in these thermal observables.

Not only the uniqueness of the phase transition but also its vast applicability make the KT theory so important. It is applied to a variety of phase transitions of 2D systems, such as magnetic phase transition of the planar-spin or XY models [7.15-21], superfluidity [7.22,23] and superconductivity [7.24-26], roughening transition of a crystal-vapor interface [7.27], metal-insulator transition of a logarithmically interacting Coulomb gas system [7.15-18,21,27-30], and melting transition of a 2D crystal [7.15,16,31-34]. It is also applicable to a one-dimensional Fermion system [7.21,30].

Owing to the uniqueness of the phase transition and to its applicability to various physical and model systems, quite an amount of Monte Carlo simulations and molecular dynamics studies [7.35-86] as well as experiments [7.87-107] have been performed. Most of the phase transitions seem to be in accordance with KT theory. For example, in a planar-spin model, vortices shown in Fig.7.1 are the corresponding topological defects [7.39]. The only exception which seems to contradict the KT theory is the melting problem. In computer experiments melting in two dimensions is found to be first order [7.50-86]. In fact, the previously listed systems can be sorted into two categories. The first class is formed by systems except 2D crystals which are mutually related through duality transformations [7.17,106,107] and reduce to the logarithmically interacting Coulomb gas system with *scalar* charges [7.15-18,28-30,108-113]. The second class is formed from 2D crystals containing dislocation *vectors* as topological defects which have orientation-dependent long-range as well as logarithmic interactions [7.114,115]. This difference in the nature of topological defects is the key point to the success and failure of KT theory. In the following we summarize Monte Carlo simulation results of a typical example of the first class, the roughening problem, and of the second class, the melting problem.

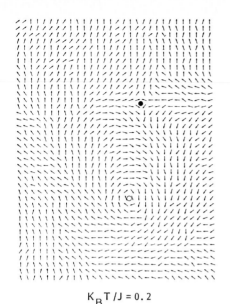

$$K_B T / J = 0.2$$

Fig.7.1. Spin configuration of isolated positive (○) and negative (●) vortices [7.39]. This configuration exists only transiently at this temperature, and the typical equilibrium configuration is the very dilute vortex-antivortex bound pairs

7.2 Roughening Transition

7.2.1 Solid-on-Solid (SOS) Model

Knowing the equilibrium structure of the crystal-vapor interface is a prerequisite for understanding and controlling the growth of crystals from vapor or dilute solution [7.114-120]. At low temperatures the interface is expected to be atomistically flat to minimize the surface energy, whereas at high temperatures it is rough and contains steps producing configurational entropy, as is shown in Fig.7.2. This change in structure is the roughening transition of the interface [7.121]. *Burton* et al. [7.117] first analyzed the roughening transition of a single-layered interface, reducing the problem to the two-dimensional Ising model, which was solved exactly [7.122] and is known to perform a second-order phase transition. This, however, turned out to be too severe a restriction of the infinite discrete degeneracy of interface position to only a two-fold degeneracy at low temperatures. Allowing a multilayer structure, in contrast, was found to extinguish the singularity of the thermodynamic quantities in the mean-field approximation [7.117] completely.

From the definition of the roughening transition, on the other hand, the interfacial width is expected to diverge at and above the roughening temperature [7.123]. Therefore one has to characterize the interface by the height or number of crystal layers h_i measured from some reference level, *which can vary from* $-\infty$ *to* $+\infty$ *in unit steps.* Crystal atoms are supposed to be located in lattice sites without allowing an empty space in the bulk of the crystal. This model is called a solid-on-solid (SOS) model, and discards vacancies, interstitial atoms, thermal vibrations of

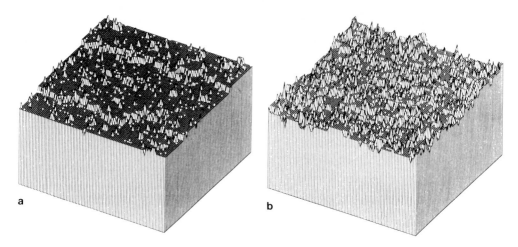

<u>Fig.7.2a,b.</u> Configurations of a crystal surface with steps produced by computer simulations at (a) low and (b) high temperature [7.116]

crystal atoms, structural reconstructions and many other effects, which are important for melting and other phase transitions. But the SOS model is generally accepted to depict the necessary minimum of configurational degrees of freedom describing the roughening transition concerned.

The degrees of freedom per site are increased from finite, for example, two of the Ising model ($h_i = 0,1$), to infinite of the SOS model ($h_i = 0, \pm1, \pm2,..., \pm\infty$). This increment of freedom induces enhancement of fluctuations, and the character of the phase transition may alter due to the fluctuation effect. The simple mean-field approximation and pair approximation [7.118-120,124-128] do not properly take the long-wavelength fluctuations into account, and thus they give no phase transition. Only self-consistent variational approximations [7.121,129] and the KT renormalization group method [7.27] treat the long-wavelength fluctuations properly, and the unique phase transition is predicted.

The energy cost due to the interfacial roughness may be represented by a Hamiltonian

$$H = J \sum_{<ij>} |h_i - h_j|^p \quad , \qquad \text{p: positive integer} \tag{7.1}$$

proportional to a height difference between the nearest-neighboring sites i and j. The model with p = 1 is usually called an SOS model or an absolute solid-on-solid (ASOS) model, and with p = 2 a discrete Gaussian (DG) model. The ASOS model was considered first and many simulations were performed [7.43,44,48], but due to its simpler analytical properties the DG model was successfully investigated theoretically [7.27,106,107,121,129] with various recent simulations [7.44-46,49]. Both models are assumed to belong to the same universality class of phase transition, along with an exactly soluble body-centered solid-on-solid (BCSOS) model [7.130].

Since the theoretically predicted essential singularities of the thermodynamic quantities are too weak to be detected, one has to use other quantities to determine the phase transition. The second and higher moments of the interface are expected to diverge at and above the roughening temperature [7.121]. However, in the system with finite size this divergence is suppressed to a finite value, proportional only to the logarithm of the system size. Monte Carlo results seem to be consistent with this behavior, but the logarithmic divergence makes it very hard to determine the roughening transition temperature T_R precisely [7.44,45].

Another possibility to determine the transition point is to use the height correlation function

$$G_{ij} = <(h_i - h_j)^2>$$ (7.2)

and study its asymptotic behavior at $r_{ij} \to \infty$. For the smooth interface below T_R the height h_j at site j will not be very different from the height h_i at site i even though their distance r_{ij} is large; therefore G_{ij} is expected to be finite even for infinite separation. The saturation value should increase on heating the system and making the height correlation weak. At the roughening temperature T_R where the correlation is lost, the asymptotic value of G_{ij} diverges as [7.27]

$$\lim_{r_{ij} \to \infty} G_{ij} = \frac{4}{\pi c} \left(\frac{T_R - T}{T_R} \right)^{-\frac{1}{2}} .$$ (7.3)

Above T_R in the rough phase, the height correlation G_{ij} is predicted to diverge logarithmically

$$G_{ij} \sim \frac{A(T)}{4\pi} \log r_{ij} ,$$ (7.4)

with coefficient A taking a universal value of $8/\pi$ at the transition point.

The Monte Carlo simulation of the ASOS and DG model seems to be consistent with the change in the asymptotic behavior of G_{ij} as the temperature is varied [7.46, 47], but the finiteness of the system size does not enable unbiased determination of the critical behavior.

7.2.2 Dual Coulomb Gas (CG) Model

A duality transformation [7.106] shows that the singularity of the free energy and therefore the character of the phase transition of the DG model are equivalent to those of the lattice Coulomb gas (CG) system, defined by the Hamiltonian

$$H_{CG} = -2\pi \sum_{i \neq j} \sum n_i n_j V_{ij} .$$ (7.5)

Here n_i are scalar integer charges, $n_i = 0, \pm 1, \ldots, \pm \infty$, under the neutrality condition

$$\sum_i n_i = 0 .$$ (7.6)

The summation in (7.5) runs over lattice points, and the interaction is given by the 2D lattice Green's function

$$V_{ij} = V(\mathbf{r}_{ij}) = \frac{1}{4N} \sum_{\mathbf{q} \neq 0} \frac{1 - \exp(i\mathbf{q} \cdot \mathbf{r}_{ij})}{2 - \cos(q_x a) - \cos(q_y a)} \quad , \tag{7.7}$$

which behaves logarithmically in the thermodynamic limit (i.e., the system size N being infinite) as

$$V_{ij} \to \frac{1}{4\pi} \ln r_{ij}/a + 1/8 \quad . \tag{7.8}$$

The temperature T of the DG model is transformed to the inverse of the temperature of the CG model, T^{CG}:

$$K = T/J = 1/\pi T^{CG} \quad . \tag{7.9}$$

The fundamental argument of the KT theory in fact starts from this CG model [7.15, 16]. Kosterlitz and Thouless found that at low temperatures T^{CG}, scalar charges form bound neutral pairs to minimize the energy cost, whereas at high T^{CG} they dissociate

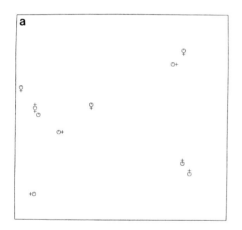

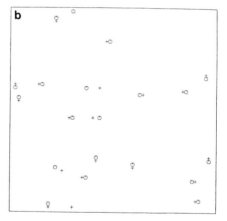

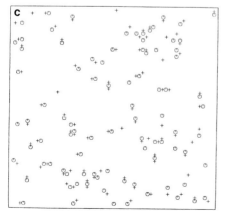

Fig.7.3a-c. Snap-shot configurations of positive (+) and negative (o) scalar charges of a simulated Coulomb gas system [7.49]. Temperatures, $K = T/J = 1/\pi T^{CG}$, are (a) 1.6, (b) 1.4 and (c) 1.2

to free charges by gaining entropy. Snap-shot configurations [7.49] of positive (+) and negative (0) charges shown in Fig.7.3 show the dissociation process of charge pairs on increasing the temperature T^{CG} (i.e., lowering the DG temperature T or K). Dissociation appears to take place at $K \lesssim 1.4$ as predicted [7.15,16,27]. Similar dissociation of vortex pairs is found in magnetic systems [7.42].

The duality transformation furthermore relates the height correlation function G_{ij} with the charge correlation function $<n_i n_j>_{CG}$

$$G_{ij} = 2KV_{ij} - (2\pi K)^2 \sum_k \sum_\ell <n_k n_\ell>_{CG} (V_{ik} - V_{jk})(V_{i\ell} - V_{j\ell}) \quad . \tag{7.10}$$

As represented by (7.9), the high temperatures of the DG system correspond to the low temperatures of the CG system, where very few charges are excited. For $T \to \infty$ or $T^{CG} \to 0$, no charge exists $n_i = 0$, and the height correlation reduces to $G_{ij} \to 2KV_{ij}$, which is proportional to the logarithm of the distance in the thermodynamic limit (7.8): The long-ranged and diverging fluctuation in the DG system is already built in the vacuum state of the CG system. In the DG model most of the simulation time is used to alter the local and short-ranged correlation of the height variables, and to alter the long-ranged asymptotic behavior these proximity effects need to be accumulated and propagated [7.131]. On the other hand, in the CG system, the local change of charge configuration influences the asymptotic behavior of G_{ij} through the long-ranged function V_{ij} directly as presented in the second term in (7.10). One also notes from (7.10) that the finite-size effect is (partly) contained in the 2D-lattice Green's function V_{ij}, appearing instead of the Green's function of the continuous space, $\log r_{ij}$.

By plotting G_{ij} against V_{ij} as in Fig.7.4a,b, one notices that at high temperatures $K \gtrsim 1.5$ all the data lie on a straight line [7.49]. The slope A(T) defined in (7.4) agrees with the KT theory, taking the universal value $8/\pi$ at the transition point $K_R = T_R/J = 1.48$ (Fig.7.5a). At low temperatures $K \lesssim 1.3$ G(r) shows saturation, but near the transition point the relaxation length towards the saturation value becomes enormous. In fact, from Fig.7.5b we see that the point of inflection in the calculated approximation of $L^{-2}(K)$ is only about 3% below the critical value K_c. But only in this narrow region is it possible to identify the critical exponent ν of the logarithm of the correlation length, according to KT $\nu = 1/2$. To perform experiments in this critical range the system size has to be large compared to the correlation length. Due to the exponential divergence of the latter the system size thus has to exceed linear dimensions of some 10^6 lattice units, which is clearly inaccessible to computer simulations.

Apart from this particular point, however, the Monte Carlo simulation [7.49] reveals quantitatively that the roughening transition of the DG model is described by KT theory. With the help of duality transformations this conclusion holds for other 2D systems of the first class defined above.

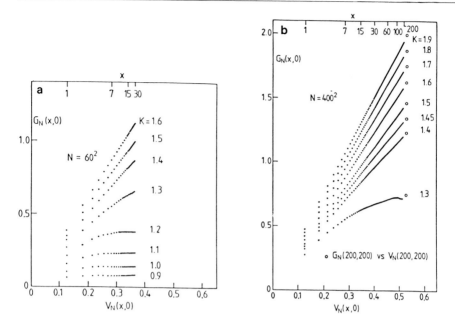

Fig.7.4a,b. Height correlation functions G in the DG model against the 2D-lattice Greens' function V for two different system sizes: (a) $N = 60^2$ and (b) $N = 400^2$ [7.49]

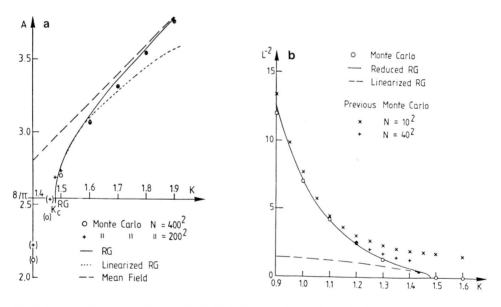

Fig.7.5. (a) Slopes A of G vs V at high temperatures $K \gtrsim 1.4$. Curves represent the results of KT *(straight)* and self-consistent theories *(dashed)*. (b) Inverse saturation value L^{-2} of the correlation functions. The broken line is the renormalization group result, the full line the low-temperature correction [7.49]

7.2.3 Step Free Energy and Crystal Morphology

The roughening transition influences the growth scheme of the crystal as a nonequilibrium property [7.116-121]. Further, it affects the equilibrium morphology of the crystal through a step free energy [7.43]. When the interface is already rough, the introduction of another step does not cost energy. Therefore the step free energy is expected to vanish above the roughening transition in the rough phase. A first indication was found by the Monte Carlo simulation of the ASOS model [7.43] (although the important extrapolation to infinite step length was not performed), confirmed by the later simulation [7.44] through system-size analysis. Mean-field and pair approximations failed to predict the vanishing of the step free energy [7.132,133] since long-wavelength fluctuations were neglected. The duality transformation can generally support that the step free energy vanishes [7.134], and the soluble BCSOS model shows this situation exactly [7.130].

The step free energy gives a singular orientation-dependent term to the surface free energy [7.135,43], which determines the equilibrium morphology of the crystal according to the Wulff theorem [7.136,137]. Below the roughening temperature, the surface free energy has a singular term and the interface is faceted with a flat area with low Miller indices. At high temperatures where the surface free energy has smooth orientation dependence, the interface is rounded. Exactly this is found in experiments for an interface between solid and superfluid ^{4}He [7.92-94], and recently explained theoretically [7.138].

Note, however, that again determination of the correlation exponent ν requires interface diameters of $\gtrsim 10^{6}$ lattice units, while gravitation limits fluctuations above $\approx 10^{8}$ lattice units. Thus on the necessarily logarithmic length scale there is not even one order of magnitude available as an interval for the determination of this exponent.

7.3 Melting Transition

7.3.1 Theoretical Predictions

Melting is one of the oldest and still unresolved examples of phase transitions. Many attempts exist to describe theoretically the melting transition in three dimensions: *Lindemann*'s semiempirical formula [7.139], the order-disorder transition model by *Lennard-Jones* and *Devonshire* [7.140], stability analysis of the solid phase under the shear stress [7.141], and various dislocation models [7.142-146] involving approximations whose validity is difficult to assess. In two dimensions similar order-disorder models [7.147] or polygon packing models [7.148] are proposed. But here one has to note that the true translational LRO is absent at finite temperatures due to the long-wavelength fluctuation [7.1,2,5,6]. The *solid* phase at low temperatures in two dimensions is characterized by the *quasi* LRO, which supports the shear stress and the transversal phonon mode. The extinction of the (infinite-range) transversal phonon mode can be used as a criterion for solid-phase melting [7.149,150]. Here we describe melting in two dimensions in terms of an elastic medium with additional excitations of topological defects, namely dislocations [7.15, 16,31-34,151].

Edge dislocations in two dimensions interact mutually through the Hamiltonian [7.114,115]

$$H_{DL} = -\frac{1}{2} J \sum_{i \neq j} \left[\mathbf{b}_i \cdot \mathbf{b}_j \log r_{ij}/a - (\mathbf{b}_i \cdot \mathbf{r}_{ij})(\mathbf{b}_j \cdot \mathbf{r}_{ij})/r_{ij}^2 \right] + E_c \sum_i \mathbf{b}_i^2 \quad , \quad (7.11)$$

where $\mathbf{b}_i$ is the Burger's vector of a dislocation at position $\mathbf{r}_i$. The coupling J is given by the bare Lamé coefficients λ and μ by $J = \mu(\mu + \lambda)[\pi(2\mu + \lambda)]^{-1}$, and dislocation core radius and core energy are represented by a and E_c respectively. Since the residual Burger's vector costs large energy proportional to the logarithm of the system size, the total amount of Burger's vectors should vanish

$$\sum_i \mathbf{b}_i = 0 \quad . \tag{7.12}$$

Since the fundamental equations (7.11,12) are quite similar to (7.5-8), *Kosterlitz* and *Thouless* concluded that melting of a 2D crystal is caused by the unbinding of dislocation pairs [7.15,16]. At low temperatures thermally created dislocations are all bound to *neutral* pairs to minimize the energy. On increasing temperature the bound pairs are dissociated, and free dislocations are allowed. These free dislocations destroy the quasi-translational order of the crystal. Furthermore, the shear stress causes these free dislocations to slide without resistance, thus renormalizing the shear modulus μ to zero. This melting transition is predicted to be of continuous order with essential singularities in thermodynamic quantities. In the low-temperature solid phase the Lamé coefficients are renormalized to μ_R and λ_R by dislocations, and the combination

$$K_R = 4a_0^2 \mu_R(\mu_R + \lambda_R)/k_B T(2\mu_R + \lambda_R) \tag{7.13}$$

is predicted to take the universal value 16π at the melting temperature, irrespective of the microscopic nature of the atomic systems.

Dissociation of dislocation pairs obviously leads to destruction of orientational order, but only in the weak sense, as it decays algebraically towards infinity. There is still *quasi*-long-range orientational order persistent, which was predicted by *Halperin* and *Nelson* (HN) as the "hexatic" phase [7.31,32]. Each single dislocation can be parametrized further as consisting of two tightly bound disclinations [7.114] of opposite "sign." The HN prediction was that at an even higher temperature these disclination pairs also dissociate, leading to a second, KT-type transition from hexatic to fluid. However, an alternative possibility to form networks by dislocations and a first-order transition by a grain-boundary mechanism was suggested by *Chui* [7.152,153].

7.3.2 Computer Experiments on Atomistic Systems

The possibility of two consecutive, continuous phase transitions with essential singularities and of an intermediate hexactic phase occurring as predicted by KT and HN is so unusual that various computer experiments on various atomistic systems were performed with substantial effort.

Frenkel and *McTague* performed molecular dynamics calculations of the Lennard-Jones system, and found two consecutive phase-transition temperatures [7.62]. At the lower one, the resistance to shear is lost and at the higher one the orientational order is lost. A similar finding is also reported by *Tobochnik* and *Chester* [7.75]. However, detailed later investigations [7.65-74] have found that the Lennard-Jones system in reality undergoes a single first-order phase transition (Fig.7.6a), and the observation of two transitions can be attributed to crossing the two-phase coexistence region in these constant-density simulations. For other hard-core [7.50-52] and soft-core [7.57,58] systems a single first-order melting transition is observed. For the r^{-1} Coulomb gas system [7.77-81] the situation is not so clear. *Hockney* and *Brown* [7.77] obtained a second-order phase transition, but later calculations [7.78,81] produce a phase transition with hysteresis.

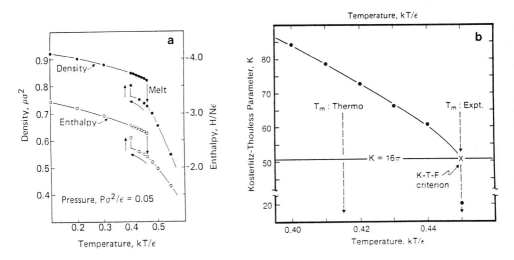

Fig.7.6. (a) The equilibrium density and enthalpy per atom as a function of temperature. One clearly notes discontinuity and hysteresis of the first-order melting transition. (b) The KT renormalized coupling constant K_R (denoted by K) as a function of temperature. T_m:Expt. is the stability limit of a solid phase and T_m:Thermo is the thermodynamic melting temperature [7.74]

A remark on computational procedures here seems pertinent. Generally, it is easy to understand that computer simulations of phase transitions should always be performed using intensive quantities ("fields") as fixed control parameters, while extensive quantities should play the role of observables. The coexistence

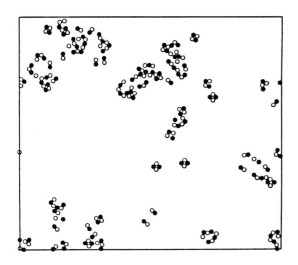

Fig.7.7. Disclinations in the Lennard-Jones system in the metastable solid phase. Atoms with coordination numbers 5 and 7 correspond to negative (o) and positive (•) disclinations [7.66]

lines in the intensive variables appear as coexistence *regions* in the extensive variables. Keeping, for example, the expectation value of the order parameter fixed (as an ensemble average), one has to struggle with all the long-time problems of phase separation. Unfortunately in molecular dynamics simulation one usually fixes the number of particles and the energy rather than chemical potential and temperature. Hence the precise location and characterization of a phase transition is a considerably demanding task as, for example, in a constant-pressure ensemble [7.70-74].

Microscopic investigations of topological defects have now revealed that on melting, the disclination pairs (= dislocations) are not dissociated but they form closed loops, Fig.7.7 [7.66]. Investigations of the renormalized coupling constant K_R (7.13) indicate that it attains the KT universal value 16π at the solid-phase stability limit, but the real melting transition takes place at a lower temperature, Fig.7.6b [7.74].

How can one interpret this discrepancy between theory and experiment? One possibility is the failure of the computer experiments. For example, the finite-size and finite simulation time restrictions might have obscured the KT phase transition [7.66,67]. Another possibility is that the true melting is caused by some other mechanism than dislocation, such as vacancies, anharmonic lattice vibrations and so on. Still a third possibility is that the dislocation model for melting is valid, but the KT and HN theories are insufficient to explain the mechanism. In fact, *Chui* has claimed that dislocation vectors form grain boundaries, and that the development of unbound pairs of grain boundaries is responsible for the first-order melting transition [7.152,153]. The phase transition of the original dislocation vector model (7.11) can be understood easily by simulating this system directly.

7.3.3 Dislocation Vector System

By Monte Carlo simulations of dislocation vector systems (7.11) two types of phase transitions are found, depending on the dislocation core energy E_c [7.85,86]. On heating a system with a large core energy, bound dislocation pairs dissociate to induce a continuous phase transition as shown in Fig.7.8. The change of microscopic configurations in Fig.7.8 is quite similar to that of the scalar Coulomb gas system in Fig.7.3. The internal energy and the specific heat show no observable singularity Fig.7.9a. The renormalized coupling K_R becomes about 16π when the dislocation pairs dissociate. The whole picture agrees with the KT theory.

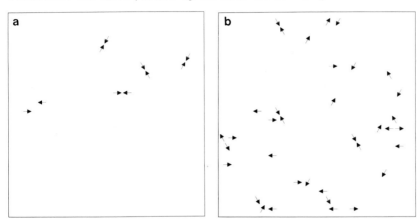

Fig.7.8a,b. Configurations of dislocation vectors for a system with a large core energy at temperatures (a) T = 0.22 and (b) T = 0.25 [7.85,86]

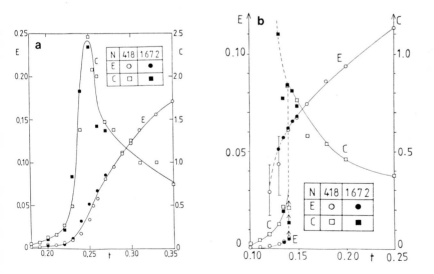

Fig.7.9a,b. Internal energy and specific heat of dislocation vector systems (a) with a large core energy E_c = 0.82 J and (b) with a small core energy E_c = 0.57 J. The melting transition is apparently continuous order for a) and first order for b) [7.85,86]

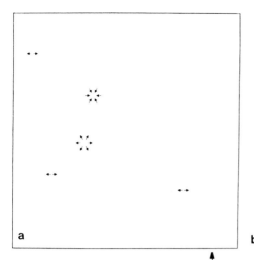

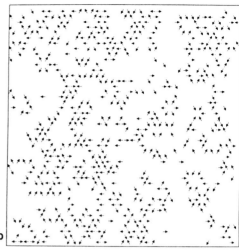

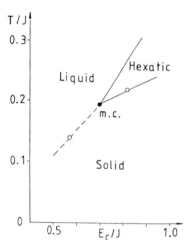

Fig.7.10a,b. Configurations of dislocation vectors for a system with a small core energy at temperatures (a) T = 0.135 and (b) T = 0.145 [7.85,86]

Fig.7.11. Schematic phase diagram of the dislocation vector system at temperature T/J and core energy E_c/J space. *Straight* lines represent continuous phase transitions, and a *dashed* line a first-order transition. A multicritical point (m.c.) is expected

However, for the system with lower core energy the melting transition turns out to be first order with discontinuity and hysteresis of the internal energy, Fig.7.9b.

Configurations of dislocation vectors in Fig.7.10 reveal their connected loop structure, forming grain boundary loops. Note the resemblance of Fig.7.10 to Fig. 7.7 depicting the Lennard-Jones system.

Sizes of the dislocation vector systems are measured in units of dislocation core radius, and therefore may correspond to large sized atomistic systems. Still the simulated system size may be small to predict the precise critical behavior. However, changes of the mechanism and the order of the phase transition are apparent for these small sizes.

From these findings we may draw a phase diagram of the dislocation vector systems at temperature T/J and core energy E_c/J space, Fig.7.11. The dislocation-unbinding mechanism postulated by KT and HN rules the melting transition occurring to the right of the multicritical point (m.c.), while the grain boundary mechanism rules the left. In the atomistic systems, the coupling J and the core free energy E_c may depend on the temperature, and therefore each system may be represented by a curved line in T/J-E_c/J phase space. The precise curve and its crossing point with the phase boundaries may be an interesting problem for future analysis.

Experiments on two-dimensional melting and freezing have been performed for various systems; electrons on liquid He [7.95-99], absorbed systems on graphite substrates [7.100-102], monolayer film on water [7.103] and substrate free films [7.104,105].

Some systems undergo a melting transition apparently consistent with KT theory [7.99-101], others show a first-order transition [7.104,105], while still others change the order of phase transition by changing a control parameter [7.102]. As yet there is no unifying picture for these various results.

References

7.1   L.D. Landau: Phys. Z. Sov. **11**, 26 (1937)
7.2   R.E. Peierls: Ann. Inst. Henri Poincaré **5**, 177 (1935)
7.3   N.D. Mermin, H. Wagner: Phys. Rev. Lett. **17**, 1133 (1966)
7.4   P.C. Hohenberg: Phys. Rev. **158**, 383 (1967)
7.5   N.D. Mermin: J. Math. Phys. **8**, 1061 (1967)
7.6   N.D. Mermin: Phys. Rev. **176**, 250 (1968)
7.7   A.M. Polyakov: Phys. Lett. B**59**, 79 (1975)
7.8   A.A. Migdal: Sov. Phys. JETP **42**, 743 (1976)
7.9   E. Brezin, J. Zinn-Justin: Phys. Rev. B**14**, 3110 (1976)
7.10 F. Wegner: Z. Phys. **206**, 465 (1967)
7.11 B. Jancovici: Phys. Rev. Lett. **19**, 20 (1967)
7.12 V.L. Berezinskii: Sov. Phys. JETP **32**, 493 (1981); **34**, 610 (1972)
7.13 H.E. Stanley, T.A. Kaplan: Phys. Rev. Lett. **17**, 913 (1966)
7.14 H.E. Stanley: Phys. Rev. Lett. **20**, 589 (1968)
7.15 J.M. Kosterlitz, D.J. Thouless: J. Phys. C**6**, 1181 (1973)
7.16 J.M. Kosterlitz: J. Phys. C**7**, 1046 (1974)
7.17 J.V. José, L.P. Kadanoff, S. Kirkpatrick, D.R. Nelson: Phys. Rev. B**16**, 1217 (1977)
7.18 D.J. Amit, Y.Y. Goldschmidt, G. Grinstein: J. Phys. A**13**, 585 (1980)
7.19 J. Villain: J. de Physique **36**, 581 (1975)
7.20 W.J. Camp, J.P. Dyke: J. Phys. C**8**, 336 (1975)
7.21 P.B. Wiegmann: J. Phys. C**11**, 1583 (1978)
7.22 D.R. Nelson, J.M. Kosterlitz: Phys. Rev. Lett. **39**, 1201 (1977)
7.23 V. Ambegaokar, B.I. Halperin, D.R. Nelson, E.D. Siggia: Phys. Rev. B**21**, 1806 (1980)
7.24 S. Doniach, B. Huberman: Phys. Rev. Lett. **42**, 1169 (1979)
7.25 L.A. Turkevich: J. Phys. C**12**, L385 (1979)
7.26 B.I. Halperin, D.R. Nelson: J. Low Temp. Phys. **36**, 599 (1979)
7.27 T. Ohta, K. Kawasaki: Prog. theor. Phys. **60**, 365 (1978)
7.28 P. Minnhagen, A. Rosengren, G. Grinstein: Phys. Rev. B**18**, 1356 (1978)
7.29 P. Minnhagen: Solid State Commun. **36**, 805 (1980)
7.30 S.T. Chui, P.A. Lee: Phys. Rev. Lett. **35**, 315 (1975)

7.31 B.I. Halperin, D.R. Nelson: Phys. Rev. Lett. **41**, 121 (1978)
7.32 D.R. Nelson, B.I. Halperin: Phys. Rev. B**19**, 2457 (1979)
7.33 A.P. Young: Phys. Rev. B**19**, 1855 (1979)
7.34 D.R. Nelson: Phys. Rev. B**26**, 269 (1982)
7.35 C. Kawabata, K. Binder: Solid State Commun. **22**, 705 (1977)
7.36 M. Suzuki, S. Miyashita, A. Kuroda, C. Kawabata: Phys. Lett. **60**A, 478 (1977)
7.37 M. Suzuki, S. Miyashita, A. Kuroda: Prog.theor. Phys. **58**, 701 (1977)
7.38 S. Miyashita, H. Nishimori, A. Kuroda, M. Suzuki: Prog. theor. Phys. **60**, 1669 (1978)
7.39 S. Miyashita: Prog. theor. Phys. **63**, 797 (1981); **65**, 1595 (1981)
7.40 W.L. McMillan: Unpublished
7.41 J. Tobochnik, G.V. Chester: Phys. Rev. B**20**, 3761 (1979)
7.42 D.P. Landau, K. Binder: Phys. Rev. B**24**, 1391 (1981)
7.43 H.J. Leamy, G.H. Gilmer: J. Crystal Growth **24/25**, 499 (1974)
7.44 R.H. Swendsen: Phys. Rev. B**15**, 5421 (1977)
7.45 R.H. Swendsen: Phys. Rev. B**18**, 492 (1978)
7.46 W.J. Shugard, J.D. Weeks, G.H. Gilmer: Phys. Rev. Lett. **41**, 1399 (1978)
7.47 W.J. Shugard, J.D. Weeks, G.H. Gilmer: Phys. Rev. B**21**, 5309 (1980)
7.48 W.J.P. van Enckebort, J.P. van der Eerden: J. Crystal Growth **47**, 501 (1979)
7.49 Y. Saito, H. Müller-Krumbhaar: Phys. Rev. B**23**, 308 (1981)
7.50 B.J. Alder, T.E. Wainwright: Phys. Rev. **127**, 359 (1962)
7.51 W.G. Hoover, B.J. Alder: J. Chem. Phys. **46**, 686 (1967)
7.52 B.J. Alder, W.R. Gardner, J.K. Hoffer, N.E. Phillips, D.A. Young: Phys. Rev. Lett. **21**, 732 (1968)
7.53 B.J. Alder, W.G. Hoover: In *Physics of Simple Liquids*, ed. by H.N.V. Temperley, J.S. Rowlinson, G.S. Rushbrooke (North-Holland, Amsterdam 1968) p.79
7.54 W.W. Wood: In *Physics of Simple Liquids*, ed. by H.N.V. Temperley, J.S. Rowlinson, G.S. Rushbrooke (North-Holland, Amsterdam 1968) p.117
7.55 W.W. Wood: J. Chem. Phys. **52**, 729 (1970)
7.56 D.A. Young, B.J. Alder: J. Chem. Phys. **60**, 1254 (1974)
7.57 J.N. Cape: J. Chem. Society Faraday II**76**, 1646 (1980)
7.58 J.Q. Broughton, G.H. Gilmer, J.D. Weeks: J. Chem. Phys. **75**, 5128 (1981); Phys. Rev. B**25**, 4651 (1981)
7.59 F.W. de Wette, R.E. Allen, D.S. Hughes, R. Rahman: Phys. Lett. **29**A, 548 (1969)
7.60 R.M.J. Cotterill, L.B. Pederson: Solid State Commun. **10**, 439 (1972)
7.61 F. Tsien, J.P. Valleau: Molecular Phys. **27**, 177 (1974)
7.62 D. Frenkel, J.P. McTague: Phys. Rev. Lett. **42**, 1632 (1979)
7.63 J.P. McTague, D. Frenkel, M.P. Allen: In *Ordering in Two Dimensions*, ed. by S.K. Sinha (North-Holland, Amsterdam 1980) p.147
7.64 S. Toxvaerd: J. Chem. Phys. **69**, 4750 (1978)
7.65 S. Toxvaerd: Phys. Rev. Lett. **44**, 1002 (1980)
7.66 S. Toxvaerd: Phys. Rev. A**24**, 2735 (1981)
7.67 S. Toxvaerd: Phys. Rev. Lett. **51**, 1971 (1983)
7.68 F. van Swol, L.V. Woodcock, J.N. Cape: J. Chem. Phys. **73**, 913 (1980)
7.69 D. Henderson: Molecular Phys. **34**, 301 (1977)
7.70 F.F. Abraham: Phys. Rev. Lett. **44**, 463 (1980)
7.71 F.F. Abraham: In *Ordering in Two Dimensions*, ed. by S.K. Sinha (North-Holland, Amsterdam 1980) p.115
7.72 F.F. Abraham: Phys. Rev. B**23**, 6145 (1981)
7.73 J.A. Barker, D. Henderson, F.F. Abraham: Physica **106**A, 226 (1981)
7.74 F.F. Abraham: Phys. Rep. **80**, 339 (1981)
7.75 J. Tobochnik, G.V. Chester: In *Ordering in Two Dimensions*, ed. by S.K. Sinha (North-Holland, Amsterdam 1980) p.339
7.76 J.M. Phillips, L.W. Bruch, R.D. Murphy: J. Chem. Phys. **75**, 5097 (1981)
7.77 R.W. Hockney, T.R. Brown: J. Phys. C**8**, 1813 (1975)
7.78 R.C. Gann, S. Chakravarty, G.V. Chester: Phys. Rev. B**20**, 326 (1979)
7.79 D.S. Fisher, B.I. Halperin, R. Morf: Phys. Rev. B**20**, 4692 (1979)
7.80 R.H. Morf: Phys. Rev. Lett. **43**, 931 (1979)
7.81 R.K. Kalia, P. Vashishta, S.W. deLeeuw: Phys. Rev. B**23**, 4794 (1981)
7.82 R.K. Kalia, P. Vashishta: J. Phys. C**14**, L643 (1981)
7.83 F.H. Stillinger, T.A. Weber: J. Chem. Phys. **74**, 4015 (1981)

7.84 T.A. Weber, F.H. Stillinger: J. Chem. Phys. **74**, 4020 (1981)
7.85 Y. Saito: Phys. Rev. Lett. **48**, 1114 (1982)
7.86 Y. Saito: Phys. Rev. B**26**, 6239 (1982)
7.87 K. Hirakawa, H. Yoshizawa, K. Ubukoshi: J. Phys. Soc. Japan **51**, 2151 (1982)
7.88 L.P. Regnault, J. Rossat-Mignod, J.Y. Henry, L.J. De Jongh: J. Magn. Mag. Mat. **31-34**, 1205 (1983)
7.89 D.J. Bishop, J.D. Reppy: Phys. Rev. Lett. **40**, 1727 (1978)
7.90 A.F. Hebard, A.T. Fiory: Phys. Rev. Lett. **44**, 291 (1980)
7.91 A. Parlovska, D. Nenow: J. Crystal Growth **39**, 346 (1977)
7.92 J. Landau, S.G. Lipson, L.M. Määttänen, L.S. Balfour, D.O. Edwards: Phys. Rev. Lett. **45**, 31 (1980)
7.93 J.E. Avron, L.S. Balfour, C.G. Kuper, J. Landau, S.G. Lipson, L.S. Schulman: Phys. Rev. Lett. **45**, 814 (1980)
7.94 K.O. Keshishev, A.Y. Parshin, A.V. Babkin: JETP-Lett. **30**, 55 (1979); Sov. Phys. JETP **53**, 362 (1981)
7.95 C.C. Grimes, G. Adams: Phys. Rev. Lett. **42**, 795 (1979)
7.96 D.S. Fisher, B.I. Halperin, P.M. Platzman: Phys. Rev. Lett. **42**, 798 (1979)
7.97 D.J. Thouless: J. Phys. C**11**, L189 (1978)
7.98 C.C. Grimes: Physica **106**A, 102 (1981)
7.99 F. Gallet, G. Gallet, A. Valdes, F.I.B. Williams: Phys. Rev. Lett. **49**, 212 (1982)
7.100 A. Widom, J.R. Owers-Broadley, M.G. Richards: Phys. Rev. Lett. **43**, 1340 (1979)
7.101 R. Feile, H. Wiechert, H.J. Lauter: Phys. Rev. B**25**, 3410 (1982)
7.102 P.A. Heiney, R.J. Birgeneau, G.S. Brown, P.M. Horn, D.E. Moncton, P.W. Stephens: Phys. Rev. Lett. **48**, 104 (1982)
7.103 B.M. Abraham, K. Miyans, S.Q. Xu, J.B. Ketterson: Phys. Rev. Lett. **49**, 1643 (1982)
7.104 D.J. Bishop, W.O. Sprenger, R. Pindak, M.E. Neubert: Phys. Rev. Lett. **49**, 1861 (1982)
7.105 D.E. Moncton, R. Pindak, S.C. Davey, G.S. Brown: Phys. Rev. Lett. **49**, 1865 (1982)
7.106 S.T. Chui, J.D. Weeks: Phys. Rev. B**14**, 4978 (1976)
7.107 H.J. Knops: Phys. Rev. Lett. **39**, 766 (1977)
7.108 E.H. Hauge, P.C. Hemmer: Physica Norvegica **5**, 209 (1971)
7.109 J. Zittarz: Z. Phys. B**23**, 55, 63 (1975); B**31**, 63, 79, 89 (1978)
7.110 J. Zittarz, B.A. Huberman: Solid State Commun. **18**, 1373 (1976)
7.111 H. Gutfreund, B.A. Huberman: J. Phys. C**10**, L225 (1977)
7.112 H.U. Everts, W. Koch: Z. Physik B**28**, 117 (1977)
7.113 L.P. Kadanoff: J. Phys. A**11**, 1399 (1978)
7.114 F.R.N. Nabbaro: *Theory of Dislocations* (Clarendon, Oxford 1967)
7.115 R. de Wit: Solid State Phys. **10**, 249 (1960)
7.116 For a general introduction to the theory and computer studies of crystal growth, see the following review articles:
      a) G.H. Gilmer, K.A. Jackson: In *1976 Crystal Growth and Materials*, ed. by K. Kaldis, H.J. Scheel (North-Holland, Amsterdam 1977) p.80
      b) H. Müller-Krumbhaar: In *Current Topics in Materials Science*, Vol.1, ed. by E. Kaldis (North-Holland, Amsterdam 1978) p.1
      c) H. Müller-Krumbhaar: In *Monte Carlo Methods in Statistical Physics*, Topics Current Phys., Vol.7, ed. by K. Binder (Springer, Berlin Heidelberg New York 1979) p.261
7.117 W.K. Burton, N. Cabrera, F.C. Frank: Phil. Trans. Roy. Soc. London A**243**, 299 (1951)
7.118 D.E. Temkin: Sov. Phys.-Crystal **14**, 344 (1969)
7.119 G.H. Gilmer, H.J. Leamy, K.A. Jackson: J. Crystal Growth **24/25**, 495 (1974)
7.120 J.D. Weeks, G.H. Gilmer, K.A. Jackson: J. Chem. Phys. **65**, 712 (1976)
7.121 For a general introduction to the roughening transition, see the following review articles:
      a) H. Müller-Krumbhaar: In *1976 Crystal Growth and Materials*, ed. by E. Kaldis, H.J. Scheel (North-Holland, Amsterdam 1977) p.116
      b) J.D. Weeks: In *Ordering in Strongly Fluctuating Condensed Matter Systems*, ed. by T. Riste (Plenum, New York 1980) p.293

7.122 L. Onsager: Phys. Rev. **65**, 117 (1944)
7.123 J.D. Weeks, G.H. Gilmer, H.J. Leamy: Phys. Rev. Lett. **31**, 549 (1973)
7.124 H.J. Leamy, K.A. Jackson: J. Appl. Phys. **42**, 2121 (1971)
7.125 J.D. Weeks, G.H. Gilmer: J. Chem. Phys. **63**, 3136 (1975)
7.126 J.D. Weeks, G.H. Gilmer: J. Crystal Growth **33**, 21 (1976)
7.127 J.P. van der Eerden: Phys. Rev. B**13**, 4942 (1976)
7.128 R.H. Swendsen: Phys. Rev. B**15**, 689 (1977)
7.129 Y. Saito: Z. Phys. B**32**, 75 (1978); in *Ordering in Strongly-Fluctuating Condensed Matter Systems*, ed. by T. Riste (Plenum, New York 1980) p.319
7.130 H. van Beijeren: Phys. Rev. Lett. **38**, 993 (1977)
7.131 R.H. Swendsen: Phys. Rev. B**25**, 2019 (1982)
7.132 J.D. Weeks, G.H. Gilmer: J. Crystal Growth **43**, 385 (1978)
7.133 G.H. Gilmer, J.D. Weeks: J. Chem. Phys. **68**, 950 (1978)
7.134 R.H. Swendsen: Phys. Rev. B**17**, 3710 (1978)
7.135 L.D. Landau: In *Collected Papers*, ed. by T. Haar (Pergamon, Oxford 1965) p. 540
7.136 G. Wulff: Z. Kristallogr. **34**, 449 (1901)
7.137 L.D. Landau, E.M. Lifshitz: *Statistical Physics* (Pergamon, New York 1980) p.520
7.138 N. Cabrera, N. Garcia: Phys. Rev. B**25**, 6057 (1982)
7.139 F.A. Lindemann: Phys. Z. **11**, 609 (1910)
7.140 J.E. Lennard-Jones, A.F. Devonshire: Proc. Roy. Soc. A**169**, 317 (1939)
7.141 M. Born: J. Chem. Phys. **7**, 591 (1939)
7.142 S. Mizushima: J. Phys. Soc. Japan **15**, 70 (1960)
7.143 A. Ookawa: J. Phys. Soc. Japan **15**, 2191 (1960)
7.144 P. Kuhlmann-Wilsdorf: Phys. Rev. **140**, A1599 (1965)
7.145 T. Ninomiya: J. Phys. Soc. Japan **44**, 263 (1978)
7.146 S.F. Edwards, M. Warner: Phil. Mag. A**40**, 257 (1979)
7.147 R. Kikuchi, J.W. Cahn: Phys. Rev. B**21**, 1893 (1980)
7.148 H. Kawamura: Prog. theor. Phys. **61**, 1584 (1979); **63**, 24 (1980)
7.149 H. Fukuyama, P.M. Platzmann: Solid State Commun. **15**, 677 (1974)
7.150 P.M. Platzman, H. Fukuyama: Phys. Rev. B**10**, 3150 (1974)
7.151 B.I. Halperin: In *Physics of Low-Dimensional Systems*, ed. by Y. Nagaoka, S. Hikami (Prog. theor. Phys., Kyoto 1979) p.53
7.152 S.T. Chui: Phys. Rev. Lett. **48**, 933 (1982)
7.153 S.T. Chui: Phys. Rev. B**28**, 178 (1983)

8. Monte Carlo Studies of "Random" Systems

K. Binder and D. Stauffer

With 5 Figures

Recent years have seen enormous activity on "random" systems: on the theoretical side, where models containing random parameters in the Hamiltonian are considered (for a magnetic system, one commonly considers randomness in the exchange interaction, in the anisotropy, in the field coupling to the order parameter, etc.); on the experimental side, systems with structural disorder (amorphous systems) as well as various sorts of randomly mixed crystals have been produced and extensively studied. A large number of new phenomena have been identified, some of which are quite different from the more "ideal" systems without such disorder, and many basic questions are still open. At the same time the theoretical methods for describing such systems are much less simple and well developed than for ideal systems, e.g., for systems such as spin glasses, even the proper formulation of a mean-field theory for infinite-range interactions has been a formidable problem, which still lacks a complete solution.

Under these circumstances, simulations are a particularly valuable tool: one can check whether a particular model faithfully describes the phenomena under study, or to what extent a model represents a particular real system, or particular approximate steps in an analytical theory; one can try to answer questions which have not yet been tackled within analytical theories at all, etc. Thus it is no surprise that a very large number of simulations on "random" systems have been made recently. We cannot attempt here to give an exhaustive description, so we concentrate on typical examples of recent work which we consider particularly illuminating. But it must be clearly emphasized that a lot of other papers, mentioned here only briefly, also contain very interesting results, and should be consulted by the interested reader. Older work on parts of this subject has been reviewed in [8.1-5], so we try to keep the overlap with these reviews as small as possible.

8.1 General Introduction

In a "random" system, two types of averages have to be taken: first the thermal average has to be performed in the presence of some "quenched" (frozen-in) disorder, such as a random distribution of magnetic and nonmagnetic atoms over the lattice sites of a model for a diluted magnet, etc., second, the average over the randomness has to be performed (one is not interested in properties related to one particular configuration of the random variables, but rather in average properties only). In the percolation problem (Sect.8.4) the situation is somewhat simpler, of course, since there are no thermal degrees of freedom admitted, and one is interested only in the geometrical properties of the configurations representing random disorder.

A typical example of this double averaging procedure is provided by the Edwards-Anderson spin glass model [8.6], Sects.8.2.1,2. Assume a lattice where on each site there is a spin $S_i = \pm 1$ (Ising case) or $\underline{S}_i = (S_i^x, S_i^y, S_i^z)$ (Heisenberg case), respectively. These spins interact with a magnetic field H (in suitable units) and with pairwise exchange variables J_{ij}. Then the energy (Hamiltonian) $\mathcal{H}$ is

$$\mathcal{H}\{J_{ij}\} = -\sum_{<i,j>} J_{ij}\underline{S}_i\underline{S}_j - H\sum_i S_i^z \quad , \qquad \text{Heisenberg case} \quad , \tag{8.1a}$$

$$\mathcal{H}\{J_{ij}\} = \sum_{<i,j>} J_{ij}S_iS_j - H\sum_i S_i \quad , \qquad \text{Ising case} \quad . \tag{8.1b}$$

The only difference to the standard Ising-Heisenberg models of magnetism is that the exchange variables $\{J_{ij}\}$ are randomly drawn from a probability distribution,

$$P(J_{ij}) = (\sqrt{2}\pi\Delta J)^{-1} \exp[-(J_{ij} - J_0)^2/2(\Delta J)^2] \quad , \qquad \text{Gaussian model} \tag{8.2a}$$

$$P(J_{ij}) = (1 - x)\delta(J_{ij} - J) + x\delta(J_{ij} + J) \quad , \qquad \begin{array}{l}\pm J \text{ model ("frustration} \\ \text{model")} \quad , \end{array} \tag{8.2b}$$

where δ is the Dirac delta function.

Thus what one needs to calculate is the free energy (and its derivatives) for a fixed set $\{J_{ij}\}$

$$F_{\{J_{ij}\}} = -k_BT \ln Z_{\{J_{ij}\}} = -k_BT \ln[\text{Tr}\{\exp(-\mathcal{H}_{\{J_{ij}\}}/k_BT)\}] \quad , \tag{8.3}$$

and *afterwards* the result (8.3) must be averaged with the distribution (8.2). This holds for all other properties as well, e.g., spin correlations $<S_iS_j>_{T,\{J_{ij}\}}$, which are not translationally invariant before the average with (8.2) is taken, are, however, translationally invariant again after the averaging (i.e., depend on the relative distance $\underline{R} = \underline{r}_i - \underline{r}_j$ rather on the sites $\underline{r}_i, \underline{r}_j$ separately)

$$g(\underline{R}) = [<S_iS_j>_{T,\{J_{ij}\}}]_{av} \quad . \tag{8.4}$$

Throughout this chapter, we shall denote by $[\dots]_{av}$ the averaging over the distribution over the disorder variables. This averaging is rather delicate, since in general the distribution function of any variable $A_{\{J_{ij}\}}$ is not symmetrically distributed around the mean value $[A_{\{J_{ij}\}}]_{av}$, and thus the mean value does not coincide with the "most probable value" of $A_{\{J_{ij}\}}$ [8.7]. This problem deserves consideration when one approximates the average $[\dots]_{av}$ by an arithmetic average over n re-

presentative realizations $\{J_{ij}\}$ drawn from distribution (8.2). Typically $n \approx 10^2$
is needed to perform averages such as (8.4) accurately [8.8].

From this discussion it is easy to guess where Monte Carlo methods will be useful for a problem such as (8.4): (i) random numbers are used to generate n sample
distributions $\{J_{ij}\}$ on which the average $[...]_{av}$ is based; (ii) for one particular
sample $\{J_{ij}\}$, one can treat the problem of estimating a thermal average $<A>_{T,\{J_{ij}\}}$
in the standard way (see Chap.1 and [8.9]). Unlike many standard methods of statistical physics, the Metropolis importance sampling method [8.9] does not need any
"homogeneity" in the interactions present in the system, no translational invariance
is required. We must emphasize a caveat, however: this importance sampling really
is a sort of *time averaging* of a dynamic model, which usually is defined in terms
of a master equation [8.9] and can be associated with the Markov process by which
configurations of the "thermal" degrees of freedom [spin configurations in the case
of (8.1)] are generated. Now random systems typically do not have only one characteristic relaxation time during which thermal fluctuations decay, but rather a *broad
spectrum of relaxation times*. Consequently, often very large observation times are
required in a Monte Carlo sampling in order to record true equilibrium properties.
For some questions, such as the behavior of the magnetic susceptibility near the
"freezing transition" at T_f in spin glasses, some relaxation times are clearly so
large that the standard Monte Carlo method [8.9] has been unable to disentagle
transient effects from true equilibrium [8.1,5].

Under these circumstances, progress has been achieved by treating only problem
(i) with Monte Carlo methods, while the thermal averaging (ii) has been treated by
another method, namely exact calculation of the partition function for small lattices [8.10-21] (in this context we also mention work emphasizing ground-state
properties only [8.22-27]). Unfortunately, this approach is restricted to Ising
problems, but it is nevertheless very interesting because one can obtain both the
exact static properties (from the partition function calculation) and the associated
dynamics (from a standard Monte Carlo simulation of exactly the same model), and comparing such static and dynamic averages is particularly revealing for such "glassy"
states.

Of course, using Monte Carlo methods for nothing else but generating a random
distribution of some disorder variables has been a common procedure for studying
low-lying excitations in disordered systems, e.g., in a disordered magnet one likes
to know the spin-wave spectrum (including the linewidth induced by the disorder),
density of states, etc. This problem can be treated either by numerical diagonaliz-
ation of the resulting "dynamical matrix" [8.28-30], or by numerically integrating
equations of motion of suitable Green's functions [8.31,32], or continued fraction
methods [8.33,34], exact enumeration of eigenstates [8.35], etc. Such techniques
have been extremely valuable because they allow the accuracy of approximate analytical techniques to be checked, such as the coherent potential approximation (CPA).

For example, in systems such as $Mn_{1-x}Co_xF_2$ and $Rb_2Mn_{0.54}Mg_{0.46}F_4$ it has been demonstrated [8.36] that such numerical methods (of the *Alben-Thorpe* type [8.31]) give a significantly better account of the observed neutron scattering spectra, which often exhibit a rich structure with many peaks [8.36], than CPA calculations, although these are also successful in some cases [8.37]. Of course, such numerical techniques have not been applied only to diluted ferromagnets [8.28-31] and antiferromagnets [8.36], amorphous ferromagnets [8.34], various spin glass models [8.29,30,32,33,35], but also to problems of disorder in lattice dynamics, the problem of electron localization in random potentials (see [8.38-44] for some numerical studies of this "Anderson transition" [8.45], which will not be discussed further here). Most percolation studies fall into this classification, since they do not involve thermal averages, as mentioned above.

As a final disclaimer, we emphasize that we are disregarding more or less completely in this review problems of structural disorder occurring in "window glasses" or amorphous systems. Of course, in the study of these systems simulation techniques are also a very helpful tool. (For an example where the structure of a metallic glass was simulated, see [8.46]; for simulations of the random network structure of glasses, see [8.47-49]; for "electron glass", see [8.50], etc.) Monte Carlo methods in the study of the lattice dynamics of disordered crystals are reviewed in [8.51].

8.2 Spin Glasses

Spin glasses are disordered magnetic materials which do not show conventional long-range order (ferromagnetism, antiferromagnetism, etc.) but do show a "freezing transition" to some new type of local order, the nature of which is still heavily debated [8.52-56]. Most theoretical work focused attention on grossly simplified models, such as given in (8.1,2), but in spite of their crudeness they are not tractable analytically. Simulations are hence very useful to study the properties of these models, Sects.8.2.1,2 . These models can resemble real systems at best in a qualitative way. It has also been achieved to simulate certain real spin glass systems faithfully, such as $Eu_xSr_{1-x}S$ and $Cu_{1-x}Mn_x$, Sect.8.2.3. Even in cases where some properties are accessible to a reliable analytical treatment, such as the infinite-range version of (8.1,2) or the one-dimensional version, simulations have significantly broadened our understanding of these models (Sects.8.2.3,4).

8.2.1 Short-Range Edwards-Anderson Ising Spin Glasses

The model considered is precisely that of (8.1b), with the sum <i,j> restricted over pairs of nearest-neighbor sites only, longer-range interactions are assumed to be zero. This model was studied extensively both by conventional Monte Carlo techniques (both thermal and disorder averaging) [8.57-85] and by the method mentioned above, where the partition function is calculated exactly [8.10-13,15-17], and only the disorder average is done by Monte Carlo methods. (Of course, if this averaging is omitted and one arbitrarily selects particular configurations as done in [8.86], one may obtain very atypical and hence misleading results.)

Although the model has been studied extensively, it is still incompletely understood and some properties are still debated. We summarize here the main findings very briefly. We begin with the two-dimensional symmetric Gaussian model [$J_0 = 0$ in (8.2b)], which is the most widely studied model. Already in [8.57] it was shown that the specific heat has a broad Schottky-like peak, while the susceptibility has a somewhat sharper cusp, in qualitative similarity to the experiment described in [8.54]. Also remament magnetization M_r was found for temperatures below freezing. This remanence decays with time slowly to zero, apparently with a nonexponential decay law [$M_r \propto t^{-a(T)}$], which is again qualitatively similar to experimental observations. This irreversible behavior of the model was studied in more detail by *Kinzel* [8.69,84], who simulated "sample cooling procedures" in close analogy with experiment. For example, by decreasing the temperature linearly with increasing time in an applied field one obtains at low temperatures a characteristic plateau in the magnetization versus temperature diagram, similar to experiment. While it is now thought that this field-cooled behavior may lead to true thermal equilibrium [8.15,84], one observes nonequilibrium behavior when in the field-cooled state the field is switched off. The resulting "thermoremanent" magnetization (TRM) differs from the "isothermal remanent magnetization" (IRM) obtained by applying a field of the same strength to a system cooled without field, which is then switched off again. It is found that the exponent a(T) defined above also depends on the strength of the field, and is different for IRM and TRM [8.69]. Many features of this behavior have close analogies in experimental observations [8.69,84]. Particularly intriguing is the observation that the critical field $H_c(T,t)$ below which IRM and TRM differ substantially follows a curve $H_c(T,t)/T_f(t) \propto [1 - T/T_f(t)]^{3/2}$ in nearly quantitative agreement with the Almeida-Thouless instability encountered in the mean-field treatment (see [8.53] for a review), although the time-dependnet "freezing temperature" $T_f (t \to \infty) \to 0$, as discussed below.

This irreversible behavior also hampers the study of susceptibility for $T < T_f$, and it turned out that the peaks found from graphically differentiating M versus H curves and from time averages of magnetization fluctuations are systematically different [8.57]. Nevertheless, it was first believed [8.1] that the susceptibility peak was due to the onset of local order as described by the Edwards-Anderson order parameter [8.6],

$$\chi = (1/k_B T)(1 - q_{EA}) \quad , \qquad q_{EA} \equiv \frac{1}{N} \sum_i [<S_i>_T^2]_{av} \quad , \tag{8.5}$$

which follows from the fluctuation-dissipation relation between susceptibility and correlations:

$$\chi = (1/k_B T) \frac{1}{N} \sum_{i,j} [<S_i S_j>_T - <S_i>_T <S_j>_T]_{av} \quad , \tag{8.6}$$

putting off-diagonal terms equal to zero, because they should vanish for a symmetric distribution $P(J_{ij}) = P(-J_{ij})$. If T_f is a phase-transition temperature, $q_{EA} \propto (1 - T/T_f)^\beta$, where β is some exponent, and this onset of order below T_f would yield a cusp in χ [8.6]. While it was first assumed that the Monte Carlo data are consistent with this idea [8.57,60], it was subsequently suggested by *Bray* and *Moore* [8.63-65] that in thermal equilibrium there would not be a transition, the susceptibility χ at zero field being just $\chi = 1/k_B T$, while the observed peak is a transient finite-observation-time effect. Since

$$\chi(t) = (1/k_B T)[1 - q(t)] \quad , \quad q(t) = \frac{1}{N} \sum_i \left[\left\{ \int_0^t S_i(t')dt'/t \right\}^2 \right]_{av} \quad , \tag{8.7}$$

a slow decay of $q(t)$ with observation time t would produce a peak in the (time-dependent) susceptibility $\chi(t)$ although there is none in thermal equilibrium.

This problem was subsequently studied [8.66,68] by performing simulations which extended up to five decades in observation time, and comparing systems with dimensionalities ranging from $d = 2$ to $d = 5$. It turned out that all these systems behave similarly, with the slow relaxation not yet settled down to equilibrium even at the longest time scales studied. Approximate analytic treatments [8.87,88] showed that such behavior can indeed be expected considering overturning single spins (and small clusters) against their "effective field" from surrounding spins, and since there is a broad distribution of these fields a broad spectrum of relaxation times arises. This work hence illustrates the general caveat made in the introduction, that in random systems broad spectra of relaxation times must be expected, which make the application of the importance-sampling Monte Carlo method very difficult. In recent and careful work on the ±J model, *Young* [8.82] used the very effective "Distributed Array Processor" (DAP) to study accurately the decay of $q(t)$ near the apparent transition temperature of the previous simulations. His work suggests that there is no transition at finite temperature, in agreement with the work described below.

The question whether for $d = 2$ an Edwards-Anderson transition in equilibrium exists was resolved by the partition function method [8.10,12,17]. Most convincing evidence against such a transition came from a study of the correlation function

$$g_{EA}(R_{ij}) = [<S_i S_j>_T^2]_{av} \quad , \quad \chi_{EA} = \frac{1}{k_B T} \sum_i g_{EA}(R_{ij}) \quad . \tag{8.8}$$

While in the case of a transition at T_f one would expect that both the correlation length $\varepsilon_{EA} [g_{EA}(R_{ij} \propto \exp(-R_{ij}/\varepsilon_{EA})]$ and the susceptibility χ_{EA} diverge at T_f, and for $T < T_f$ one has $\lim_{R_{ij} \to \infty} g_{EA}(R_{ij}) = q_{EA}^2 > 0$, it was found [8.10] that g_{EA} decreases to zero with distance at all temperatures. At the value of T_f where the *dynamic simulations of exactly the same model* indicate the onset of freezing, the correlation length ξ_{EA} is only about two lattice spacings [8.10]. Although the partition function method is restricted to rather small lattices (linear dimension 18 lattice spacings or less), finite-size effects *near* T_f should not affect the behavior of $g_{EA}(R_{ij})$ significantly. This conclusion is corroborated by a careful Monte Carlo study of $g_{EA}(R_{ij})$ for very large lattices and $T \geq T_f$ [8.82], and the observation that the onset of freezing is due to a finite fraction of spins becoming very slow at T_f [8.77,79]. At this point, one advantage of simulations clearly emerges: one can study the relaxational behavior of any spin individually, and correlate it with the local configuration of the random bonds J_{ij}. *Kinzel* [8.77] found that at T_f slow spins occur in very small clusters well isolated from each other, but which steadily grow in size as the temperature is lowered. The gradual onset of a static correlation χ_{EA} hence has a dynamic counterpart in the size of these slow spin clusters.

The partition function calculations suggest that both the specific heat C and the entropy S vary linearly with T at low temperatures, and $S(T = 0) = 0$. However, in the dynamic simulations where the system is frozen in a certain "valley" in configuration space there is residual entropy [8.83] as in structural glasses. For the ±J model, on the other hand, there is a finite ground-state entropy $S(0) > 0$ even in thermal equilibrium [8.10,61]. It turns out that many properties of the ±J model are qualitatively similar to those of the Gaussian model, such as the temperature variation of ξ_{EA} [8.10,82]. But the ±J model has a $t^{-\frac{1}{2}}$ relaxation towards "equilibrium" [8.61], rather than the nearly logarithmic relaxation seen in the Gaussian model, and even at $T = 0$ $g_{EA}(R_{ij})$ decays to zero, while in the Gaussian model $g_{EA} \equiv 1$ at $T = 0$.

Thus a phase transition where χ_{EA} diverges occurs in this model at $T \to 0$ only. But this zero-temperature transition leads to a rather rapid temperature variation of the nonlinear susceptibility χ_{nl} near T_f already [8.78]. Here χ_{nl} is defined by

$$M = [<S_i>_T]_{av} = \chi H - \chi_{nl} H^3 + \cdots \quad , \tag{8.9}$$

and for symmetric bond distributions is simply related to χ_{EA}: $\chi_{nl} \approx (k_B T)^{-2} \chi_{EA}$. The variation of M/H with field H is found [8.78] to be rather similar to corresponding experimental data [8.89]; for $T \to 0$ a crossover from (8.9) to a singular variation $M \propto H^x$, with $x \approx 0.27$ [8.84] is found, which explains the "plateaus" observed in the field-cooled magnetization.

The picture of a spin glass at low temperatures as a system which can be frozen into one of many more or less equivalent "valleys" in configuration space has been substantiated by studies of ground-state properties [8.10,12,16,17,22,27]. To go from one valley to the next, a "packet of solidary spins" has to be overturned together; the size of this "packet" is estimated at about 13×13 lattice spacings [8.17]. *Morgenstern* [8.17] suggests that the resulting *free*-energy barrier becomes small at about the freezing temperature of the dynamic simulations, and thus shows the self-consistency of this description.

Much less is known about the asymmetric Gaussian model ($J_0 \neq 0$) in (8.2a) or asymmetric $\pm J$ model ($x \neq 1/2$) in (8.2b). A critical value of $J_0/\Delta J$ exists above which the system is ferromagnetic, which, however, is not yet known accurately [8.75]. The critical concentration x_c for the disappearence of ferromagnetism in the $\pm J$ model was obtained by several studies [8.10,24,27,61,72] which all fall into the range $x_c \cong 0.12 \pm 0.02$. But there is still disagreement [8.16,26] whether inbetween the ferromagnetic phase and the spin glass at $T = 0$ a "random antiphase state" also occurs. An interesting problem in connection with these ground-state properties is also the percolation between frustrated or unfrustrated "plaquettes" [8.73,74].

The Monte Carlo results for three- and higher-dimensional spin glasses [8.57,64, 65,68,70] are qualitatively rather similar to the results for $d = 2$. One finds [8.68] that the peak in specific heat becomes perhaps somewhat more sharp as d increases, consistent with the expectation that for $d \geq 6$ there is a cusp in C at T_f as predicted by mean-field theory.

Again the slow dynamics established as $T \rightarrow T_f$ (which above T_f is consistent with an ordinary critical slowing down [8.70]) makes a reliable study of the equilibrium properties at low temperatures difficult. For $d = 3$ the partition function method has again been applied [8.11]. Calculating (8.8) for $4 \times 4 \times 10$ cubes where $\mathbf{R}_{ij}$ is parallel to the long direction of the cube, one finds again that ξ_{EA} is very small near T_f. Of course, this statement is much less well founded than for $d = 2$, since two linear dimensions are also rather small, but apparently the dependence of the apparent order parameter $q(t)$, $t < 2000$ MCS/spin, on temperature is quite independent of size, and can already be studied reliably with a system as small as $4 \times 4 \times 4$ [8.11]. Recent careful Monte Carlo work [8.82] is consistent with these findings but leaves open the possibility that χ_{EA} and ξ_{EA} diverge at a much lower but nonzero temperature.

8.2.2 Short-Range Edwards-Anderson Heisenberg Spin Glasses

Of course, no exact partition function calculations are available for short-range Edwards-Anderson Heisenberg spin glasses (8.1b), and also standard Monte Carlo simulations are much more time-consuming than for the Ising model. Early work [8.60, 90] observed a broad peak in the specific heat and a somewhat sharper "cusp" of χ in this model as well, but it soon became clear that this peak of χ is a nonequilibrium effect. In a more detailed study, *Stauffer* and *Binder* [8.91] studied the relaxation of $q(t)$ for a model where $\underline{\underline{S}}_i$ is an n-component spin for spatial dimensionalities $d = 2$ to $d = 6$, Fig.8.1. Again it was found that for short observation times spin-glass order exists at very low temperatures, and the behavior depends on

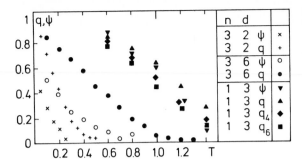

Fig.8.1. Order parameters q,ψ versus temperature for a nearest-neighbor Gaussian Edwards-Anderson n-vector model, for various n and d. Here ψ is an order parameter measuring the alignment to a particular ground state [8.91]

n	d		
3	2	ψ	×
3	2	q	+
3	6	ψ	○
3	6	q	●
1	3	ψ	▼
1	3	q	▲
1	3	q_4	◆
1	3	q_6	■

the spin dimensionality n only weakly (for $n > 1$). For larger observation times the order parameter $q(t)$ was found to decay proportionally to $\ln t$, as in the Ising case, for several decades of time. A theoretical explanation for this behavior was subsequently presented by *Bray* and *Moore* [8.92], who suggested that a typical local energy minimum in configuration space has sufficiently many directions for which the energy surface is locally flat. This assumption was also confirmed by simulations for $n = 2$ and $d = 2$, while for $d = 3$ the situation is less clear [8.93]. While in the Ising case it is obvious that thermally activated processes dominate the dynamics, it is not clear how important they are for the present model. Some evidence for the existence of low-lying metastable states in Heisenberg spin glasses was provided by *Reed* [8.94], and they might well influence the relaxation of $q(t)$ at very long times. A study of this relaxation by Monte Carlo methods is rather difficult, since for $n > 1$ no spin direction is preferred and hence in a finite system there may be rotational diffusion of the direction of the resulting magnetic moment. Thus a careful study of size dependence is needed to exclude this effect [8.95].

It is also worth mentioning that for Heisenberg systems the Monte Carlo method does not provide a faithful description of the real system dynamics (Chap.1). But the ground-state spin configurations found by Monte Carlo methods can be used to find the spin-wave spectrum [8.29,93,96,97]. It now appears that for $d = 3$, $n = 2$ the density of excited states with excitation energy λ vanishes as $\lambda^{\frac{1}{2}}$, and the spin-wave stiffness constant is nonzero [8.93].

A question which can be linked to that of the lower critical dimension (above which a freezing transition would occur in equilibrium at nonzero temperature) is the sensitivity of the free energy to boundary conditions [8.98]. Considering a cylindrical domain of length L and cross-sectional area A, at the end either periodic (P) or antiperiodic (AP) boundary conditions are applied. Defining the free-energy difference per spin Δf as $\Delta f = (F_{AP} - F_P)/N$, the quantity of interest is the configurational average $\gamma \equiv [(\Delta f)^2]_{av}$, since $[\Delta f]_{av} = 0$ for a spin glass. It is then proposed [8.98] that in a spin glass γ decays algebraically with $L(\gamma \propto A^{-r}L^{-p})$, while in a paramagnetic phase γ would decay exponentially with L. For the $d = 3$ Heisenberg spin glass at $T = 0$, $p = 3$, $r = \frac{1}{2}$ [8.98], and it is suggested that these results imply that the system is below its lower critical dimension at d = 3 [8.98]. For short observation times, however, where the system has not fully adjusted to a change of boundary condition, one obtains p = 1 instead [8.99], which implies that then the system behaves as if it were on its lower critical dimension.

Again all this work refers to symmetrical distributions only. Only very recently was the crossover studied from spin-glass behavior to ferromagnetism in a model with asymmetric bond distribution [8.100].

8.2.3 Site-Disorder Models

The simplest site-disorder models result when one considers random binary mixtures $A_x B_{1-x}$ with exchange constants $J_{AA} \neq J_{BB} \neq J_{AB}$ between the Ising spins. For $J_{AB} = -|J_{AA}| = -|J_{BB}|$ and $x = 0.5$, the resulting "*Mattis* spin glass" [8.101] can be mapped via gauge transformations to an (Ising) ferromagnet. For $x \neq 0.5$ or other choices of interaction parameters, however, the behavior of the system is not trivial. The phase diagram of this model was studied by *Tatsumi* [8.102]. Recently the nonlinear susceptibility was also studied but only for the trivial symmetric case [8.103].

More interesting is the class of models where a magnet containing competing interactions is diluted. A simple cubic Ising system with nearest-neighbor ferromagnetic exchange J_1 and next-nearest-neighbor exchange J_2 which is antiferromagnetic in one lattice direction ("ANNNI model" [8.104]) was studied as a function of concentration of magnetic atoms [8.105]. It was found that the wave vector describing the modulated phase changes with concentration and order breaks down before the percolation threshold is reached.

A square lattice with nearest-neighbor exchange ferromagnetic, next-nearest-neighbor exchange antiferromagnetic was studied with the partition function method [8.13]. This model was predicted to have spin-glass behavior in a broad concentration range [8.106]. The numerical results [8.13] are consistent with this prediction and suggest behavior of the correlation function $g_{EA}(R_{ij})$ qualitatively similar to the ±J model.

A similar model for the fcc lattice but with Heisenberg rather than Ising spins is a rather realistic model of $Eu_x Sr_{1-x} S$, a classic spin-glass material studied extensively experimentally [8.107]. Monte Carlo work on this model could explain [8.106] the decrease of the ferromagnetic Curie temperature $T_c(x)$ with decreasing x nicely, without any adjustable parameter. But for $x < 0.5$, where experiment shows spin-glass behavior, $q(t)$ is found to decay to zero as a function of time rather rapidly, there is nearly no evidence for a well-defined T_f. The latter is obtained only if dipolar anisotropy is included [8.56]. The same conclusion, that anisotropy is needed to stabilize spin-glass behavior, was also reached in simulations of metallic spin glasses [8.108]. A model similar to the above, but on the bcc lattice and with Ising spins, was studied in [8.109] with reference to $Fe_{1-x} Al_x$ alloys.

Grest and *Gabl* [8.110] considered diluted Ising antiferromagnets on triangular and fcc lattices. While in all previous models spin-glass behavior arises in competition between ferro- and antiferromagnetic bonds, in the present "fully frustrated" models there are no ferromagnetic bonds. Nevertheless, spin-glass behavior is found here also [8.110-112]. In the case of a diluted classical Heisenberg fcc antiferromagnet no clear evidence of spin-glass behavior was obtained, however [8.111].

Of course, models of metallic spin glasses, where the randomly located dilute magnetic atoms interact via the long-range Rudermann-Kittel-Kasuya-Yoshida (RKKY) interaction, would be particularly interesting from the experimental point of view. The partition function method is restricted to the one-dimensional Ising version of this model [8.113] and systems containing at most $N = 16$ spins. The results [8.113] indicate behavior similar to the infinite-range model [8.114], i.e., the order-parameter square increases with the size of the system at low temperatures, in contrast to the short-range case where the opposite tendency occurs [8.10]. Note, however, that in one dimension the RKKY interaction falls off with distance R as slowly as R^{-1}. For three-dimensional RKKY systems attention has been focused on the Heisenberg case [8.30,108,115-117]. Early work [8.115] was restricted to a discussion of the "distribution of effective fields" at low temperatures (for a related simulation for dipolar rather than RKKY interactions, see [8.118]). Due to the intrinsic difficulty of performing Monte Carlo simulations with long-range potentials, studies of thermodynamic properties are restricted to rather small systems ($N = 96$ in [8.116,117]). Very recently it has become possible [8.108] to study the thermodynamic properties of RKKY systems containing between 500 and 4928 spins. A typical example (appropriate for CuMn at an Mn concentration of 0,9%) is shown in Fig.8.2. It is seen that without anisotropy the order parameter falls off towards zero at very low temperatures, while rather small anisotropies are sufficient to stabilize spin-glass order for the time scales accessible in Monte Carlo computer experiments. Unfortunately, the freezing temperatures so obtained still depend rather sensitively on the size of the system studied, at least for small anisotropy. Thus a quantitative comparison with experiment in the vicinity of the freezing temperature cannot yet be done. *Walker* and *Walstedt* [8.108] also studied the behavior of the model in an applied magnetic field, and found some evidence for freezing of the transverse spin component as predicted in mean-field calculations [8.119].

These computer model studies of a metallic spin glass [8.30,108] are a beautiful example of how far one can push Monte Carlo simulations of rather complicated model systems at present. These calculations have contributed significantly to the understanding of real spin-glass materials. Particularly interesting has also been the result [8.30] that most of the experimental specific heat (which is nearly linear in temperature at low temperatures) can be accounted for by the elementary excitations of the model; hence processes involving hopping over energy barriers

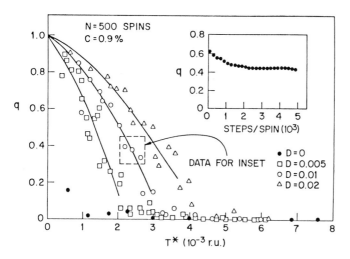

Fig.8.2. Spin-glass order parameter q versus temperature for an RKKY spin-glass model. The parameter D measures the strength of the pseudo-dipolar anisotropy. Inset shows the time variation of q in a typical run [8.108]

do not seem to be important for the linear specific heat, contrary to theoretical suggestions [8.120] and views on "window" glasses ("two-level systems" [8.120]). Also the spin-wave stiffness constant has been estimated for this model [8.121].

Finally we mention that preliminary simulation results are also available for a spin glass with RKKY and additional Dzyaloshinski-Moriya interactions [8.122], but this study also uses only N = 96 spins.

8. 2. 4 The Infinite-Range Model

We here return to the model of (8.1,2), but now allow for interactions between *arbitrary* pairs of spins <i,j>, the exchange constants J_{ij} always being drawn from the same distribution (8.2) irrespective of distance. This model was introduced [8.114] as an attempt to make the mean-field approximation of [8.6] precise. While it is generally accepted that this model has a freezing transition at T_f below which order develops, it now is clear that a single-order parameter as introduced in (8.5) does not tell the whole story. Simulations have contributed to the understanding of this model in many respects. *Thouless* et al. [8.123] based some of their assumptions on properties of the simulated distribution function of effective fields [8.124], studied further in [8.125,126] including the Heisenberg case [8.125] or magnetic field [8.126]. There are now two main approaches to the understanding of the mean-field theory of spin glasses: (i) *Parisi*'s replica symmetry breaking [8.114] (ii) *Sompolinsky*'s theory [8.127] based on nonergodic dynamic behavior (the phase space has many valleys with barriers inbetween, with barrier heights tending to infinity as N → ∞: averages over one valley or many valleys yield different results). Some specific predictions of both theories have been tested by Monte Carlo studies [8.128-131,18]. *Parisi* [8.128] studied the field dependence of magnetization, energy and one of his order parameters and found agreement with his replica theory [8.114], though the statistical accuracy of his data is rather limited. *Ma* and *Payne* [8.129] studied the entropy S as a function of magnetic field

252

H, applying the new method of estimating the entropy from "coincidence counting" [8.132]. They found that S(H) is (nearly) independent of field, consistent with the *Parisi-Toulouse* so-called projection hypothesis [8.133]. One now knows that this hypothesis is not exact [8.134], but numerical evidence suggests that it is a fairly accurate approximation. *Young* and *Kirkpatrick* [8.18] applied the method of calculating partition functions of systems containing up to 20 spins exactly, carefully studying the size dependence. They found that elementary excitations involve overturning $N^{\frac{1}{2}}$ spins relative to the ground state, and that in the infinite-range model the off-diagonal terms in (8.6) must not be neglected, and hence the relation (8.5) is not valid in the infinite-range case [8.135]. *MacKenzie* and *Young* [8.131] studied the relaxation times needed for going from one low-lying state to the next, and confirmed the qualitative aspects of *Sompolinsky*'s theory [8.127]. *Dasgupta* and *Sompolinsky* [8.130] confirmed their analytic prediction that the square of the "staggered magnetization" associated with the largest eigenvalues of the interaction matrix scales with size as $N^{5/6}$, by Monte Carlo simulations. They presented further evidence that short-time properties (system stays in one "valley" in phase space) correspond to an individual solution of the "TAP equation" [8.123], as suggested previously [8.30,136]. For such nonergodic systems it is hence crucial to remember that standard Monte Carlo sampling yields time averages rather than ensemble averages, which would correspond to averaging over all solutions rather than an individual solution.

8.2.5 One-Dimensional Models

The one-dimensional random-bond Ising model has been studied by various numerical transfer-matrix methods [8.137-139], ordinary Monte Carlo methods [8.140-142], and by extending *Glauber*'s solution [8.143] for calculating dynamic correlations to inhomogeneous systems [8.144], where then again only the average over the bond distribution is performed by Monte Carlo techniques. Apart from their possible application to biopolymers [8.137,138], the one-dimensional models are an interesting testing ground for various methods, as it is rigorously established that there cannot be any finite transition temperature, for instance. While the static properties, apart from anomalous behavior of entropy and magnetization as a function of field at low temperatures [8.139], are not so interesting, it turns out that remanent magnetization can be observed with a logarithmic decay over severval decades in time [8.141,142,144], just as in the higher-dimensional models. This model is thus a simple example of spin-glass behavior arising entirely from dynamic effects, even though there are not even any "frustrated" bonds in the ground state (for H = 0). Similar relaxation is also found in modified Mattis models [8.145], which are also not frustrated but have a phase transition. This observation again corroborates our general conclusion that due to the broad distribution of relaxation times in random systems, their equilibrium behavior is rather hard to obtain from the standard "dynamic" Monte Carlo sampling, and a careful study of observation time effects in always needed.

8.3 Other Systems with Random Interactions

In contrast to the enormous activity in spin-glass simulations, work on systems involving other kinds of randomness, e.g., random fields, random anisotropies, is relatively sparse. The simplest randomness concerns nonmagnetic impurities in a ferromagnet with no competing interactions [8.1]. Recent progress with this problem was made by *Novotny* and *Landau* [8.146], who showed that in the dilute Baxter-Wu model the critical exponents of its second-order transition differ from those of the undiluted model, as theoretically expected. In the Ising model no such clear evidence is found [8.1,147].

More dramatic effects are predicted for systems where impurities act like a random field [8.148]. The nearest-neighbor Ising model in a random field $h_i = \pm h$

$$\mathcal{H} = -J \sum_{<i,j>} S_i S_j - \sum_i h_i S_i$$

has been studied both by ordinary Monte Carlo (MC) methods [8.149], by Monte Carlo renormalization group methods (MCRG) [8.150] and by the partition function method [8.21]. Both MC and MCRG are consistent with a transition to a ferromagnetic state in weak random fields in three dimensions, but not in two dimensions where order is destroyed [8.21,148]. It has been shown that in weak fields there is a "smeared transition" (the specific heat peak is progressively rounded with increasing strength of the random field, Fig.8.3) to a state consisting of large ordered domains [8.21]. This observation is qualitatively similar to corresponding experimental results [8.151], which refer to *three*-dimensional systems, however. A controversy exists whether there should be an instability of ferromagnetic order to weak random fields [8.152], which is not found in [8.149,150]. Since the problem can be related to the problem of the random-field-induced width of interfaces, an attempt has been made to study the latter problem directly by MC methods as well [8.153]. Unfortunately, such problems are also rather difficult to study since large system sizes are required, and again the simulations are hampered by large relaxation time effects.

The same caveat applies to the problem of isotropic ferromagnets with random anisotropies, where theories imply absence of ferromagnetic long-range order [8.154], and again simulations have had difficulties to note this instability [8.155], but more recent work treating the Ising limit of infinitely strong random anisotropy shows the absence of ferromagnetic long-range order [8.156]. In this limit, however, from the very beginning the model is identical to an Ising-type spin glass.

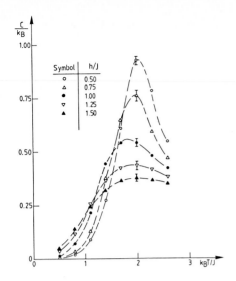

Specific heat of two-dimensional Ising models versus temperature for various values of the random magnetic field. Data are from an exact partition function calculation for systems of size 12×12, averaged over 30 samples of random field configurations [8.21]

8.4 Percolation Theory

More than a hundred papers on percolation are published every year, and many of them use Monte Carlo simulation. As for spin glasses we cannot possibly review all work, so restrict ourselves to a few and perhaps particularly simple and fundamental aspects. More details and references are given in specialized reviews [8.3, 157,158] and, in particular, in the collection of articles on percolation structures and processes [8.159]. We also refer to these reviews for the large variety of generalizations, modifications and applications of percolation theory; the present overview concentrates on the usual, i.e., random, percolation of sites in periodic lattices.

8.4.1 Cluster Numbers

As explained in Chap.1, in the usual site percolation process each lattice site is occupied randomly with probability p and left empty with probability 1-p, independent of what its neighbors are doing. A "cluster" is a group of neighboring occupied sites. For p larger than some threshold p_c one infinite cluster appears besides many finite clusters, whereas for p below p_c all clusters are finite in the thermodynamic limit. For bond percolation, instead of occupied sites we work with occupied bonds between lattice sites; for simplicity we restrict ourselves here to site percolation.

The average number per lattice site of clusters containing s sites each is called n_s; right at the critical point $p = p_c$, n_s decays with a power law

$$n_s \propto s^{-\tau} \quad (s \to \infty) \quad .$$

(8.10)

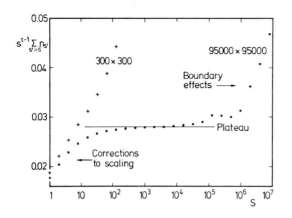

Fig.8.4. Variation of the scaled number of clusters larger than s with cluster size in the triangular lattice at $p = p_c = 1/2$ [8.161]

Monte Carlo evidence for this simple behavior was already given in [8.1,3] and the references therein. Scaling arguments [8.3] relate τ to the other exponents through

$$1/(\tau - 2) = \delta = 1 + \gamma/\beta \quad ,$$

where β is the exponent for the size of the infinite cluster ("spontaneous magnetization") and γ that for the "mean cluster size" or "susceptibility"; δ is the critical exponent describing the percolation analog of the critical isotherm. In two dimensions, $\tau = 187/91$ presumably is exact [8.160] whereas $\tau \simeq 2.2$ in three dimensions, and $\tau = 5/2$ "classically", i.e., for high dimensions above six. Therefore in two dimensions we may multiply n_s with s^τ to see if the result is independent of s, as predicted by (8.10). Figure 8.4 shows that this is not the case for two reasons: (8.10) is valid only for s going to infinity, and in the thermodynamic limit of lattice size going to infinity; the clusters must be large, but still much smaller than the whole lattice. Therefore only for intermediate cluster sizes s can one expect $s^\tau n_s$ to be independent of s. Figure 8.4 shows no such region of constant $s^\tau n_s$ for a 300×300 triangular lattice, a size already quite large compared with many simulations in Ising models or spin glasses. But for a 95000×95000 lattice, containing 10^5 times more sites, a plateau is clearly visible [8.161] near s = 1000, as required by (8.10). And for that larger lattice size it is easier to see the two reasons for deviations from the plateau: corrections to scaling for small s, and finite-size effects for large s. Thus for small s, (8.10) is not valid since critical exponents are defined only asymptotically. And for large s near 10^5 the effects from the boundary of the lattice give systematic corrections to the cluster numbers (periodic boundary conditions are not necessarily useful [8.162]). For smaller lattices the two effects merge together, prevent the formation of a clear plateau, and make an accurate determination of the parameters describing $n_s(p_c)$ impossible.

Thus one should not rely too much on small lattices if two-dimensional percolation is concerned; sometimes in smaller systems violations of theories like universality are observed which later [8.163] vanish in larger systems. The single simulation for 130000×130000 presumably was one of the largest systems ever dealt with on a computer (but see [8.164]).

For very large systems one may try to find from the deviations for small s from the above-mentioned plateau the scaling corrections to the leading term:

$$n_s \propto s^{-\tau}(1 + \text{const } s^{-\Omega} + \ldots) \quad . \tag{8.11}$$

The data of Fig.8.4 suggest a correction exponent $\Omega \simeq 0.6$. From a phenomenological assumption of *Aharony* and *Fisher* [8.165] one may derive that this exponent Ω is not necessarily the true correction exponent from an irrelevant operator, but instead signals a correction due to the fact that the true scaling fields are not just $p-p_c$ and $1/s$ but analytic combinations thereof. Quantitatively, one then expects [8.161] $\Omega = 55/91 = 0.6044$ in two dimensions, compatible with the Monte Carlo data; this result seems to be the first numerical confirmation of the phenomenological assumptions of [8.165].

Had we simulated 10^5 times a 300×300 lattice we would have used about as much computer time as for one 95000×95000 simulation. But for large s we still would not have achieved the desired plateau; instead we would have got the boundary effects more accurately. Thus it is not always better to simulate many systems of intermediate sizes than one large system, as we suggested in Chap.1. The reason is also clear. In the thermodynamic limit, quantities like the magnetization of the susceptibility have fluctuations diverging at the critical point, whereas [8.166] cluster numbers n_s for a fixed large s have a mean square fluctuation equal to n_s apart from trivial normalization factors. These fluctuations at fixed s do not diverge at the critical point, just as the susceptibility in a fixed magnetic field does not diverge at $T = T_c$. Therefore right at $p = p_c$ the fluctuations in n_s and thus the statistical erros in n_s behave as usual if the system size is increased, whereas the errors in the spontaneous magnetization contain an additional factor diverging for system size $\to \infty$ at $T = T_c$. Similarly, investigation of the size of the largest cluster for percolation, as in Fig.1.1, is strongly hampered by fluctuations containing a factor [8.167] diverging at p_c; and accurate determination of the related critical exponent thus requires many simulations on intermediate system sizes rather than one simulation on a very large system. The three-dimensional data there are thus based on at least a thousand runs for each system size [8.168].

Another general lesson can also be learned from the same percolation example in Fig.1.1. Simulations of large lattices allow corrections to the asymptotic scaling behavior to be analysed, like the scaling corrections in (8.11). One is no longer restricted, as was the case when [8.1] was written, to plot data on log-log paper and to draw a straight line through them to determine the exponent from the slope of this line. Instead, for large enough systems and good enough statistics the data are often accurate enough to show some curvature which now should no longer be neglected. This curvature is due to a correction term as in (8.11). Thus a suitable analysis of this curvature gives the leading exponent as well as the first correction exponent. Unfortunately this larger number of parameters also means more errors for each: a hundred times more computer effort does not give ten times smaller error bars than analyses published before that ignoring the correction term. That does not mean that the new analysis is less efficient than the old one; only the old error bars were overly optimistic since they neglected systematic deviations due to curvature.

For p not exactly equal to p_c but close to it, and for large s, the cluster numbers $n_s(p)$ seem to obey, for dimensionality up to six, the scaling law

$$n_s(p) \propto s^{-\tau} f((p - p_c)s^{\sigma}) \quad , \tag{8.12}$$

as was confirmed by *Nakanishi* and *Stanley* in two to six dimensions by Monte Carlo analysis [8.169]. The scaling function $f(z)$ with $z = (p-p_c)s^{\sigma}$ seems to be a simple Gaussian in six dimensions with $\sigma = 1$. This simple form, $\log f \propto -(p-p_c)^2 s$, agrees with the exact solution on the Bethe lattice [8.157]. The lower the dimensionality d, the more asymmetric is f about the critical point $z = 0$: the scaling function $f(z)$ then develops a maximum at negative z, i.e., below p_c. This maximum increases relative to the value at the threshold $z = 0$; the ratio of the maximum value to the value at $z = 0$ is supposed to be universal and approaches infinity in one dimension where $f(z) = z^2 e^z$ exactly, with $\sigma = 1$, $\tau = 2$, and $p_c = 1$ [8.3].

8.4.2 Computational Techniques

For experts wishing to make their own Monte Carlo simulations of random percolation we now review some details; other readers may jump to Sect.8.4.3. The easy task of the usual simulation technique is the production of a sample: we simply determine one random number for each of the L^d lattice sites and occupy that site if the random number is smaller than p. The complication arises in the analysis of the configuration, for example in the counting of the clusters to determine n_s.

The algorithm of *Hoshen* and *Kopelman* [8.170] is particularly suited for large systems since only a part of the system needs to be stored in the memory at any given time. In two dimensions, only one or two lines are needed, in three dimensions only one or two planes, etc. The appendix of [8.170] gives a complete Fortran program to produce and count three-dimensional clusters for intermediate system sizes. Table 8.1 gives a somewhat faster but much longer program enabling an $L \times L$ triangular lattice to be simulated up to $L = 131070$. But the user should be aware that it blocks a CDC Cyber 76 computer for more than a day to simulate that lattice.

The classification subroutine of *Hoshen* and *Kopelman* [8.170], called KLASS in [8.171] (and in a similar form called LASS in [8.158]) is written in Table 8.1 into the main program three times to make it run faster. (This and other tricks

Table 8.1. Cluster counting program for the triangular lattice

```
      PROGRAM PERC(OUTPUT,TAPE6=OUTPUT)
      DIMENSION LEVEL(131071),N(95000),NS(7),NSUM(32)
      LEVEL 2, LEVEL $ COMMON/HGB/LEVEL
      LOGICAL TOP,LEFT,BACK
      DATA NSUM/32*0/,NS/7*0/
C     TRIANGULAR SITE PERCOLATION WITH NAKANISHI RECYCLING
      CALL RANSET(100001)
      L=131070
      P=0.5
      TAUM1=96./91.
      ALOG2=1.0000001/ALOG(2.0)
      LARGE=1048576
      MAX=95000
      MAX3=MAX*3
      N(MAX)=MAX
      LP1=L+1
      LIMIT=MAX-L*0.20
      INDEX=IREC=0 $ CHI=0. $ IF(L.GT.131070) STOP 1
      DO 1 I=1,LP1
1     LEVEL(I)=MAX
      DO 3 K=1,L
      IF(INDEX.LT.LIMIT) GOTO 20
      IREC=IREC+1
      IF(K.LE.3) STOP 2
      J=INDEX
      DO 21 I=2,LP1
      LEV=LEVEL(I)
      IF(LEV.EQ.MAX) GOTO 21
C     HOSHEN-KOPELMAN RECLASSIFICATION OF CURRENT LINE
C     EMPTY PLACES HAVE LEVEL = MAX
      IF(LEVEL(I-1).NE.MAX) GOTO 24
      LABEL=LEV
      IF(N(LEV).GE.0) GOTO 27
      MS=N(LEV)
22    LABEL=-MS
      MS=N(LABEL)
      IF(MS.LT.0) GOTO 22
      N(LEV)=-LABEL
      IF(LABEL.GT.INDEX) GOTO 25
27    J=J+1
      N(J)=N(LABEL)
      N(LABEL)=-J
      LEVEL(I)=J-INDEX
      GOTO 21
24    LEVEL(I)=LEVEL(I-1)
      GOTO 21
25    LEVEL(I)=LABEL-INDEX
21    CONTINUE
      IF(J.GE.MAX) STOP 3
      IF(IREC.EQ.1) WRITE(6,2) P,J
C     END OF RECYCLING; NOW ANALYSIS OF FINISHED CLUSTERS
      DO 26 IS=1,INDEX
      NIS=N(IS)
      IF(NIS.LE.0) GOTO 26
      FNIS=NIS
      CHI=CHI+FNIS*FNIS
      IF(NIS.GE.8) GOTO 261
      NS(NIS)=NS(NIS)+1
      GOTO 26
```

Table 8.1 (cont.)

```
C       STORE SIZES IN BINS WITH UPPER LIMIT = 2 * LOWER LIMIT
261     INTEG=ALOG(FNIS)*ALOG2
        NSUM(INTEG)=NSUM(INTEG)+1
        IF(NIS.GE.LARGE) WRITE(6,97) NIS
26      CONTINUE
        INDEX1=J-INDEX
        IF(INDEX1.LE.O) STOP 5
        DO 23 IND=1,INDEX1
23      N(IND)=N(IND+INDEX)
        INDEX=INDEX1
        IF(IREC.NE.(IREC/500)*500) GOTO 20
        WRITE(6,2) P,INDEX,J,K
        WRITE(6,96) NS,J
        WRITE(6,93) NSUM
20      CONTINUE
        MOLD=MAX
        DO 3 I=2,LP1
        LBACK=MBACK=MOLD
        MOLD=LEVEL(I)
        IF(RANF(I).GT.P) GOTO 9
        MLEFT=LEVEL(I-1)
        MTOP=LTOP=MOLD
        IF(MLEFT+MTOP+MBACK.EQ.MAX3) GOTO 4
        LEFT=MLEFT.LT.MAX $ TOP=MTOP.LT.MAX $ BACK=MBACK.LT.MAX
C       FIRST HOSHEN-KOPELMAN CLASSIFICATION OF TOP NEIGHBORS
        IF(.NOT.TOP.OR.N(LTOP).GE.O) GOTO 12
        MS=N(LTOP)
13      MTOP=-MS
        MS=N(MTOP)
        IF(MS.LT.O) GOTO 13
        N(LTOP)=-MTOP
C       NOW COMES THE BACK NEIGHBOR (LEFT OF TOP)
12      IF(.NOT.BACK.OR.N(LBACK).GE.O) GOTO 11
        MS=N(LBACK)
14      MBACK=-MS
        MS=N(MBACK)
        IF(MS.LT.O) GOTO 14
        N(LBACK)=-MBACK
C       LEFT NEIGHBOR NEEDS NO RECLASSIFICATION
11      LEVEL(I)=MNEW=MINO(MTOP,MBACK,MLEFT)
        ICI=1
        IF(TOP) ICI=ICI+N(MTOP)
        IF(LEFT.AND.MTOP.NE.MLEFT) ICI=ICI+N(MLEFT)
        IF(BACK.AND.MBACK.NE.MLEFT.AND.MBACK.NE.MTOP)
     1  ICI=ICI+N(MBACK)
        N(MNEW)=ICI
C       ICI IS THE SIZE OF THE CLUSTER AT THIS STAGE
C       WATCH OUT HERE FOR INTEGER OVERFLOW IF 32 BIT COMPUTERS ARE USED
        IF(TOP .AND.MTOP .NE.MNEW) N(MTOP )=-MNEW
        IF(LEFT.AND.MLEFT.NE.MNEW) N(MLEFT)=-MNEW
        IF(BACK.AND.MBACK.NE.MNEW) N(MBACK)=-MNEW
        GOTO 3
4       LEVEL(I)=INDEX=INDEX+1
C       START OF NEW CLUSTER
        N(INDEX)=1
        GOTO 3
9       LEVEL(I)=MAX
3       CONTINUE
```

Table 8.1 (cont.)

```
C      NOW FINAL ANALYSIS
       IF(INDEX.EQ.0) GOTO 35
       DO 6 IS=1,INDEX
       NIS=N(IS)
       IF(NIS.LT.0) GOTO 6
       FNIS=NIS
       CHI=CHI+FNIS*FNIS
       IF(NIS.GE.8) GOTO 61
       NS(NIS)=NS(NIS)+1
       GOTO 6
. 61   INTEG=ALOG(FNIS)*ALOG2
       NSUM(INTEG)=NSUM(INTEG)+1
       IF(NIS.   GE.LARGE) WRITE(6.97) NIS
6      CONTINUE
35     WRITE(6,96) NS
96     FORMAT(" NS:",I9,7I8)
97     FORMAT(" CLUSTER OF SIZE ",I12)
93     FORMAT(" BINS:",4(/,8I9))
2      FORMAT(F15.8,3I10,F20.4)
       WRITE(6,93) NSUM
       CHI=(CHI/L)/L
       WRITE(6,2)P,L,INDEX,IREC,CHI
       I=2**32
       ISUM=0
       DO 905 INDEX=3,32
       INTEG=35-INDEX
       ISUM=ISUM+NSUM(INTEG)
       CHI=(ISUM*FLOAT(I)**TAUM1/L)/L
       IF(ISUM.GT.0) WRITE(6,2) CHI,ISUM,I,INTEG
905    I=I/2
       DO 906 INDEX=1,7
       NIS=8-INDEX
       ISUM=ISUM+NS(NIS)
       CHI=(ISUM*FLOAT(NIS)**TAUM1/L)/L
906    WRITE(6,2) CHI,ISUM,NIS
       STOP $ END
*EOF
```

speeded up the program of [8.158] by a factor of 3, down to 2.2 microseconds per site on Cyber 76 [8.172] in a *Fortran* program.) The labels N of already analysed clusters are recycled (as in [8.169]) in regular intervals in the part starting with "If (index.lt.limit) Go to 20" and ending with label 20. If label N is positive it gives the size of the cluster to which it refers whereas a negative N points to another label which is either positive and thus the cluster size or negative and thus a pointer to another label [8.170]. This recycling method is due to *Nakanishi* [8.169] and differs from that of [8.170]. A program very similar to that in Table 8.1 was used by *Margolina* et al. [8.161] for L = 50000 (9 hours on IBM 370/168), L = 70000 (12 hours on IBM 3081) and L = 130000 (26 hours on CDC Cyber 76). The program gives the number of clusters in the bins 8 to 15, 16 to 31, etc., for cluster size s. It multiplies the final result with the appropriate power of s to give a plateau asymptotically. To reduce fluctuations we actually work with a number of clusters larger than a given size s in Fig.8.4, which is a quantity decaying as $s^{1-\tau}$ instead of $s^{-\tau}$.

This method of cluster counting has, of course, been applied to many problems involving n_s, not only to their values at p_c. An entirely different way to simulate clusters, which works best below p_c, is *Leath*'s cluster growth method [8.173]. One starts from one occupied site and then lets the cluster grow by investigating successive neighbor shells of this central site. For each site in each shell one determines once and for all whether it is occupied or empty, with probability p and 1-p, respectively. The growth stops if all sites in one shell are empty or not connected with the central site. This method was improved further by *Alexandrowitz* [8.174] and *Pike* and *Stanley* [8.175]. A third method, which does not give cluster numbers but only the cluster structure, is mentioned in Sect.8.4.3.

8.4.3 Cluster Structure

One of the simplest quantities characterizing the structure of a cluster is its radius R_s, which may be defined as the radius of gyration [8.173] or the maximum linear extent [8.176]. Scaling theory [8.3,157] suggests that R_s at $p = p_c$ varies with the cluster mass s as

$$R_s \propto s^{(\tau-1)/d} \tag{8.13}$$

in d dimensions. One may interpret the reciprocal exponent $(d/(\tau - 1))$ as an effective or fractal dimension [8.177]; we refer to [8.178] for more recent literature and a discussion of self-similarity.

The largest cluster at the critical point in a finite system will have a radius of the order of the linear dimension L of the system. Thus the size S_∞ of this largest cluster will vary with L as $S_\infty \propto L^{d/(\tau-1)}$ according to (8.13). In this sense Fig.1.1 gives the effective dimension of percolation clusters at p_c through the slope of the log-log plot, which is compatible with 91/48, the theoretical expectation in two dimensions, and is extrapolated to about 2.53 in three dimensions [8.168]. Finite-size scaling, on the other hand, gives $S_\infty/L^d \propto L^{-\beta/\nu}$ as the analog of spontaneous magnetization, which is the same result if hyperscaling is used but may seem more plausible to readers familiar with thermodynamic phase transitions. Thus $\beta/\nu \approx 0.47$ is the three-dimensional Monte Carlo estimate.

Another quantity of interest for the cluster structure is the cluster perimeter t, which is the number of empty sites neighboring a cluster site. It is similar to the number of occupied-empty neighbor pairs, which is the energy in a lattice gas. A detailed discussion of perimeters is found in [8.3,157]. The perimeter t is proportional to the cluster size s on the average and should not be interpreted as a

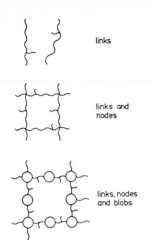

links

links and
nodes

links, nodes
and blobs

Fig.8.5. Historical development of a simplified picture
the infinite cluster [8.183]

measure of the cluster surface. Only the "excess" perimeter $t - s \cdot (1-p)/p$ may be
proportional to a surface area [8.3]. A detailed Monte Carlo investigation of the
surface structure in two and three dimensions is given in [8.124]; these cluster
surfaces are always rough in the thermodynamic limit, i.e., their thickness increases to infinity if $s \to \infty$. A surface tension analogous to the interface free energy
(domain wall energy) of Ising models has also been defined and calculated [8.179]
for large clusters and p above p_c. In three dimensions, corrections proportional
to the first, second and third inverse power of the cluster radius were estimated
for this surface tension [8.180].

Finally, let us look at a more quantitative question of cluster structure: How
does the infinite cluster look at or very near p_c? Early attempts, as reviewed by
Kirkpatrick [8.181] approximated poorly the infinite network by one-dimensional
channels of occupied sites ("links"). Then came the "links and nodes" picture
[8.182] where the one-dimensional links cross each other at nodes separated by the
correlation length (typical cluster radius) ξ. This picture lead to difficulties
in the interpretation of two-dimensional conductivity data. Therefore it was superseded by the "links, nodes and blobs" picture [8.183], where the one-dimensional
channels between the nodes were also allowed to contain smaller d-dimensional regions of higher density which are multiply connected, Fig.8.5. In addition, in all
these three pictures there are dangling ends, that means sections of the cluster
which would not carry any current if the cluster is thought of as an electrical
conductor to which a voltage is applied. Visual inspection of computer printouts
for infinite clusters [8.184] does not indubitably confirm this picture.

More quantitatively, all sites of the infinite cluster can be divided into three
groups: the "yellow" parts of the cluster are the dangling or dead ends; the rest
of the cluster, i.e., the current-carrying part, forms the backbone [8.184]. The
backbone in turn is divided into "red" parts which cannot be removed without split-
ting the cluster, and "blue" parts, which are less crucial, since removal of a

263

single blue site does not split the cluster. The blue parts form loops and thus over their short length a network of parallel conductors. Roughly one may identify the blue parts with the blobs and the red parts with the links. This concept can also be applied to finite clusters slightly below the threshold. It seems [8.183] that most of the material is "yellow"; of the nonyellow backbone, most of the material is "blue." Thus the crucial "red" parts form the smallest contribution to the total cluster mass. Critical exponents for the red and blue backbone parts were calculated by Monte Carlo simulation (see [8.175] for precise definitions of these critical exponents) and confirmed theoretical predictions [8.183] on red cluster parts. We refer to *Coniglio*'s paper [8.183] for more details in this field.

A technical point: if one is interested in the structure of clusters much larger than the correlation length, and not in their average number n_s, one can apply the shape fluctuation algorithm [8.3]. Starting from a single isolated cluster, one tries to exchange one occupied cluster site with one empty perimeter site, both selected randomly with the restriction that the exchange is not allowed to cut the cluster into parts. Then one calculates the difference Δt between the new and old perimeter of the cluster, and actually takes into account the attempted exchange with probability $(1-p)^{\Delta t}$ only. This method is not exact but works for large clusters [8.185]. It is the analog of thermal Monte Carlo simulations in Kawasaki's kinetic Ising model (conserved magnetization, fluctuating energy), in the sense that Δt corresponds to the energy change and $1-p$ to the probability exp(-const/T) to break a bond between Ising spins. All investigations of cluster surfaces in [8.179] used this algorithm; it can also easily be applied to the "lattice animal" limit $p \to 0$ where all clusters of a given size s appear with the same statistical weight, independent of their perimeter t [8.186].

8.4.4 Large-Cell Monte Carlo Renormalization

Percolation theory gives a particularly simple explanation of the technique of real space (or position space) renormalization group, with the help of Monte Carlo simulations of large cells. *Reynolds* et al. [8.187] have shown that this method they developed is equivalent to finite-size scaling [8.188] for sufficiently large lattices. Thus this method is somewhat analogous to the renormalization of subsystem magnetizations in Ising models [8.189] discussed in Sect.1.2.1, which also is equivalent to finite-size scaling.

To determine the percolation threshold p_c in a finite system of L^d sites by a Monte Carlo simulation, we may fill up the lattice by occupying more and more sites until at a concentration p_c of occupied sites for the first time a cluster connects top and bottom of the system. (In practice one does not want to repeat this analysis after every addition of a site. Thus one starts with a trial value for p_c, say 1/2. If then the cluster percolates, one reduces p_c to 1/4; otherwise one increases p_c to 3/4. With the changed p_c one again checks if the system perco-

lates from top to bottom; this bisection is repeated until p_c is determined with
the desired accuracy. Of course, one has to use the same sequence of random numbers
for one sequence of bisections to get a definite estimate for p_c in one such sequence. For a more detailed discussion of the optimization of such bisection methods
we refer to [8.172]). By repeating this Monte Carlo experiment again and again for
different sequences of random numbers, we get a set of different thresholds p_c and
may average over all experiments to get $<p_c>$.

For sufficiently large lattices the probability distribution $\Pi(p_c)$ for the thresholds p_c seems to approach a Gaussian [8.190] whose width $W = (<p_c^2> - <p_c>^2)^{\frac{1}{2}}$ decreases with increasing system size:

$$\Pi(p_c) = (2\pi W^2)^{-\frac{1}{2}} \exp[-(p_c - <p_c>)^2/2W^2] \quad . \tag{8.14}$$

Finite-size scaling [8.188] predicts $W \propto L^{-1/\nu}$ in systems with linear dimension L;
here ν is the critical exponent of the correlation length $\xi \propto |p - p_c|^{-\nu}$. Thus

$$1/\nu = - \log W/\log L + const/\log L \tag{8.15}$$

for large L, where the constant comes from the proportionality factor for W. In a
plot of $-\log(W)/\log(L)$ versus $1/\log(L)$ a straight line should therefore fit the
data asymptotically (but see below). In this way *Levinshtein* et al. [8.186] determined quite accurately the exponent ν in two and three dimensions from moderately
large lattices, without any reference to renormalization ideas.

On the other hand, the renormalization technique [8.187] transforms cells of
size L^d with occupation probability p into one supersite each, with occupation
probability $p' = R(p)$; and one has to calculate the behavior of the renormalization
function R close to its fixed point

$$R(p^*) = p^* \quad .$$

The renormalized supersite is occupied if the corresponding L^d cell percolates
from top to bottom [8.187]. Thus the renormalized occupation probability $p' = R(p)$
for the supersite equals the probability that the threshold p_c for this L^d cell is
smaller than p':

$$R(p) = \int_0^p \Pi(p_c)dp_c \tag{8.16a}$$

or

$$\Pi(p) = dR(p)/dp \quad . \tag{8.16b}$$

Expanding linearly about the fixed point

$$p' - p^* = \Lambda \cdot (p - p^*) \quad \text{with} \quad \Lambda \equiv \Pi(p^*) \quad . \tag{8.17}$$

For sufficiently large cells, p^* approaches the true threshold p_c^{true}. Therefore
$\Lambda \propto 1/W$ from (8.14); factors of order unity can be absorbed into the proportionality
constant. Renormalization technique identifies the old and new correlation lengths

$$L|p' - p_c^{true}|^{-\nu} = |p - p_c^{true}|^{-\nu} \quad . \tag{8.18}$$

Therefore in general

$$1/\nu = \log(\Lambda)/\log(L) \quad . \tag{8.19}$$

Here, with the above result for Λ inserted, we get

$$1/\nu = - \log(W)/\log(L) + const'/\log(L) \quad ,$$

in agreement with finite-size scaling (8.15). Thus for large systems the simple be-havior for the probability distribution in (8.14) enables renormalization and fi-nite-size scaling to give the same final result.

But with finite-size scaling we see better what the next term is in the expan-sion (8.15). Obviously,

$$W \propto L^{-1/\nu}(1 + const" \cdot L^{-\omega} + ...) \tag{8.20a}$$

is a plausible form with some phenomenological correction exponent $\omega > 0$. Then

$$-\log(W)/\log(L) = 1/\nu - const/\log(L) - const" L^{-\omega}/\log(L) + ... \tag{8.20b}$$

similar to (1.24). In other words, if we denote $-\log(W)/\log(L)$ by y and $1/\log(L)$ by x, then in a plot of y versus x a straight line $y = 1/\nu - const \cdot x$ describes the asymptotic behavior only, and the next correction term is quite complicated and not simply proportional to x^2.

This result that one should not simply fit a parabola through the y versus x plot was overlooked in [8.190] and in some later papers following that bad example. But it agrees with an exact result $y = 1/\nu(1 - const \cdot x + O(L^{-d}))$ derived [8.191] from a simple "majority rule" approximation. Thus instead of using a parabolic fit for y versus x, one should fit the data directly on (8.20) with a usual correction-to-scaling factor.

Numerically, these fine details turned out to be irrelevant in an investigation of triangular site percolation: Monte Carlo analysis [8.190,192] confirmed $\nu = 4/3$ with an accuracy of about 1%, using 6000 samples for 1000×1000 and 40 for 10000×10000 lattices. Later the transfer matrix approach [8.193] gave even more accurate confirmation.

On the other hand, that transfer matrix approach is very difficult in three-dimensional percolation [8.194] thus making Monte Carlo methods more useful here. The above bisection method gave $p_c = 0.311_7$ and $\nu \simeq 0.88$ in the simple cubic lattice [8.195]. From this exponent and the above estimate $\beta/\nu \simeq 0.47$ from Fig.1.1 and [8.168] we get via scaling the other critical exponents of three-dimensional perco-lation

$$2 - \alpha = 2\beta + \gamma = 2\beta + (2 - \eta)\nu = \beta(\delta + 1) = d\nu \quad . \tag{8.21}$$

For example, the correlation exponent η is about -0.06 here [8.168] from a Monte Carlo analysis which included corrections to scaling, compared with a field theoretical estimate of about -0.13 [8.196] and a series estimate of 0.03 ± 0.03 [8.197]. These minor differences will hopefully be resolved in the future. For higher dimensions not much seems to be known yet from the Monte Carlo renormalization technique, so standard Monte Carlo methods have to be relied upon [8.3,157,169]. We already mentioned that two-dimensional critical exponents are believed to be known exactly:

$$\alpha = -2/3, \quad \beta = 5/36, \quad \gamma = 43/18, \quad \delta = 18\tfrac{1}{5}, \quad \nu = 4/3, \quad \eta = 5/24 \quad .$$

8.4.5 Other Aspects

The large number of variants of percolation theory, and the large number of questions asked even for normal (random) percolation, make it impossible to mention all aspects here; thus we shortly review only a few points.

a) *Conductivity*. One may imagine that each site is electrically conducting if occupied and electrically insulating if empty. Current flows between neighboring conductors only. The conductivity Σ of a large sample between two bus bars is nonzero only for p above p_c, and varies close to the threshold as

$$\Sigma \propto (p - p_c)^\mu \quad . \tag{8.22}$$

Earlier hopes [8.1,182] to relate the conductivity exponent μ to the other "static" exponents of percolation through the scaling laws

$$\mu = 1 + (d - 2)\nu \quad \text{or} \quad \mu = (d - 1)\nu \tag{8.23}$$

failed to materialize. In two dimensions with $\nu = 4/3$ (thus $\mu = 1$ or $= 4/3$), the conductivity exponent seems to be reliably larger than unity [8.198-202] and probably smaller than 4/3. In three dimensions is seems to be close to 2 [8.203].

This Monte Carlo estimate was obtained by looking at the conductivity right at p_c as a function of system size, employing finite-size scaling:

$$\Sigma (p_c) \propto L^{-\mu/\nu} \quad . \tag{8.24}$$

Early simulations in two dimensions [8.198,199] suggested $\mu \simeq \nu$, but later simulations by the Harvard group [8.198] gave $\mu/\nu \simeq 0.97$. Independently, *Mitescu* et al. [8.200] found 0.91 for this ratio, and the transfer matrix approach [8.201] gave 0.96. The conventional analysis as a function of $p-p_c$ gave $\mu = 1.31 \pm 0.04$ from 800×800 lattices [8.204] and $\mu = 1.23 \pm 0.04$ from 1400×1400 lattices [8.205]. (In these last two simulations as well as in the recent Harvard work [8.198] a special method restricted to two dimensions was used, which gives the conductivity of a finite lattice exactly, and not only after a time-consuming iterative solution of Kirchhoff's equation [8.181]. The average of these determinations is $\mu/\nu = 0.96 \pm 0.02$, or $\mu = 1.28 \pm 0.02$.

A new hypothesis [8.202]

$$2\mu = \nu(3d - 4) - \beta \tag{8.25}$$

seems at present the only surviving hope for a conductivity scaling law. It gives $\mu = 91/72 = 1.264$ in two dimensions, and about 2 in three dimensions, in excellent agreement with the above Monte Carlo estimates. Also in higher dimensions the behavior of (8.25) is reasonable [8.202]. Of course, it is also possible that there is no relation at all between conductivity and the other exponents.

The conductivity exponent is related to that of a diffusion process in the percolation clusters. We refer the interested reader to the review by *Mitescu* and *Roussenq* in [8.159]. Dielectric constants are discussed, e.g., by *Gefen* et al. [8.206]. See the Note Added in Proof for a further discussion of (8.25).

b) *Nonrandom Percolation*. Whereas in random percolation all sites are occupied or empty independently of their neighbors, in reality correlations often exist between neighbors. For example, a site in a simple cubic lattice might not be allowed to have more than m of its six neighbors occupied [8.207]. Or two neighbors feel an attraction between them in thermal equilibrium for the lattice gas (Ising model). A more detailed review of interacting percolation clusters is given by *Kertész* et al. in [8.159]. We mention here only two conclusions [8.207,208]: (1) As long as these interactions have finite range or produce correlations over finite distances only, the critical exponents remain those of random percolation. Only for infinite correlation lengths at the critical point [8.207] or infinite interaction ranges [8.208] does the universality class change. (2) Clusters in the lattice gas diverge at the correct critical point only if they are defined as groups of up-spins connected by lines drawn randomly with probability [8.209]:

$$p_B = 1 - \exp(-E/k_B T) \quad , \tag{8.26}$$

where E is the energy to break the interaction bond between the two up-spins.

c) *Kinetic Aspects*. One may discuss the diffusion of a probe ("ant") on a fixed percolation cluster ("labyrinth") or lattice animal ("parasite problem" [8.210]). Various ant species have already been found, as reviewed by *Mitescu* and *Roussenq* (Chap.6 and [8.159]).

The shape fluctuation algorithm [8.3] allows a Monte Carlo study of the diffusivity of percolation clusters and lattice animals [8.186]. Leath's growth process [8.173], on the other hand, can also be interpreted kinetically [8.174,211], and cluster growth times near the threshold have been found. This process also serves as a model of epidemic spread with immunity effects [8.212a]; for long times the equilibrium distribution is that of random percolation.

Results different from random percolation are found even for long times if every site repeatedly has the chance to become occupied due to some diffusion process. This is the case in a model of polymer gelation where a nonuniversal amplitude ratio was found [8.212b]. Also in contrast to random percolation, the cluster size distribution n_s may show a peak [8.213] at some characteristic size which grows with increasing time and concentration. This model is quite complicated and consists of lattice sites with zero, two and four bonds to their neighbors, and initiators whose diffusion through the lattice causes these bonds to become permanently occupied [8.214].

Chemical reactions are, by definition, kinetic phenomena. But some of them can be described by "oriented" or "directed" percolation, a static concept. This rapidly growing field of directed percolation has been reviewed by *Kinzel* and *Redner* in [8.159].

8.5 Conclusion

It is evident from this brief review that Monte Carlo methods for "random systems" have played a central role in the development of this field, and have significantly contributed both to the understanding of real materials and to checking theoretical concepts. In "random systems," where there is locally no translational invariance, the usual experimental diffraction methods give a much less complete description of the "order" or structure than in "ideal" systems such as periodic crystals; also many theoretical tools are either much more complicated (such as the mean-field theory in the case of spin glasses) or apply to dimensionalities far from the physical one. The renormalization group expansions in $\varepsilon = d_u - d$, where d is the spatial dimensionality and d_u the upper critical dimensionality, have $d_u = 6$ for percolation, spin glasses, random field and random anisotropy problems, rather than $d_u = 4$ as for ordinary critical phenomena, polymer statistics, etc. Thus all these methods are somewhat less useful, and reliable information from numerical techniques is very valuable.

While initially the broad spectrum of relaxation times has hampered interpretation of simulations of spin glasses (as well as systems with random fields, random anisotropies, etc.), this problem has been overcome in several ways. For low-dimensional Ising systems, a combination of Monte Carlo and transfer matrix techniques was useful, as well as brute-force calculations on parallel processors [8.82]; for Heisenberg spin glasses with RKKY interactions, the availability of large-scale computations on very fast machines such as CRAY-1 promises similar progress [8.108]; under these circumstances it is conceivable that "special purpose processors" dedicated exclusively to the simulation of random Ising systems will be constructed [8.215] and provide an even more significant step forward.

With the simulation of percolation problems, also very valuable progress in the technical aspects of simulations was obtained, and now it is possible to simulate extremely large lattices to count percolation clusters in a very efficient way, etc. The estimates for critical properties thus obtained are competitive with the best renormalization group methods.

Finally we emphasize again that we somewhat arbitrarily focused attention on magnetic systems, but we think that similar progress is conceivable if one considers other related systems as well. For instance, a problem currently of interest is the occurrence of "quadrupolar glasses" in diluted systems which can show orientational order (e.g., mixtures of ortho- and para-hydrogen or mixed N_2-Ar crystals). This problem in a sense is intermediate between spin glasses and structural glasses. After earlier work studying models of pure molecular crystals both in three [8.216-218] and two [8.219-221] dimensions, diluted lattices of interacting quadrupoles have also attracted attention [8.222-224]. Indeed evidence for a glass-like orientational order has been found, and the results help to clarify related experimental findings. In view of the progress achieved with enormous effort for the spin-glass problem, it is likely that the field of quadropolar glasses may see much simulation activity in the future as well.

References

8.1　K. Binder, D. Stauffer: In *Monte Carlo Methods in Statistical Physics*, Topics Current Phys., Vol.7, ed. by K. Binder (Springer, Berlin Heidelberg New York 1979) p.301
8.2　K. Binder: J. Phys. (Paris) **39**, C6-1527 (1978)
8.3　D. Stauffer: Phys. Repts. **54**, 1 (1979)
8.4　N.K. Jaggi: Unpublished
8.5　K. Binder: In *Ordering in Strongly Fluctuating Condensed Matter Systems*, ed. by T. Riste (Plenum, New York 1980) p.343, 423
8.6　S.F. Edwards, P.W. Anderson: J. Phys. F**5**, 965 (1975)
8.7　B. Derrida, H. Hilhorst: J. Phys. C**14**, L539 (1981)
8.8　J.L. van Hemmen, I. Morgenstern: J. Phys. C**15**, 4353 (1982)
8.9　K. Binder (ed.): *Monte Carlo Methods in Statistical Physics*, Topics Current Phys., Vol.7 (Springer, Berlin Heidelberg New York 1979) Chap.1
8.10　I. Morgenstern, K. Binder: Phys. Rev. Lett. **43**, 1615; Phys. Rev. B**22**, 288 (1980)
8.11　I. Morgenstern, K. Binder: Z. Phys. B**39**, 227 (1980)
8.12　I. Morgenstern, H. Horner: Phys. Rev. B**25**, 504 (1982)
8.13　I. Morgenstern: Z. Phys. B**42**, 23 (1981)
8.14　I. Morgenstern: Z. Phys. B**43**, 33 (1981)
8.15　I. Morgenstern: Phys. Rev. B**25**, 6067 (1982)
8.16　I. Morgenstern: Phys. Rev. B**25**, 6071 (1982)
8.17　I. Morgenstern: Phys. Rev. B**27**, 4522 (1983); J. Appl. Phys. **53**, 7682 (1982)
8.18　A.P. Young, S. Kirkpatrick: Phys. Rev. B**25**, 440 (1982); S. Kirkpatrick, A.P. Young: J. Appl. Phys. **52**, 1712 (1981)
8.19　B. Derrida, J. Vannimenus: Phys. Rev. B**27**, 4401 (1983)
8.20　J.F. Fernandez: Phys. Rev. B**25**, 417 (1981) [Note that the conclusions reached in this work are misleading, as pointed out in K. Binder, I. Morgenstern: Phys. Rev. B**27**, 5826 (1983)]
8.21　I. Morgenstern, K. Binder, R.M. Hornreich: Phys. Rev. B**23**, 287 (1981)
8.22　J. Vannimenus, J.M. Maillard, L. de Séze: J. Phys. C**12**, 4523 (1979)
8.23　R. Rammal, R. Suchail, R. Maynard: Solid State Commun. **32**, 487 (1979)
8.24　I. Bieche, R. Maynard, R. Rammal, J.P. Uhry: J. Phys. A**13**, 2253 (1980)
8.25　F. Barahona, R. Maynard, R. Rammal, J.P. Uhry: J. Phys. A**15**, 673 (1982)
8.26　R. Maynard, R. Rammal: J. Phys. **43**, L347 (1982)
8.27　J. Vannimenus, G. Toulouse: J. Phys. C**10**, L537 (1977); J. Vannimenus, L. de Séze: J. Appl. Phys. **50**, 7342 (1979)
8.28　S. Kirkpatrick, A.B. Harris: Phys. Rev. B**12**, 4980 (1975)
8.29　W.Y. Ching, D.L. Huber, K.M. Leung: Phys. Rev. B**21**, 3708 (1980); B**23**, 6126 (1981); W.Y. Ching, D.L. Huber: Phys. Rev. B**26**, 6164 (1982)

8.30   L.R. Walker, R.E. Walstedt: Phys. Rev. Lett. **38**, 514 (1977); Phys. Rev. B**22**, 3816 (1980)
8.31   R. Alben, M.F. Thorpe: J. Phys. C**8**, L275 (1975); M.F. Thorpe, R. Alben: J. Phys. C**9**, 2555 (1976)
8.32   W.Y. Ching, K.M. Leung, D.L. Huber: Phys. Rev. Lett. **39**, 725 (1977)
8.33   U. Krey: Z. Phys. B**38**, 243 (1980); B**42**, 231 (1981); J. Magn. Mag. Mater. **28**, 231 (1982)
8.34   U. Krey: Z. Phys. B**30**, 367 (1978); B**31**, 247 (1978)
8.35   J. Richter, S. Kobe: J. Phys. C**15**, 2193 (1982)
8.36   R.A. Cowley, R.J. Birgenau, G. Shirane: In *Ordering in Strongly Fluctuating Condensed Matter Systems*, ed. by T. Riste (Plenum, New York 1980) p.157
8.37   W.J.L. Buyers, D.E. Pepper, R.J. Elliott: J. Phys. C**6**, 1933 (1973)
8.38   D.C. Licciardello, D.J. Thouless: J. Phys. C**8**, 4157 (1975)
8.39   D. Weaire, A.R. Williams: J. Phys. C**10**, 1239 (1977)
8.40   K. Schönhammer, W. Brenig: Phys. Lett. **42**A, 447 (1973)
8.41   S. Yoshino, M. Okazaki: J. Phys. Soc. Japan **43**, 415 (1977)
8.42   B. Kramer, D. Weaire: J. Phys. C**11**, L5 (1978)
8.43   P. Prelovsek: Phys. Rev. B**18**, 3657 (1978)
8.44   J. Stein, U. Krey: Z. Phys. B**37**, 13 (1980)
8.45   P.W. Anderson: Phys. Rev. **109**, 1492 (1958)
8.46   D.S. Boudreaux, J.M. Gregor: J. Appl. Phys. **48**, 152 (1977)
8.47   P. Steinhardt, R. Allen, D. Weaire: J. Non-Crystalline Solids **15**, 199 (1974)
8.48   M.G. Duffy, D.S. Boudreaux, D.E. Polk: J. Non-Crystalline Solids **15**, 435 (1974)
8.49   D.A. Smith: Phys. Rev. Lett. **42**, 729 (1979)
8.50   J.H. Davies, P.A. Lee, T.M. Rice: Phys. Rev. Lett. **49**, 758 (1982)
8.51   W.M. Visscher, J.E. Gubernatis: In *The Dynamical Properties of Solids*, Vol.III, ed. by G.K. Horton, A.A. Maradudin (North Holland, Amsterdam 1980)
8.52   K. Binder: In *Fundamental Problems in Statistical Mechanics V*, ed. by E.G.D. Cohen (North-Holland, Amsterdam 1980) p.21
8.53   K.H. Fischer: Phys. Status Solidi (b) **116**, 357 (1983)
8.54   J.A. Mydosh: *Lecture Notes in Physics*, Vol.149, ed. by C. di Castro, L. Peliti (Springer, Berlin Heidelberg New York 1981)
8.55   P.A. Beck: In *Liquid and Amorphous Metals*, p. 545, ed. by E. Lüscher, H. Coufal (Sijthoff and Noordhoff, Alphen van den Rijn, Holland 1980)
8.56   K. Binder, W. Kinzel: J. Phys. Soc. Japan Suppl. **52**, 209 (1983)
8.57   K. Binder, K. Schröder: Phys. Rev. B**14**, 2142 (1976); Solid State Commun. **18**, 1361 (1976)
8.58   K. Binder, D. Stauffer: Phys. Lett. A**57**, 177 (1976)
8.59   I. Ono: J. Phys. Soc. Japan **41**, 345 (1976); I. Ono, Y. Matsuoka: J. Phys. Soc. Japan **41**, 1425, 1427 (1976)
8.60   K. Binder, D. Stauffer: Z. Phys. B**26**, 339 (1977); Physica B**86-88**, 871 (1977)
8.61   S. Kirkpatrick: Phys. Rev. B**16**, 4630 (1977)
8.62   M. Sakata, F. Matsubara, Y. Abe, S. Katsura: J. Phys. C**10**, 2887 (1977)
8.63   A.J. Bray, M.A. Moore: J. Phys. F**7**, L333 (1977)
8.64   A.J. Bray, M.A. Moore, P. Reed: J. Phys. C**11**, 1187 (1978)
8.65   P. Reed, M.A. Moore, A.J. Bray: J. Phys. C**11**, L139 (1978)
8.66   D. Stauffer, K. Binder: Z. Phys. B**30**, 313 (1978)
8.67   D.C. Rapaport: J. Phys. C**11**, L111 (1178)
8.68   D. Stauffer, K. Binder: Z. Phys. B**34**, 97 (1979)
8.69   W. Kinzel: Phys. Rev. B**19**, 4595 (1979); J. Physique **39**, C6-905 (1978)
8.70   S. Kirkpatrick: In *Ordering in Strongly Fluctuating Condensed Matter Systems*, ed. by T. Riste (Plenum, New York 1980) p.459
8.71   N.K. Jaggi: J. Phys. C**13**, L623 (1980)
8.72   N.K. Jaggi: J. Phys. C**13**, L177 (1980)
8.73   A.J. Kolan, R.G. Palmer: J. Phys. C**13**, L575 (1980); see also E. Domany: J. Phys. C**12**, L119 (1979)
8.74   A. Sadiq, R.A. Tahir-Kheli, M. Wortis, N.A. Bhatti: Phys. Lett. **84**A, 439 (1981)
8.75   H. Takayama, S. Takase: J. Phys. Soc. Jpn. **50**, 3555 (1981)
8.76   S. Takase, H. Takayama: J. Phys. Soc. Jpn. **50**, 1075 (1981)

8.77   W. Kinzel: Z. Phys. B46, 59 (1982); Phys. Rev. B26, 6303 (1982)
8.78   K. Binder: Z. Phys. B48, 319 (1982);
       K. Binder, W. Kinzel: J. Magn. Mag. Mat. 31-34, 1309 (1983)
8.79   K. Nemodo, H. Matsukawa, H. Tayakama: J. Phys. Soc. Jpn. 51, 3126 (1982);
       H. Tayakama, K. Nemoto, H. Matsukava: J. Magn. Mag. Mat. 31-34, 1303 (1983)
8.80   H. Betsuyaku: J. Magn. Mag. Mat. 31-34, 1311 (1983)
8.81   O. Nagai, M. Toyonaga, D. The-Hung: J. Magn. Mag. Mat. 31-34, 1311 (1983)
8.82   A.P. Young: Phys. Rev. Lett. 50, 917 (1983); J. Phys. C17, L517 (1984)
8.83   J. Jäckle, W. Kinzel: J. Phys. A16, L163 (1983)
8.84   W. Kinzel, K. Binder: Phys. Rev. Lett. 50, 1509 (1983)
8.85   I. Ono, T. Oguchi: Unpublished
8.86   H. Betsuyaku: J. Magn. Mag. Mat. 31-34, 1311 (1983)
8.87   A.J. Bray, M.A. Moore: J. Phys. C12, L477 (1979)
8.88   C. Dasgupta, S.-K. Ma, C.-K. Hu: Phys. Rev. B20, 3837 (1979)
8.89   P. Monod, H. Bouchiat: J. Phys. Lett. (Paris) 43, L45 (1982);
       B. Barbara, A.P. Malozemoff, Y. Imry: Phys. Rev. Lett. 47, 1852 (1981)
8.90   W.Y. Ching, D.L. Huber: Phys. Lett. A59, 383 (1977); AIP Conf. Proc. 34,
       370 (1977)
8.91   D. Stauffer, K. Binder: Z. Phys. B41, 237 (1981)
8.92   A.J. Bray, M.A. Moore: J. Phys. C15, 2417 (1982); C14, 2629 (1981); Phys.
       Rev. Lett. 47, 120 (1981)
8.93   R.B. Grzonka, M.A. Moore: J. Phys. C16, 1109 (1983)
8.94   P. Reed: J. Phys. C12, L859 (1979)
8.95   C. Kawabata, J.V. Jose, S. Kirkpatrick: J. Phys. C14, 3015 (1981)
8.96a P. Reed: J. Phys. C12, L475 (1979)
8.96b D.L. Huber, W.Y. Ching, M. Fibich: J. Phys. C12, 3535 (1979)
8.97   D.L. Huber, W.Y. Ching: J. Phys. C13, 5579 (1980)
8.98   J.R. Banavar, M. Cieplak: Phys. Rev. Lett. 48, 832 (1982)
8.99   J.R. Banavar, M. Cieplak: Phys. Rev. B29, 469 (1984)
8.100 M.Z. Cieplak, M. Cieplak, J.R. Banavar: Phys. Rev. B26, 2482 (1982)
       M. Cieplak, J.R. Banavar: Phys. Rev. B27, 293 (1983)
8.101 D.C. Mattis: Phys. Lett. A56, 421 (1976)
8.102 T. Tatsumi: Progr. Theor. Phys. 59, 1428, 1437 (1978); 57, 1799 (1977)
8.103 T. Oguchi, T. Ishikawa: J. Phys. Soc. Japan 50, 2180 (1981)
8.104 W. Selke, M.E. Fisher: Phys. Rev. B20, 257 (1979)
8.105 T. Kawasaki: J. Phys. Soc. Japan Suppl. 52, S-239 (1983); J. Magn. Mag.
       31-34, 1469 (1983)
8.106 K. Binder, W. Kinzel, D. Stauffer: Z. Phys. B36, 161 (1979);
       W. Kinzel, K. Binder: Phys. Rev. B24, 2701 (1981)
8.107 H. Maletta, W. Felsch: Phys. Rev. B20, 1245 (1979)
8.108 L.R. Walker, R.E. Walstedt: Phys. Rev. Lett. 47, 1624 (1981); J. Magn. Mag.
       Mat. 31-34 (1983); J. Appl. Phys. 53, 7985 (1982)
8.109 G.S. Grest: Phys. Rev. B21, 165 (1980)
8.110 G.S. Grest, E.F. Gabl: Phys. Rev. Lett. 43, 1182 (1979)
8.111 W.Y. Ching, D.L. Huber: J. Appl. Phys. 52, 1715 (1981)
8.112 C.Z. Andêrico, J.F. Ferñandez, T.S. Streit: Phys. Rev. B26, 3824 (1982);
       J.F. Ferñandez, C.Z. Anderico, T.S. Streit: J. Appl. Phys. 53, 7991 (1982)
8.113 D. Ariosa, M. Droz, A. Malaspinas: Helv. Phys. Acta 55, 29 (1982)
8.114 D. Sherrington, S. Kirkpatrick: Phys. Rev. Lett. 35, 1792 (1975);
       G. Parisi: Phys. Repts. 67, 25 (1980)
8.115 F.A. de Rozario, D.A. Smith, C.H.J. Johnson: Physica B86-88, 861 (1977)
8.116 A. Freudenhammer: J. Magn. Mag. Mat. 9, 46 (1978)
8.117 W.Y. Ching, D.L. Huber: J. Phys. F8, L63 (1978)
8.118 M. Fähnle: J. Magn. Mag. Mat. 15-18, 133 (1980); Appl. Phys. 23, 267 (1980)
8.119 G. Toulouse, M. Gabay: J. Phys. Lett. (Paris) 42, L163 (1981);
       M. Gabay, G. Toulouse: Phys. Rev. Lett. 47, 201 (1981)
8.120 B.I. Halperin, P.W. Anderson, C.M. Varma: Phil. Mag. 25, 1 (1972);
       S.K. Ma: Phys. Rev. B22, 4484 (1980)
8.121 R.W. Walstedt: Phys. Rev. B24, 1524 (1981)
8.122 C.G. Morgan-Pond: Physica 108B, 767 (1981); Phys. Rev. Lett. 51, 490 (1983)
8.123 D.J. Thouless, P.W. Anderson, R.G. Palmer: Phil. Mag. 35, 593 (1977)

8.124 S. Kirkpatrick, D. Sherrington: Phys. Rev. B**17**, 4384 (1978)
8.125 R.G. Palmer, C.M. Pond: J. Phys. F**9**, 1451 (1981)
8.126 F.T. Bantilan, Jr., R.G. Palmer: J. Phys. F**11**, 261 (1980)
8.127 H. Sompolinsky: Phys. Rev. Lett. **47**, 935 (1981)
8.128 G. Parisi: Phil. Mag. B**41**, 677 (1980)
8.129 S.-K. Ma, M. Payne: Phys. Rev. B**24**, 3984 (1981)
8.130 C. Dasgupta, H. Sompolinsky: Phys. Rev. B**27**, 4511 (1983)
8.131 N.D. MacKenzie, A.P. Young: Phys. Rev. Lett. **49**, 301 (1982)
8.132 S.-K. Ma: J. Stat. Phys. **29**, 717 (1981)
8.133 G. Parisi, G. Toulouse: J. Phys. Lett. **41**, L361 (1980)
8.134 G. Toulouse, M. Gabay, T.C. Lubensky, J. Vannimenus: J. Phys. Lett. **43**, L109 (1982)
8.135 K.H. Fischer: Solid State Commun. **18**, 1515 (1976)
8.136 A.P. Young: J. Phys. C**14**, L1085 (1981)
8.137 I. Morgenstern, K. Binder, A. Baumgärtner: J. Chem. Phys. **69**, 253 (1978), and references therein
8.138 T.R. Funk, D.M. Crothers: Biopolymers **6**, 893 (1968)
8.139 M. Puma, J.F. Fernandez: Phys. Rev. B**18**, 1391 (1978);
       J.F. Fernandez: Phys. Rev. B**16**, 5125 (1977);
       S.-K. Ma, H.H. Chen: J. Stat. Phys. **29**, 717 (1982)
8.140 D.P. Landau, M. Blume: Phys. Rev. B**13**, 287 (1976); B**16**, 598 (1977)
8.141 J.F. Fernandez, M. Medina: Phys. Rev. B**19**, 3561 (1979)
8.142 D. Kumar, J. Stein: J. Magn. Mag. Mat. B**15-18**, 225 (1980); J. Phys. C**13**, 3011 (1980), see also
       J.V. José et al.: Phys. Rev. B**27**, 334 (1983)
8.143 R.J. Glauber: J. Math. Phys. **4**, 294 (1963)
8.144 J.D. Reger, K. Binder: Z. Physik B**60**, 137 (1985)
8.145 R. Medina, J.F. Fernandez, D. Sherrington: Phys. Rev. B**21**, 2915 (1980)
8.146 M.A. Novotny, D.P. Landau: Phys. Rev. B**24**, 1468 (1981)
8.147 D.P. Landau: Phys. Rev. B**22**, 2450 (1980)
8.148 Y. Imry, S.-K. Ma: Phys. Rev. Lett. **35**, 1399 (1975)
8.149 D.P. Landau, H.H. Lee, W. Kao: J. Appl. Phys. **49**, 1356 (1878);
       D. Stauffer, C. Hartzstein, K. Binder, A. Aharony: Z. Phys. B**55**, 324 (1984);
       D. Chowdhury, D. Stauffer: Z. Physik B**60**, 249 (1985);
       D. Andelman, H. Orland, L.C.R. Wijewardhana: Phys. Rev. Lett. **52**, 145 (1984)
8.150 E.B. Rasmussen, M.A. Novotny, D.P. Landau: J. Appl. Phys. **53**, 1925 (1982)
8.151 H. Rohrer, H.J. Scheel: Phys. Rev. Lett. **44**, 876 (1980)
8.152 K. Binder, Y. Imry, E. Pytte: Phys. Rev. B**24**, 6736 (1981);
       G. Grinstein, S.-K. Ma: Phys. Rev. Lett. **49**, 685 (1981)
8.153 J.F. Fernandez, G. Grinstein, Y. Imry, S. Kirkpatrick: Phys. Rev. Lett. **51**, 203 (1983)
8.154 A. Aharony, E. Pytte: Phys. Rev. Lett. **45**, 1583 (1980), and references contained therein
8.155 M.C. Chi, R. Alben: J. Appl. Phys. **48**, 2987 (1977);
       M.C. Chi, T. Egami: J. Appl. Phys. **50**, 165 (1979);
       R. Harris, S.H. Sung: J. Phys. F**8**, L299 (1978)
8.156 C. Jayaprakash, S. Kirkpatrick: Phys. Rev. B**21**, 4072 (1980)
8.157 J.W. Essam: Repts. Progr. Phys. **43**, 843 (1980)
8.158 D. Stauffer: *Lecture Notes in Physics*, Vol.149 (Springer, Berlin Heidelberg New York 1981) p.9
8.159 G. Deutscher, R. Zallen, J. Adler: Ann. Israel. Phys. Soc., Vol.5 (1983) ("Percolation Structures and Processes")
8.160 B. Nienhuis, E.K. Riedel, M. Schick: J. Phys. A**13**, L189 (1980);
       R.P. Pearson: Phys. Rev. B**22**, 2579 (1980); see also
       M.P.M. den Nijs: J. Phys. A**12**, 1857 (1979);
       B. Nienhuis: J. Phys. A**15**, 199 (1982)
8.161 A. Margolina: Thesis, Boston University (1983)
       A. Margolina, Z. Djordjevic, H.E. Stanley, D. Stauffer: Phys. Rev. B**28**, 1652 (1983)
8.162 D.W. Heermann, D. Stauffer: Z. Physik B**40**, 133 (1980);
       but see D.C. Rapaport: J. Phys. A**18**, L175 (1985), favoring periodic boundary conditions

8.163 N. Jan, D. Stauffer: Phys. Lett. **69**A, 39 (1982)
8.164 D. Dhar, M. Barma, M.K. Phani: Phys. Rev. Lett. **47**, 1238 (1981)
8.165 A. Aharony, M.E. Fisher: Phys. Rev. B**27**, 4394 (1983)
8.166 A. Coniglio, H.E. Stanley, D. Stauffer: J. Phys. A Lett. **12**, L323 (1979)
8.167 A. Coniglio, D. Stauffer: Lett. Nuovo Cim. **28**, 33 (1980)
8.168 A. Margolina, H.J. Herrmann, D. Stauffer: Phys. Lett. **69**A, 73 (1982)
8.169 H. Nakanishi, H.E. Stanley: Phys. Rev. B**22**, 2466 (1980);
For renormalization in 4 and 5 dimensions see N. Jan, D.C. Hong, H.E. Stanley: J. Phys. A**18**, L935 (1985)
8.170 J. Hoshen, D. Stauffer, G.H. Bishop, R.J. Harrison, G.P. Quinn: J. Phys. A**12**, 1285 (1979)
J. Hoshen, R. Kopelman: Phys. Rev. B**14**, 3438 (1976)
8.171 D. Stauffer, A. Coniglio, M. Adam: Adv. Polymer Sci. **44**, 103 (1982)
8.172 T. Gebele: Staatsexamensarbeit, Cologne University (1983), and J. Phys. A**17**, L51 (1984)
8.173 P.L. Leath: Phys. Rev. Lett. **36**, 921 (1976); Phys. Rev. B**14**, 5046 (1976)
8.174 Z. Alexandrowitz: Phys. Lett. **80**A, 284 (1980)
8.175 R. Pike, H.E. Stanley: J. Phys. A**14**, L169 (1981)
8.176 G.D. Quinn, R.J. Harrison, G.H. Bishop: J. Phys. A**9**, L9 (1976)
8.177 B.B. Mandelbrot: *The Fractal Geometry of Nature* (Freeman, San Francisco 1982)
8.178 A. Kapitulnik, A. Aharony, G. Deutscher, D. Stauffer: J. Phys. A**16**, L269 (1983)
8.179 H. Franke: Z. Phys. B**40**, 61 (1981); **45**, 247 (1982); Phys. Rev. B**25**, 2040 (1982)
8.180 H. Franke, J. Kertész: Phys. Lett. **95**, 52 (1983)
8.181 S. Kirkpatrick: Rev. Mod. Phys. **45**, 574 (1973)
8.182 A.S. Skal, B.I. Shklovskii: Sov. Phys. Semicond. **8**, 1029 (1974);
P.G. de Gennes: J. Physique (Paris) **37**, L1 (1976)
8.183 H.E. Stanley: Lecture Notes in Physique, Vol.149 (Springer, Berlin Heidelberg New York 1981) p.59
A. Coniglio: J. Phys. A**15**, 3829 (1982)
8.184 S. Kirkpatrick: AIP Conf. Proc. **40**, 99 (1980)
8.185 H.P. Peters, D. Stauffer, H.P. Hölters, K. Loewenich: Z. Phys. B**34**, 399 (1979)
8.186 H. Gould, K. Holl: J. Phys. A**14** L443 (1981)
8.187 P.J. Reynolds, H.E. Stanley, W. Klein: Phys. Rev. B**21**, 1223 (1980)
8.188 M.E. Levinshtein, B.I. Shklovskii, M.S. Shur, A.L. Efros: Sov. Phys. JETP **42**, 197 (1976)
8.189 K. Binder: Phys. Rev. Lett. **47**, 693; Z. Phys. B**43**, 119 (1981)
8.190 P.D. Eschbach, D. Stauffer, H.J. Herrmann: Phys. Rev. Lett. **23**, 422 (1981)
8.191 C. Tsallis: J. Physique **43**, L471 (1982)
8.192 D. Stauffer: Phys. Lett. **83**A, 404 (1981)
8.193 H.W.J. Blöte, M.P. Nightingale, B. Derrida: J. Phys. A**14**, L45 (1981);
B. Derrida, D. Stauffer: J. Phys. (Paris) **46**, 1623 (1985)
8.194 B. Derrida, H.J. Herrmann: Private communication
8.195 D.W. Heermann, D. Stauffer: Z. Phys. B**44**, 449 (1981); see also
S. Wilke: Phys. Lett. **96**A, 344 (1983)
8.196 J.S.J. Reeve: J. Phys. A**15**, L521 (1982)
8.197 D.S. Gaunt, M.F. Sykes: J. Phys. A**16**, 783 (1983)
8.198 C.J. Lobb: J. Phys. C**12**, L827 (1979) and private communication. Later, more accurate determinations gave $\mu = 1.30$, see for example
J.G. Zabolitzky: Phys. Rev. B**30**, 4077 (1984), in disagreement with prediction (8.25)
8.199 A.K. Sarychev, A.P. Vinogradoff: J. Phys. C**14**, L487 (1981)
8.200 C.D. Mitescu, M. Allain, E. Guyon, J.P. Clerc: J. Phys. A**15**, 2523 (1982)
8.201 B. Derrida, J. Vannimenus: J. Phys. A**15**, L557 (1982); for three dimensions see B. Derrida, D. Stauffer, H.J. Herrmann, J. Vannimenus: J. Physique (Paris) **44**, L701 (1983)
8.202 S. Alexander, R. Orbach: J. Physique (Paris) **43**, L625 (1982)
For further discussion of their scaling law (8.25) see
R. Rammal, G. Toulouse: J. Physique (Paris) **44**, L13 (1983);

S. Wilke, Y. Gefen, V. Ilković, A. Aharony, D. Stauffer: J. Phys. A**17**, 647 (1984)

F. Leyvraz, H.E. Stanley: Phys. Rev. Lett. **51**, 2048 (1983)

8.203   C.M. Mitescu, M.J. Musolf: J. Physique (Paris) **44**, L679 (1983); M. Sahimi, B.D. Hughes, L.E. Scriven, H.T. Davis: J. Phys. C**16**, L521 (1983)

8.204   R. Fogelholm: J. Phys. C**13**, L571 (1980)

8.205   P.S. Li, W. Strieder: J. Phys. C**15**, 6591, L1235 (1982)

8.206   Y. Gefen, A. Aharony, S. Alexander: Phys. Rev. Lett. **50**, 77 (1983)

8.207   J. Kertesz, B.K. Chakrabarti, J.A.M.S. Duarte: J. Phys. A**15**, L13 (1982)

8.208   M. Gouker, F. Family: Phys. Rev. B**28**, 1449 (1983)

8.209   A. Coniglio, W. Klein: J. Phys. A**13**, 2775 (1980)

8.210   S. Wilke: Staatsexamensarbeit, Cologne University (1983) and in [8.202]

8.211   P. Grassberger: Math. Bioscience, in press (1983)

8.212a P. Manneville, L. de Seze: In *Numerical Methods in the Study of Critical Phenomena*, Springer Ser. Synergetics, Vol.9, ed. by J. Della Dora, J. Demongeot, B. Lacolle (Springer, Berlin Heidelberg New York 1981)

8.212b H.J. Herrmann, D.P. Landau, D. Stauffer: Phys. Rev. Lett. **49**, 412 (1982); J. Phys. A**16**, 1221 (1983); R.B. Pandey: J. Stat. Phys. **34**, 163 (1984)

8.213   N. Jan, T. Lookman, D. Stauffer: J. Phys. A**16**, L117 (1983); for other two-dimensional work see A. Rushton, F. Family, H.J. Herrmann: Unpublished; T. Lookman, R.B. Pandey, N. Jan, D. Stauffer, L. Moseley, H.E. Stanley: Phys. Rev. B**29**, 2805 (1984)

8.214   R. Bansil, H.J. Herrmann, D. Stauffer: Preprints; for other work involving zero-functional sites see D. Matthews-Morgan, D.P. Landau, H.J. Herrmann: Statistical Physics Conference (Edinbourgh 1983); Phys. Rev. B**29**, 6328 (1984)

8.215   J.H. Condon, A.T. Ogielski: Rev. Sci. Instrum. **56**, 1961 (1985)

8.216   M.H. James: Phys. Rev. **167**, 862 (1968)

8.217   M.J. Mandell: J. Chem. Phys. **60**, 4880 (1974)

8.218   M.A. Klenin, S.F. Pate: Physica **107**B, 185 (1981)

8.219   M.A. Klenin, S.F. Pate: Phys. Rev. B**26**, 3969 (1982)

8.220   S.F. O'Shea, M.L. Klein: Chem. Phys. Lett. **66**, 381 (1979)

8.221   O.G. Mouritsen, A.J. Berlinsky: Phys. Rev. Lett. **48**, 181 (1982)

8.222   M.A. Klenin: Phys. Rev. Lett. **42**, 1549 (1979)

8.223   M.A. Klenin: Solid State Commun. **33**, 631 (1980)

8.224   M.A. Klenin: Phys. Rev. B**28**, 5199 (1983)

9. Monte Carlo Calculations in Lattice Gauge Theories

C. Rebbi

With 4 Figures

9.1 Lattice Gauge Theories: Fundamental Notions

Monte Carlo simulations have been applied to the analysis of a variety of statistical systems and recently have been successfully extended to the study of lattice gauge theories. In the context of this book, where so much is said on Monte Carlo simulations, I need not dwell on the details of the method itself; more important for the reader are the answers to the following questions:

i)   What motivates the study of a quantum field theory on a space-time lattice?

ii) How is a lattice gauge theory formulated and what is the analogy with a statistical system?

iii) What results have been achieved by Monte Carlo simulations?

Points (i) and (ii) will be considered in the present section, while Sects.9.2,3 contain an exposition of the most relevant results.[1]

Quantized fields and more specifically quantized gauge fields play a fundamental role in our understanding of particle dynamics: the most widely accepted theories for particle interactions are all based on the notion of a gauge field. While there are several ways to define a quantized field theory, the formulation of greatest use for Monte Carlo simulations proceeds through the following steps:

i)   space-time is made Euclidean by a Wick rotation to an imaginary time axis, $t \to it$:

ii) the vacuum expectation values of quantum observables are defined as averages over all possible configurations of the system with measure $\exp\{-S\}$, where S is the action functional (units where $c = \hbar = 1$ are adopted throughout this chapter):

$$<\mathcal{O}> = Z^{-1} \int \mathcal{D}A\mathcal{D}\bar{\psi}\mathcal{D}\psi\, \mathcal{O}(A,\bar{\psi},\psi)\exp[-S(A,\bar{\psi},\psi;g)] \quad , \tag{9.1}$$

$$Z = \int \mathcal{D}A\mathcal{D}\bar{\psi}\mathcal{D}\psi\, \exp[-S(A,\bar{\psi},\psi;g)] \quad ; \tag{9.2}$$

---

1 A more detailed account of Monte Carlo simulations for lattice gauge theories can be found in [9.1], which has a more thorough bibliography. The very large number of papers published in the last few years on the subject of this chapter would make an exhaustive list of references excessively long, so only a few among the most relevant works will be quoted.

iii) the values of the observables are analytically continued, if necessary, back
to Minkowski space time.

The motivation for Step (i) is to have a better defined measure (exp{-S} rather
than exp{-iS}) in the functional integrations. In (9.1,2) A denotes the gauge fields;
$\bar{\psi}$ and ψ, all the possible matter fields; $\int \mathcal{D}A\mathcal{D}\bar{\psi}\mathcal{D}\psi$, a functional integration over
all field configurations. The action functional S consists of two terms: the ac-
tion of the gauge field itself, $S_G(A) = 1/4 \int d^4xF^2$, where F is the field strength
associated with A (see below) and the matter action S_M, a functional of A and the
matter fields as well. Both S_G and S_M generally depend on one or more coupling constant denoted by g; S_M may furthermore contain additional parameters, such as particle masses. The normalizing factor Z is the vacuum-to-vacuum permanence amplitude,
also referred to as partition function because of the analogy with statistical
mechanics.

Of course, a precise mathematical meaning must be given to the functional integrals in (9.1,2): this is done by regularization or renormalization. Most of the
regularization procedures are based on perturbative expansion: the measure and
functional integrals are formally expanded into powers of g and the individual terms
in the series are regularized. However, there are quantities which cannot be evaluated perturbatively, since some phenomena of particle physics are governed by
large coupling constants or, even more fundamental, some observables are given by
expressions having essential singularities for g = 0 (this can be demonstrated) and
therefore defy a perturbative expansion. The lattice formulation of a quantum theory
provides a regularization independent of any expansion in powers of g: hence it is
ideally suited to study those phenomena of particle physics which are nonperturba-
tive in nature. It is hardly surprising that the most successful applications of
lattice regularization are those made to the gauge theory of strong interactions
known as Quantum Chromo Dynamics (QCD).

Gauge invariance consists in the possibility of performing local transformations
on the matter fields:

$$\psi(x) \rightarrow \exp[ig\lambda^{\alpha}(x)\tau_{\alpha}]\psi(x) \quad , \tag{9.3}$$

where τ_{α} are the infinitesimal generators of the gauge group $\mathcal{G}$ in the representation to which ψ belongs and $\lambda^{\alpha}(x)$ are the parameters of the transformation. A scalar product containing derivatives of the fields is not gauge invariant and the role
of the gauge field is to enable definition of a covariant derivative. Rather than
considering the transformation properties of a scalar product such as $(\partial_{\mu}\bar{\psi})\psi$, let
us consider (which is equivalent)

$$s = \bar{\psi}(x + dx)\psi(x) \quad . \tag{9.4}$$

Here s is not invariant because different gauge transformations can be performed
at the neighboring points x and x + dx. To make s invariant, $\psi(x)$ must be "transpor-

ted" from x to x + dx before being contracted with $\bar{\psi}(x + dx)$. The gauge field A_μ^α (x) provides precisely the "transporter," given by the group element

$$U = \exp(igA_\mu^\alpha dx^\mu \tau_\alpha) \quad . \tag{9.5}$$

The scalar product

$$\tilde{s} = \bar{\psi}(x + dx)U\psi(x) \tag{9.6}$$

is now gauge invariant provided that U itself transforms as

$$U \to \exp[ig\lambda^\alpha(x + dx)\tau_\alpha]U \exp[-ig\lambda^\alpha(x)\tau_\alpha] \quad . \tag{9.7}$$

This gives the transformation law

$$A_\mu^\alpha \tau_\alpha \to \exp(ig\lambda^\beta \tau_\beta)A_\mu^\alpha \tau_\alpha \exp(-ig\lambda^\gamma \tau_\gamma) + [\partial_\mu \exp(ig\lambda^\beta \tau_\beta)]\exp(-ig\lambda^\gamma \tau_\gamma) \tag{9.8}$$

for A.

In the lattice regularization the space-time points are replaced by the vertices of a discrete lattice. We shall assume the lattice to be hypercubical and the symbol a denotes the lattice spacing throughout this chapter. Matter fields are assigned to sites of the lattice: in the regularization, $\psi(x)$ is replaced with ψ_i, where i denotes a generic lattice site. The gauge dynamical variables are then group elements $U_{ji} \in G$ associated with the oriented links of the lattice between neighboring lattice sites i,j [9.2,3]. A gauge transformation is defined assigning to each site a group element G_i. The matter fields transform as

$$\psi_i \to G_i\psi_i \quad ; \tag{9.9}$$

the gauge dynamical variables U_{ji} transform as

$$U_{ji} \to G_jU_{ji}G_i^{-1} \tag{9.10}$$

and act as transporters between neighboring points i and j. The scalar product $\bar{\psi}_jU_{ji}\psi_i$ is gauge invariant.

The dynamical behavior of a system is determined by the action functional. We must therefore associate a gauge action S_G and a matter action S_M to a lattice field configuration, specified by the variables U_{ij} and ψ_i. The gauge part of the action is of utmost importance.

In continuum theory from the gauge fields A_μ^α (x) one forms the field strength

$$F_{\mu\nu}^\alpha(x) = \partial_\mu A_\nu^\alpha(x) - \partial_\nu A_\mu^\alpha(x) + gf_{\beta\gamma}^\alpha A_\mu^\beta(x)A_\nu^\gamma(x) \tag{9.11}$$

(where g is the coupling constant and $f_{\beta\gamma}^\alpha$ are the structure constants of the gauge group) and defines

$$S_G = \frac{1}{4} \int d^4x F_{\mu\nu}^\alpha(x)F_\alpha^{\mu\nu}(x) \quad . \tag{9.12}$$

As $A_\mu^\alpha(x)$ parametrizes the transport between infinitesimally neighboring points and $x^\mu + dx^\mu$, so a geometrical meaning can be assigned to the field strength. Further,

$F^\alpha_{\mu\nu}$ parametrizes the transport around the closed path defined by a parallelogram of sides dx^μ, dx^ν, given by

$$U = \exp(igF^\alpha_{\mu\nu}\tau_\alpha \; dx^\mu \wedge dx^\nu) \quad . \tag{9.13}$$

In the lattice version of the theory it is therefore natural to consider the transports around the elementary squares of the lattice, also called plaquettes, of side a. Let i, j, k and ℓ be the vertices of such a square. We associate with it a plaquette transport operator

$$U_p = U_{i\ell}U_{\ell k}U_{kj}U_{ji} \tag{9.14}$$

and construct the lattice gauge action through U_p.

At this point it pays to be specific and to consider a definite gauge group, which we take to be SU(2). A generic group element can then be expressed as

$$U = \cos\theta + i\boldsymbol{\sigma}\cdot\hat{\mathbf{n}} \sin\theta \quad , \tag{9.15}$$

where σ_α are the Pauli matrices and $\hat{\mathbf{n}}$ is a unit vector. The plaquette transporters will also be of the form

$$U_p = \cos\theta_p + i\boldsymbol{\sigma}\cdot\hat{\mathbf{n}}_p \sin\theta_p \quad . \tag{9.16}$$

In the continuum theory, the action density $\frac{1}{4}|F^\alpha_{\mu\nu}|^2$ vanishes if the transport around the closed path of sides dx^μ and dx^ν gives the identity $\mathbf{I}$ in group space and is positive otherwise, growing larger the more the transport deviates from $\mathbf{I}$. Correspondingly, one can take

$$E_p = 1 - \cos\theta_p \tag{9.17}$$

as a measure of the deviation from the identity of the transport around the pla-quette. Here E_p is called the internal energy of the plaquette because of the ana-logy with statistical mechanics to be exposed later, but it should be remembered that E_p is related to the action density rather than to a physical energy. Further, E_p varies between 0 and 2 as U_p ranges over all possible values in the SU(2) group manifold. Allowing for a proportionality constant in the definition of the action, we set

$$S_p = \frac{4}{g^2} E_p \tag{9.18}$$

and define the action for the SU(2) lattice gauge theory as

$$S_G = \sum_p S_p = \frac{4}{g^2} \sum_p E_p \quad , \tag{9.19}$$

where the sum is extended over all plaquettes of the lattice. The specific way the coupling constant is inserted in (9.18) is motivated by the fact that in a conti-nuum limit, where U_{ij} is parametrized as in (9.5) with $|dx^\mu|$ formally set equal to the lattice spacing a and a is then set to zero, the lattice action (9.19) reduces to the continuum form of (9.12). Of course, the particular form we have chosen for the SU(2) lattice action is not the only one that reproduces the continuum action if the $a \rightarrow 0$ limit is taken. With any gauge group several alternative forms of the action can be considered. The action of (9.19), which can also be expressed as

$$S_G = \frac{2}{g^2} \sum_p \text{Tr}\left\{ I - \frac{U_p^+ + U_p}{2} \right\}$$

(9.20)

and which generalizes in this form to the SU(N) groups, has been the most widely used and is referred to as Wilson's lattice action.

The matter action can be easily defined on the lattice by replacing derivatives with finite difference operators and making the coupling gauge invariant, as explained before. It turns out, however, that several of the most relevant features of the interaction between gauge and matter fields are a consequence of the quantum mechanics of the gauge field itself. For this reason many investigations have focused on the properties of the gauge systems alone and here we shall also concentrate on the quantum dynamics determined by the pure gauge action $S_G(\{U_{ji}\})$.

Given any observable $\mathcal{O}(\{U_{ji}\})$ its quantum expectation value, in the vacuum state, is given by

$$<\mathcal{O}> = Z^{-1} \int \prod_{\{ij\}} dU_{ji} \, \mathcal{O}(\{U_{ji}\}) \, \exp[-S_G(\{U_{ji}\})] \quad ,$$

(9.21)

where the integrals are now ordinary (as opposed to functional) invariant integrals over the group manifold and Z is given by the integration of the measure factor alone. If one considers initially a system of finite volume V, to proceed to the thermodynamical limit $V \rightarrow \infty$ only after computation of the observables, then the integrals on the rhs of (9.18) are well-defined finite integrals and a regularization of the quantum theory has been achieved. This regularization is independent of any scheme of expansion. Rewriting (9.18) as

$$<\mathcal{O}> = Z^{-1} \int \prod_{\{ij\}} dU_{ji} \, \mathcal{O} \exp\left[-\left((\text{const}/g^2) \sum_p E_p \right) \right]$$

(9.22)

(the const being 4 for the SU(2) gauge group), the analogy with the statistical formulation of thermodynamics becomes manifest, with the correspondence

$$\frac{\text{const}}{g^2} \leftrightarrow \frac{1}{kT} \quad .$$

(9.23)

Indeed, the notation β for the coupling parameter const/g^2 in the lattice gauge action is frequently used. The fact that the lattice is four-dimensional is nevertheless a reminder that one is dealing with a quantized field theory rather than with a thermodynamical system.

The lattice regularization allows for the standard weak coupling perturbative expansion. This can be formulated in terms of Feynman diagrams, but is made rather cumbersome by the loss of rotational invariance and by the presence of couplings of higher order. It also allows strong coupling expansions [9.4], analogous to the high-temperature expansions for thermodynamical systems. However, the possibility of performing expansions for large values of the coupling constant in the regularized lattice gauge theory does not necessarily solve all problems of strong interactions. Indeed, although the lattice formulation constitutes an extremely useful intermediate step, eventually the lattice regularization must be removed by letting the lattice spacing a tend to zero. In this process, the coupling constant g, which plays the role of an unrenormalized coupling constant, cannot be kept fixed, as we shall presently demonstrate. In particular, to achieve a continuum limit, g may have to approach values well outside the range of validity of strong coupling expansions.

Given any dimensional quantity, which to fix ideas we shall take to be a length ℓ, its value will be given by an expression of the form

$$\ell = af(g) \quad , \tag{9.24}$$

where the lattice spacing enters trivially and the dynamical content of the theory is given by the dimensionless function $f(g)$ (g is also dimensionless) which measures ℓ in units of the lattice spacing. For ℓ to keep a constant value as a is sent to zero, a critical point g_{cr} must exist such that

$$f(g) \xrightarrow[g \to g_{cr}]{} \infty \quad . \tag{9.25}$$

Then one can define a continuum limit which will preserve a finite value for the physical quantity ℓ by letting $g \to g_{cr}$ as $a \to 0$, with a definite functional relationship

$$f = g(a) \quad , \quad \text{or} \quad a = a(g) \tag{9.26}$$

obtained, precisely, demanding that

$$af(g(a)) = \text{const.} \tag{9.27}$$

Moreover, the critical point must have scaling properties: given any other physical quantity q which in the lattice formulation will be expressed as a suitable power of a times another dimensionless function of g,

$$q = a^p \tilde{f}(g) \quad , \tag{9.28}$$

the same functional relation (9.25), which allows ℓ to remain constant as $a \to 0$, $g \to g_{cr}$, when inserted into (9.28) must also make q approach a finite value in the same limit. At the end of the renormalization procedure one will thus have a quantized field theory defined in continuum space-time: all reference to a and g will have disappeared from the theory; all physical quantities will be expressed in terms of one physical quantity (or a few, if the lattice theory contained more than one coupling parameter), such as ℓ above, which will set the scale.

From these considerations it is obvious that although the statistical properties of the lattice gauge theory may be interesting per se, for the quantum field theorist the most crucial questions are where the critical points of the regularized theory lie, what their scaling properties are and, eventually, how to use the scaling behavior to obtain the physical values of the observables. In the gauge theory of strong interactions and, generally, for non-Abelian gauge theories, perturbative arguments based on renormalization groups can be used to demonstrate that $g = 0$ is a possible scaling critical point (asymptotic freedom) [9.5]. Also, the same analysis shows that the functional relationship between a and g (or β) must be of the form

$$a = \frac{1}{\Lambda} \left(\frac{24\pi^2}{11g^2}\right)^{51/121} \exp(-12\pi^2/11g^2)[1 + O(g^2)]$$

$$= \frac{1}{\Lambda} \left(\frac{6\pi^2}{11}\beta\right)^{51/121} \exp[-(3\pi^2/11)\beta][1 + O(1/\beta)] \qquad (9.29)$$

for the SU(2) gauge theory ($\beta = 4/g^2$) and

$$a = \frac{1}{\Lambda} \left(\frac{16\pi^2}{11g^2}\right)^{51/121} \exp(-8\pi^2/11g^2)[1 + O(g^2)]$$

$$= \frac{1}{\Lambda} \left(\frac{8\pi^2}{33}\beta\right)^{51/121} \exp[-(4\pi^2/33)\beta][1 + O(1+\beta)] \qquad (9.30)$$

for the SU(3) gauge theory ($\beta = 6/g^2$). In these equations Λ is a dimensional scale introduced for convenience, with no direct physical significance, being a formal parameter which will keep a finite value as $a \rightarrow 0$. All physical observables of the continuum limit may then be expressed as functions of Λ, and Λ may be eliminated to yield meaningful relations between the observables themselves.

The present theoretical understanding of strong interactions is based on the fact that $g = 0$ is indeed the critical point for the SU(3) gauge theory of interacting quarks (QCD), but, at the same time, that no phase boundary prevents continuation to that limit of properties, such as quark confinement, which manifest themselves in the strong coupling domain. Strong coupling expansions (valid for small β) however, cannot be carried out to sufficiently high orders for the transition to the scaling behavior toward the continuum limit, which is recovered for $\beta \rightarrow \infty$. The great advantage of Monte Carlo simulations is that they can provide information, albeit numerical, also for intermediate values of β, beyond the domain of validity of strong coupling expansions. Thus, by Monte Carlo simulations, several theoretical conjectures about the gauge theory of strong interactions have been verified and a variety of physical quantities have been evaluated. Moreover, Monte

Carlo calculations have been a powerful tool for the analysis of lattice gauge theories as a statistical system, a quite interesting subject per se.

9.2 General Monte Carlo Results for Lattice Gauge Systems

In a pure gauge system the dynamical variables are the group elements U_{ji} associated with the links of the lattice (4N in number, for a four-dimensional lattice of N sites). These must be stored in the memory of the computer and their collection defines a configuration of the system. If the gauge group G is a Lie group, the group elements will be represented by matrices subject to suitable constraints. The number of real parameters needed to specify the variable U_{ji} is therefore generally smaller than the number of real constants in the matrix. However, when expressed in terms of the minimal number of parameters the algebra of the group normally involves transcendental operations, so it is more convenient to store in memory the full matrices, even if this implies some redundancy, and to realize the group operations by matrix algebra. In a lattice gauge theory the group does not have to be continuous, and discrete gauge groups can be considered as well. Then of course the representation of the dynamical variables U_{ji} reduces to labeling the group elements by an index and the group operation can be conveniently implemented through a call to a multiplication table, also stored in memory. For an explicit example of a Monte Carlo program for systems with discrete gauge groups, see [9.6].

In a Monte Carlo simulation the dynamical variables U_{ji} are upgraded one at a time: following the standard implementation of the *Metropolis* algorithm [9.7] a new candidate U_{ji}' is selected. Frequently U_{ji}' is obtained multiplying U_{ji} with a group element $R^{(k)}$ chosen at random among a large set of matrices $\{R^{(1)}, R^{(2)} \ldots R^{(m)}\}$ prestored in memory. The upgrading matrices $R^{(i)}$ are selected at random according to some probability distribution peaked about the identity of the group; they obey the property, necessary for detailed balance, that both $R^{(i)}$ and $R^{(i)-1}$ belong to the set. The width of the distribution (about the identity) is a parameter that one may change according to the value of the coupling constant. It is convenient to proceed with large variations in upgrading in a regime of strong coupling, by smaller changes in a regime of weak coupling. The upgrading $U_{ji} \rightarrow U_{ji}'$ is accepted if the equality

$$r < \exp[-S(U_{ji}' \ldots) - S(U_{ji} \ldots)] = \exp(-\Delta S) \qquad (9.31)$$

is satisfied, r being a pseudorandom number uniformly distributed between 0 and 1.

In what I shall call the standard Metropolis algorithm, after completing the above Monte Carlo (MC) step one proceeds to another dynamical variable, either going through the lattice in a systematic way or by random selection. When every U_{ji} has been upgraded once (exactly or on the average) 1 MC iteration has been completed. In several instances the method described above leads to quite a high rate

of rejections and is therefore inefficient. It becomes convenient then to try n
upgradings of the *same dynamical variable* before proceeding to the next. This slight
variation of the method has often been referred to as improved Metropolis procedure. While increasing n requires more computer time, the efficiency may also increase drastically, and generally an optimal value of n is found empirically. If
n tends to infinity, then the new value of the dynamical variable becomes independent of the previous one and distributed with a probability proportional to
$\exp[-S(U_{ji},...)]$, where all the other dynamical variables in S are kept fixed. This
upgrading procedure has been called "the heat-bath method." Of course, implementation of the heat-bath method by performing a very large number of upgradings of the
same variable is not computationally convenient; but there are a few cases when
the selection of the new value for U_{ji} according to the Boltzmann distribution
(all other U's being kept fixed) can be done directly, and then the heat-bath algorithm becomes quite efficient [9.8].

As has been explained in the previous section, given a lattice gauge theory,
knowing the critical points which may be used to define a continuum limit is of
the greatest importance. Thus, the earliest applications of the MC method to lattice gauge theories were devoted precisely to establishing the phase structure. The
expectation was that non-Abelian models, such as those defined over the SU(2) or
SU(3) gauge groups, would possess no critical point, apart from g = 0, $\beta = \infty$. Then
the continuum limit would be recovered by letting g and the lattice spacing a approach zero simultaneously, according to (9.29,30). The resulting theory would be
asymptotic free [9.6] (i.e., the interaction becomes weaker at short distances),
but the absence of any phase transition in the whole range $0 \leq \beta \leq \infty$ would guarantee
that properties such as quark confinement (in the application to QCD), easily de-
monstrable in the strong coupling domain, would survive in the continuum limit.
On the contrary, an Abelian gauge theory based on the U(1) group (the group of
quantum electro dynamics) should undergo a phase transition at some finite value
β_{cr} of $\beta = 1/e^2$ (the coupling constant is now the unrenormalized charge): since
electric charges are not confined in the real word, beyond the strong coupling,
confining phase, the Abelian theory must exhibit also a Coulombic phase (i.e., a
phase where the potential between two charges would assume the familiar Coulomb
form), and the continuum limit would be recovered within this phase.

Monte Carlo methods were used in [9.9] to investigate the phase structure of
Abelian lattice gauge theories. Both the system defined with the gauge group U(1)
and those defined with the finite gauge groups Z_N were considered, where Z_N can be
thought of as the group with elements $\exp(2\pi i n/N)$, n = 0, 1, ..., N - 1, the group
operation being ordinary complex multiplication. (It is of course also the additive
group of integers mod N.) As N approaches ∞, Z_N tends to U(1). This Monte Carlo
study [9.9] was carried out on lattices having of the order of 40000 dynamical variables and the phase structure was investigated performing thermal cycles (i.e., MC

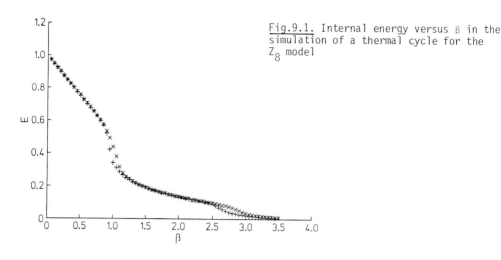

Fig.9.1. Internal energy versus β in the simulation of a thermal cycle for the Z_8 model

computations where β is changed adiabatically in the course of the simulation from
0 to some maximum value and then back to zero), and by various studies of metastability and of the response to mixed initial configurations. In thermal cycles evidence for phase transitions is obtained from hysteresis loops, which manifest themselves where a critical point implies a marked increase in the relaxation time. The
other analyses were used to refine the information obtained from thermal cycles.
Figure 9.1 displays the values measured for the average internal energy of the Z_8
model as a function of β in the course of the simulation of a thermal cycle. Two
hysteresis loops, indicating two phase transitions and a three-phase structure, are
apparent.

The outcome of the investigations of [9.9] was that the models with groups Z_2,
Z_3 and Z_4 have a two-phase structure with a single critical point. These models
are self-dual, they may be mapped into themselves through a change of dynamical
variables which leaves the functional form of action unchanged, but with β replaced
with a new $\tilde{\beta}$. The critical points coincide with the self-duality points $\beta_{cr} = (\tilde{\beta}_{cr})$.
As the order of the group increases beyond 4 a three-phase structure becomes apparent,
with critical points $\beta_{cr}' < \beta_{cr}''$. As N increases β_{cr}' is seen to remain rather stable
tending to a value $\beta_{cr}' \approx 1.005$ in the N → ∞ limit, i.e., for the U(1) lattice gauge
model. Instead β_{cr}'' approaches infinity growing like N^2. The interpretation is that
for N large enough the Abelian Z_N models have a three-phase structure. The phase for
$\beta < \beta_{cr}'$ is the strong coupling phase, where charges are confined (the energy needed
to separate a pair of opposite charges increases linearly with separation); the intermediate phase, for $\beta_{cr}' < \beta < \beta_{cr}''$, is a phase of spin-wave excitations: these are the
collective excitations of the field which become the photons in the continuum limit;
the phase for $\beta_{cr}'' < \beta$ is a highly ordered phase, with short-range correlations,
characteristic of the fact that one is dealing with a discrete group where a gap in
the action exists between the action of a plaquette in the identity configuration
$(U_p = I)$ and the action of a plaquette in the immediately higher state of excitation.

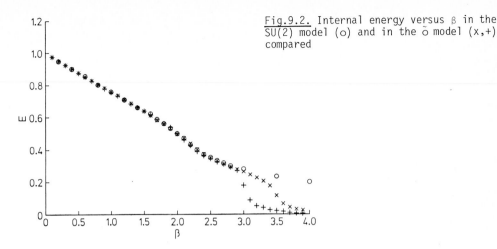

Fig.9.2. Internal energy versus β in the SU(2) model (o) and in the õ model (x,+) compared

In the U(1) system only the strong coupling phase and the photon phase are present. The transition in the U(1) model was studied in detail by finite-size scaling techniques in [9.10].

The first evidence that the SU(2) system for non-Abelian lattice gauge models has no critical point (other than g = 0) was achieved by the MC analysis of [9.11]. Whereas the five-dimensional SU(2) model (also considered in [9.11] exhibited a clear first-order transition at some value of β, with an apparent discontinuity in the function E(β), no evidence for a singularity in E(β) (not even though the appearance of hysteresis loops) was found in the four-dimensional model. Systems having as gauge groups finite non-Abelian subgroups of SU(2) were studied in [9.12, 13]. Figure 9.2 compares the simulation results of a thermal cycle for the system defined with the 48-element subgroup of SU(2) (the covering of the rotation group of the cube; curves marked by + and x) with results obtained for the SU(2) system directly (open circles). Comparison between Figs.9.1,2 is instructive, because the two discrete models have the same action gap. Correspondingly, the transition to the highly ordered phase characteristic of the discrete gauge group occurs at similar values of β. But the non-Abelian discrete system shows no indication of any other phase transition, like the one at $β_{cr}' \approx 1$ in the Z_8 model which develops into the phase transition of the U(1) system. Comparison of the results obtained with the full SU(2) group and its subgroup is also interesting. The two models give essentially identical results for the internal energy up to values of β rather close to the critical point of the discrete model. The SU(2) group has a larger finite subgroup Ỹ, the 120-element covering group of the rotation group of the icosahedron. The Ỹ and SU(2) systems cause indistinguishable results for all measurable quantities up to values of β well beyond the range of values used in practical applications. Thus, in several instances, the Ỹ model has been used as an excellent approximation to the SU(2) model, which allows substantial saving in computer time.

Later investigations in which actions depending on more than one parameter have been considered (see [9.1] for bibliography and details) have shown that even non-Abelian models [such as the SU(2) theory] have a rich phase structure. However, the presence of critical surfaces in these multidimensional parameter spaces do not, according to the best numerical evidence, spoil the fundamental result, i.e., that the lattice regularization can be used to define a continuum theory preserving the most relevant features of the strong coupling regime.

9.3 Monte Carlo Determination of Physical Observables

Once we accept from theoretical prejudices and numerical evidence the notion that the scaling critical point for a non-Abelian gauge theory is at $\beta = \infty$ ($g = 0$), we can attempt to evaluate the observables of the continuum theory. A quantity q, having physical dimension a^p (a denotes the lattice spacing), will be calculated at progressively larger values of β, q being given by

$$q = a^p f(\beta) \tag{9.32}$$

(9.28), where the dimensionless function $f(\beta)$ represents the result of the calculation of q in the regularized lattice theory. As β tends to infinity, $f(\beta)$ must behave as

$$f(\beta) \xrightarrow[\beta \to \infty]{} c[\Lambda a(\beta)]^{-p} \ , \tag{9.33}$$

the function $\Lambda a(\beta)$ being given by (9.29,30) for the SU(2) and SU(3) theories respectively. Finding (from computation) behavior as in (9.33) verifies the scaling properties of the model. The continuum value of q will then be given by

$$q = c\Lambda^{-p} \ . \tag{9.34}$$

This equation expresses q in terms of a scale parameter Λ, which has no direct physical meaning; indeed, it can also be thought of as a definition of Λ in terms of q. However, once expressions of the form of (9.34) become available for several quantities, Λ can be eliminated among them, and one obtains physically meaningful relations between observables of the theory.

From the above discussion it is clear that to recover properties of the continuum theory it is not really necessary to proceed to the limit $\beta \to \infty$, but it suffices to extrapolate the results to values of β large enough that evidence for scaling is found. In this respect strong coupling expansions are inadequate; extrapolations of strong coupling series do not show any clear signal for scaling. Vice versa, Monte Carlo simulations for lattice gauge theories are so successful because they can be carried out at intermediate values of β (beyond the range of validity of the strong coupling expansions), where evidence for scaling is seen. [Note that in principle Monte Carlo computations can be performed at any value of β, no matter how large. However, (9.29,30) imply a relationship between β and lattice spacing: hence, if a lattice of finite size is considered, β is also related to the physical volume of the lattice. Since computational constraints do not allow simulations of four-dimensional systems extending for much more than eight or ten sites in each direction, the size limitations also limit the largest β which can be meaningfully considered.] The most interesting applications have focused on calculating observables of quantum chromo dynamics. Correspondingly, the simulations involve the SU(3) lattice gauge theory. Several interesting computations (indeed, probably the majority of the simulations) have however been performed for the SU(2) lattice gauge theory with the assumption that results can be extrapolated without incurring substantial errors from the computationally simpler SU(2) model to the more appropriate SU(3) system.

This section gives a perforce concise review of applications of Monte Carlo simulations to the determination of observables of QCD. It is useful to say a few words on the measurable quantities, defined over the lattice, which form the basis for such computations. Any measurable quantity must be gauge invariant. In a pure gauge theory, where the only dynamical variables are the link variables U_{ji}, one can associate a gauge invariant quantity with each closed path γ in the lattice in the following way. If the vertices along   are i, i_2, ..., i_N, one constructs first the transporter along γ

$$U_\gamma = U_{i_1 i_N}, \; \ldots, \; U_{i_3 i_2} U_{i_2 i_1} \; .$$

(9.35)

In a gauge transformation

$$U_\gamma \rightarrow U_\gamma ' = G_{i_1} U_\gamma G_{i_1}^{-1} \; .$$

(9.36)

It follows that the quantity

$$W_\gamma = Tr \; U_\gamma$$

(9.37)

is gauge invariant. Further, W_γ is called the Wilson factor associated with the loop γ. Expectation values of Wilson loop factors are of paramount importance for lattice theory. Of particular relevance are the factors associated with rectangular paths extending for m and n links, denoted by $W_{m,n}$. Also, in most applications lattices with periodic boundary conditions are considered. Specifically, let us assume the system to be periodic in time with period n_t. It follows that a path consisting of n_t time links all emerging from a common space coordinate $\mathbf{x}$ and successive in time will be closed by virtue of the periodic boundary conditions. The corresponding Wilson loop factor is also very important. For reasons which will become clear later, this is called a thermal loop factor, denoted by $W_{\mathbf{x}}$.

The rate of decrease of $<W_\gamma>$ as the area A enclosed by the loop increases plays the role of order parameter for the lattice gauge theory. Since an entirely random averaging over any of the U_{ji} in (9.35) produces $<W_\gamma> = 0$ and since the correlation among the link variables in U_γ decreases as γ becomes larger, a falloff of $<W_\gamma>$ for increasing loop size is expected. The crucial question is whether the falloff follows an area law

$$<W_\gamma> \sim e^{-\sigma A} \qquad (9.38)$$

or happens at a slower rate. Behavior as in (9.38) is found in strong coupling expansions and indicates linear growth of the potential between two charges (confinement). To see this last point notice that $<W_\gamma>$ can also be thought of as the ratio between the partition function Z_γ of the theory, where the action has been modified by the term induced by a charge circulating along γ and the vacuum-to-vacuum permanence amplitude, the ordinary partition function Z:

$$<W_\gamma> = \frac{Z_\gamma}{Z} \quad . \qquad (9.39)$$

[To understand better the relation between W_γ and a current along γ, the reader may recall that in the continuum limit and in the Abelian case W_γ takes the form $\exp(ig \int_\gamma A_\mu dx^\mu)$]. If γ is a rectangular loop extending for m sites along a space axis (say the x axis) and n sites along the time direction, (9.39) implies

$$<W_\gamma> \xrightarrow[t \to \infty]{} \exp[-V(r)t] \quad , \qquad (9.40)$$

where $r = ma$ and $t = na$ are the space and time extent of the lattice and $V(r)$ represents the potential energy of two static opposite charges at separation r. Comparing (9.38,40) we see that area law behavior implies a linearly growing potential

$$V(r) \xrightarrow[r \to \infty]{} \sigma r \quad . \qquad (9.41)$$

The coefficient σ is commonly referred to as string tension.

In a lattice computation an area-law falloff will manifest itself through

$$<W_{mn}> \sim \exp[-K(\beta)mn] \quad . \qquad (9.42)$$

From $m = r/a$, $n = t/a$ one infers that the dimensionless function $K(\beta)$ is related to σ by $K(\beta) = \sigma a^2$ and therefore that $K(\beta)$ should scale as $[\Lambda a(\beta)]^2$. The first Monte Carlo evidence for nonvanishing string tension in the SU(2) theory, with the correct scaling behavior, was presented in [9.14,15]. Figures 9.3a,b illustrate the results obtained in [9.16], where a 16^4 lattice was considered (approximating the SU(2) gauge group with its icosahedral subgroup $\tilde{Y}$, Sect.9.2). In an actual computation it is impossible to evaluate W_γ for loops large enough for the leading term in the exponential falloff to be identified with certainty. One tries therefore to isolate the coefficient of the area-law term by considering ratios of $<W_\gamma>$ for loops geometrically similar and differing only in area. Thus, in Figs.9.3a,b effective string tensions $K(I,\beta)$ are plotted, where

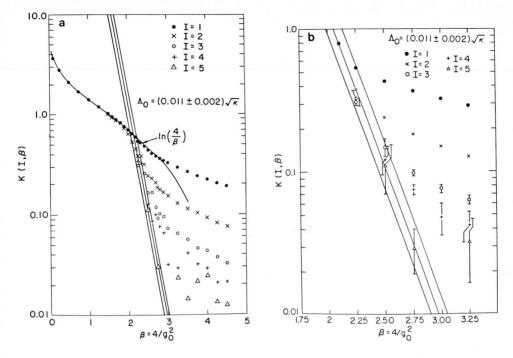

Fig.9.3. (a) Monte Carlo determination of string tension. (b) Detail of Fig.9.3a, with statistical error bars

$$K(I,\beta) = - \ln \frac{<W_{I,I}>}{<W_{I+1,I-1}>} \quad . \tag{9.43}$$

The square loop of sides I and the rectangular loop of sides $I + 1$ and $I - 1$ differ by one lattice unit of area. Therefore, for $I \to \infty$, $K(I,\beta)$ reduces to the area-law coefficient $K(\beta)$ in (9.42). In Fig.9.3a the Monte Carlo results for $K(I,\beta)$ are seen to follow the lowest-order strong coupling result (the curve originating at $\beta \approx 0$) for small values of β. At a value of $\beta \approx 2$ they depart from the strong coupling curve and the tendency of the MC data to envelope an asymptotic curve $K(\beta)$ as I increases is clear. The three parallel lines represent the expected scaling be- havior (including errors), with a coefficient found through a fit to the points which seem to lie better on the asymptotic curve (see Fig.9.3b, where error bars due to statistical fluctuations are also included). Thus one determines the rela- tion between the physical string tension σ and the scale parameter Λ. Assuming from the phenomenology of strong interactions $\sigma \approx (450 \text{ MeV})^2$, one finds [9.14-16] $\Lambda \approx 6$ MeV for the SU(2) theory and [9.17] $\Lambda \approx 4$ MeV for the SU(3) theory. The relation between the lattice scale Λ and other scale parameters used in perturbative studies of QCD (in domains of energies and momentum transfer where perturbative expansions are justified) can be evaluated analytically [9.18,19] and reasonably good agreement be- tween these independent determinations of the scale parameters is found.

The expectation value $<W_x>$ of the thermal loop factor can be related to the free energy of a single static source with space coordinate **x**. Indeed the partition function of a system at physical temperature T can be reformulated in terms of a functional integral, where the system extends for a length 1/T in imaginary time (units where Boltzmann's constant k equals 1 are used) with periodic boundary conditions. Thus, if one considers systems periodic in Euclidean space-time extending for n_t sites in the time direction, one can identify the appropriate functional integrals with partition functions at finite temperature $T = 1/n_t a$. An extent n_s of the system in the space directions much larger than n_t is required (at least in principle) for the above interpretation. Also notice that these considerations imply that some caution should be exerted in associating measurements performed on periodic lattices of small volume with the values of physical observables in the ground state (at zero physical temperature).

Coming back to $<W_x>$, this quantity is the ratio between the partition function of the system in the presence of a fixed source at **x** and the partition function of the unperturbed system. Hence the free energy of a static source (relative to the vacuum) is

$$F = -T \ln<W_x> = -(1/n_t a) \ln<W_x> \quad . \tag{9.44}$$

If charges are confined and a static source cannot exist in isolation, one expects $F = \infty$ and therefore $<W_x> = 0$. However, finite temperature effects may cause a Debye-screening phenomenon, by which F can become finite even if the system possesses confining properties at lower temperature. Thus, a deconfining transition would manifest itself with a critical temperature T_c such that $<W_x>$ acquires a nonvanishing expectation value for $T > T_c$.

It is interesting to observe that gauge systems possess a symmetry which would insure $<W_x> = 0$ if not spontaneously broken. This symmetry consists in multiplying all the variables U_{ji} associated with the time links originating at a fixed time coordinate by an element of the center of the group (these considerations apply to such groups as U(1), SU(N) having a nontrivial center, i.e., set of elements commuting with all elements of the group, and for sources transforming nontrivially under the center of the group). The above operation leaves the action of the system invariant (because each plaquette contains either zero or two of the transformed link variables with opposite orientation), but modifies the value of $<W_x>$ (because every thermal loop factor contains just one of the transformed variables). Hence one expects $<W_x> = 0$ if the symmetry is realized at the level of expectation values.

A numerical investigation of the thermal properties of the SU(2) system was first presented in [9.20,21]. From a study of $<W_x>$ it was inferred that the system undergoes a phase transition at a temperature $T_c \approx 0.39 \sqrt{\sigma}$. Notice that the temperature $T = (n_t a)^{-1}$ can be modified by varying both n_t and the effective lattice spacing a through a change of β. Systems with different n_t develop a nonvanishing expec-

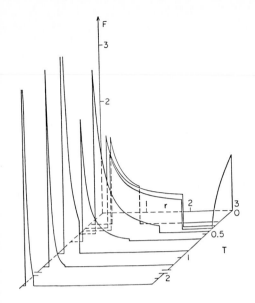

Fig.9.4. Behavior of the force between
two static sources in the fundamental
representation in SU(2) theory, as functions of physical temperature T and separation r (all quantities are expressed
in units of appropriate powers of string
tension). F is displayed only in the
ranges of separation where it is measured.
The curve to the right represents the
temperature-dependence behavior
$\sigma = (1 - T/T_c)^{\frac{1}{2}}$ of $\sigma(t)$. As r increases F
is seen to approach constant values for
$T < T_{cr} \approx 0.39$, to vanish exponentially
(Debye screening) for $T > T_{cr}$

tation value for W_x at different critical values β_{cr} of β. It was found in [9.21]
that the relationship between β_{cr} and n_t is consistent with scaling toward a continuum limit.

From the expectation value of products of loop factors $<W_x W_o>$ one can extract
the free energy of two (or more) static sources. The transition from a confining
to a Debye-screened form of the potential was verified in [9.20,21]. A graph of the
force between two static sources, as a function of temperature and separation, is
displayed in Fig.9.4 (taken from [9.22], where the recovery of rotational symmetry
in the transition from the strong coupling domain to the scaling domain was studied
and found to take place).

Monte Carlo simulations have been used to estimate masses in the spectra of
quantized excitations of various models. These analyses proceed through measurements of the expectation values of products of suitable operators. Let $\mathcal{O}(\mathbf{x},t)$ be
an operator which couples the vacuum to the state under investigation ($\mathbf{x},t$ denote
lattice sites). Then, the connected Green's function

$$G(\mathbf{x},t) = <\mathcal{O}^{\dagger}(\mathbf{x},t)\,\mathcal{O}(0)> - \,|<\mathcal{O}>|^2 \qquad (9.45)$$

can be expressed in terms of a complete set of intermediate states

$$G(\mathbf{x},t) = \sum_{n \neq vacuum} |<n|\mathcal{O}|0>|^2 \exp(-E_n t + i\mathbf{P}_n \cdot \mathbf{x}) \quad . \qquad (9.46)$$

The Euclidean character of space-time implies a decreasing, rather than oscillatory, behavior of G with time, and the energy (energies) of the lowest state(s) excited by $\mathcal{O}$ can be estimated from the rate of decay of G.

Actually it is convenient to sum $\mathbf{x}$ over all space positions in (9.45), defining
a new Green's function

$$\tilde{G}(t) = \sum_{x} G(x,t) \quad . \tag{9.47}$$

This projects out of the sum over intermediate states all states except those with vanishing space lattice momentum **P**, for which $E_n = m_n$. Thus the rate of decay of $\tilde{G}$ directly produces information on masses:

$$\tilde{G}(t) = \sum_{n \neq 0} |C_n|^2 \exp(-m_n t) \quad , \tag{9.48}$$

C_n being a suitable amplitude.

This technique, coupled with a variational procedure by which a few parameters incorporated in the definition of $\mathcal{O}$ are varied to obtain the largest possible amplitude to the lowest lying state, has been used in [9.23-26] to estimate the mass gap of the SU(2) and SU(3) pure gauge theories. If these systems possess confining properties, indeed, one does not expect the presence of massless excitations and the lowest quantum state must occur at some nonvanishing physical mass m_g.

The Monte Carlo calculations produce estimates of m_g in terms of the scale parameter Λ and, from the independent measurements of the string tension, also in terms of $\sqrt{\sigma}$. With $\sqrt{\sigma} \approx 450$ MeV the various MC results (see [9.1] for an exposition of other methods to estimate m_g) are consistent with $m_g \approx 1000$ MeV for the SU(2) theory, $m_g \approx 800$ MeV for the SU(3) theory, and errors of the order of 15% to 20%.

The method to estimate physical masses outlined above can also be applied to systems containing matter fields coupled to the gauge fields. Of particular interest is the SU(3) theory with fermions in the fundamental representation, since this constitutes the lattice-regularized version of quantum chromo dynamics. It is, however, not straightforward to extend Monte Carlo simulations to systems with fermions. Basically, the problem one encounters is that fermionic degrees of freedom correspond to binary variables representing the fact that a given fermionic state can be either empty or occupied. However, because of Fermi statistics (the anti-commuting nature of fermionic fields) a reformulation of the quantum-mechanical expectation values in terms of sums over these binary variables will in general involve a measure which is not positive definite (even in Euclidean space-time) and thus not suitable for numerical simulations. Only in models with one space and one time dimension has it been possible to express expectation values involving fermions in terms of a positive definite measure and to apply MC simulations to the corresponding systems [9.27].

An alternative procedure, which is viable when the matter field action is bilinear in the fermionic fields, consists in performing explicitly the Gaussian integration over these variables. Thus the fermionic degrees of freedom are eliminated from the functional integrals, which now involve sums over gauge field configurations only. However, Gaussian integration produces a modification of the pure gauge measure $\exp\{-S_g(U)\}$ and the action is replaced by a new effective action $S_{eff}(U)$, which is nonlocal in the gauge variables (it involves coupling among all

U_{ji} no matter how far apart). Present research efforts are directed toward finding efficient computational methods to evaluate the variation of $S_{eff}(U)$ in the upgrading $U \rightarrow U + \Delta U$, as required for the implementation of the Metropolis procedure [9.28].

Interesting results have, however, been obtained by neglecting the modification $S_G \rightarrow S_{eff}$, induced by the fermionic dynamical reaction, and performing the quantum averages integrating over gauge field configurations with the pure gauge measure $\exp\{-S_G(U)\}$. Suppose, for instance, that one wishes to estimate numerically the mass of a meson consisting of a quark and an antiquark. This will be created from the vacuum by an operator of the type $\bar{\psi}(x,t)\Gamma\psi(x,t)$, where Γ is a suitable γ matrix, and, according to the discussion presented above for the mass gap, the mass of the meson can be deduced from the rate of decay of the Green's function

$$G(t) = \sum_{x} <\bar{\psi}(x,t)\Gamma\psi(x,t)\bar{\psi}(0)\Gamma\psi(0)> \quad . \tag{9.49}$$

The idea is to consider a large sample of gauge field configurations generated by the standard MC algorithm with action $S_G(U)$. Within each of the configurations the propagator of a quark is calculated by inverting the corresponding Dirac's operator (by some relaxation method) and the propagator of the meson is obtained by multiplying the quark propagators. The result is then averaged over all the configurations in the sample to approximate $G(t)$. This procedure accounts for all dynamical effects (including binding) due to the interaction of the propagating quarks and antiquarks with the gauge field and to the self-interaction of the gauge field with itself. What the approximation neglects are the effects associated with the creation from the vacuum and subsequent annihilation of quark-antiquark pairs. From general theoretical and phenomenological arguments these effects are expected to be of lesser importance in determining the spectrum of lowest lying states.

The approximation described above has been applied to numerical determination of masses in the quark model spectrum [9.29]. The results are subject to margins of uncertainty primarily due to computational limitations which preclude both consideration of large lattices (as one would like, to make sure that the results are not affected by finite-volume effects) and of large samples of configurations (desirable to produce better statistics). Yet they have been quite encouraging. When the mass of the quark is sent to zero the mass of the lowest pseudoscalar excitation (π meson) is also seen to approach zero, as one would expect from a dynamical realization of chiral symmetry. The other masses approach finite limits, with values determined entirely by the string tension. The quark masses (free parameters for QCD) may be fixed from the experimental values of the masses of the pseudoscalars (π meson, K meson, etc.) and this input, together with the string tension, allows the other masses in the spectrum to be estimated. Typical results give:

$m_\rho = (800 \pm 100)$ MeV   (exp.776),

$m_\delta = (950 \pm 150)$ MeV   (exp.981),

$$m_{A_1} = (1100 \pm 150) \text{ MeV} \qquad (\text{exp.} \sim 1100),$$

$$m_p = (1000 \pm 150) \text{ MeV} \qquad (\text{exp.938}),$$

$$m_\Delta = (1300 \pm 150) \text{ MeV} \qquad (\text{exp.1236}),$$

for the masses of a few selected mesons (ρ, δ, A_1) and baryons (proton, Δ resonance).

In conclusion, applications of Monte Carlo methods to lattice gauge theories have produced remarkable results for particle physics, especially for the theory of strong interaction: this notwithstanding the fact that the four-dimensional nature of the systems severely limits the extent of the lattices. Probably, part of the success of the simulations depends on the actual physical properties of hadrons. Experiments indeed reveal an abrupt change in dynamical behavior of the quarks as the scale of distances is reduced by a rather small factor: from confined constituents at separation of the order of hadronic sizes, they become almost free objects (asymptotic freedom) at shorter distances. Correspondingly, the crossover from the strong coupling regime to the scaling regime is also abrupt in lattice computations, and this is certainly one of the main reasons why MC simulations have been successful. In any event the results obtained up to now are far from complete and remarkable progress can be expected both from the advent of more powerful computational facilities and from the intelligent coupling of analytical insight with numerical techniques.

References

9.1 M. Creutz, L. Jacobs, C. Rebbi: Phys. Repts. C**95**, 201 (1983)
9.2 K. Wilson: Phys. Rev. D**10**, 2445 (1974)
9.3 F. Wegner: J. Math. Phys. **12**, 2259 (1971)
9.4 R. Balian, J. Drouffe, C. Itzykson: Phys. Rev. D**10**, 3376 (1974); D**11**, 2098, 2104 (1975)
9.5 D. Gross, F. Wilczek: Phys. Rev. Lett. **30**, 1343 (1973);
D. Politzer: Phys. Rev. Lett. **30**, 1346 (1973)
9.6 G. Bhanot, C. Lang, C. Rebbi: Computer Phys. Comm. **25**, 275 (1982)
9.7 N. Metropolis, A.W. Rosenbluth, M.N. Rosenbluth, A.H. Teller, E. Teller: J. Chem Phys. **21**, 1087 (1953)
9.8 M. Creutz: Phys. Rev. D**21**, 2308 (1980)
9.9 M. Creutz, L. Jacobs, C. Rebbi: Phys. Rev. Lett. **42**, 1390 (1979); Phys. Rev. D**20**, 1915 (1979)
9.10 B. Lautrup, M. Nauenberg: Phys. Lett. **95**B, 63 (1980)
9.11 M. Creutz: Phys. Rev. Lett. **43**, 553 (1979)
9.12 C. Rebbi: Phys. Rev. D**21**, 3350 (1980)
9.13 D. Petcher, D. Weingarten: Phys. Rev. D**22**, 2465 (1980)
9.14 M. Creutz: Phys. Rev. D**21**, 2308 (1980)
9.15 M. Creutz: Phys. Rev. Lett. **45**, 313 (1980)
9.16 G. Bhanot, C. Rebbi: Nucl. Phys. B**180**, 469 (1981)
9.17 E. Pietarinen: Nucl. Phys. B**190**, 349 (1981)
9.18 A. Hasenfratz, P. Hasenfratz: Phys. Lett. **93**B, 165 (1980)
9.19 R. Dashen, D. Gross: Phys. Rev. D**23**, 2340 (1981)
9.20 L. McLerran, B. Svetisky: Phys. Lett. **98**B, 195 (1981)
9.21 J. Kuti, J. Polonyi, K. Szlachanyi: Phys. Lett. **98**B, 199 (1981)
9.22 C. Lang, C. Rebbi: Phys. Lett. **115**B, 137 (1982)

9.23 B. Berg, B. Billoire, C. Rebbi: Ann. Physics **142**, 185 (1982)
9.24 M. Falcioni, E. Marinari, M.L. Paciello, G. Parisi, F. Rapuano, B. Taglienti, Zhang Yi-Cheng: Phys. Lett. **110**B, 295 (1982)
9.25 B. Berg, A. Billoire: Phys. Lett. **113**B, 65 (1982)
9.26 K. Ishikawa, M. Teper, G. Schierholtz: Phys. Lett. **116**B, 429 (1982)
9.27 R. Blankenbeckler, J. Hirsch, D. Scalapino, R. Sugar: Phys. Rev. Lett. **47**, 1628 (1982)
9.28 F. Fucito, E. Marinari, G. Parisi, C. Rebbi: Nucl. Phys. B**180**, 369 (1981);
     E. Marinari, G. Parisi, C. Rebbi: Nucl. Phys. B**190**, 266 (1981);
     D. Scalapino, R. Sugar: Phys. Rev. Lett. **46**, 519 (1981);
     D. Weingarten, D. Petcher: Phys. Lett. **99**B, 333 (1981);
     H. Hamber: Phys. Rev. D**24**, 951 (1981);
     J. Kuti: Phys. Rev. Lett. **49**, 183 (1982)
9.29 H. Hamber, G. Parisi: Phys. Rev. Lett. **47**, 1792 (1981);
     E. Marinari, G. Parisi, C. Rebbi: Phys. Rev. Lett. **47**, 1798 (1981);
     H. Hamber, E. Marinari, G. Parisi, C. Rebbi: Phys. Lett. **108**B, 314 (1982);
     D. Weingarten: Phys. Lett. **109**B, 57 (1982);
     A. Hasenfratz, P. Hasenfratz, Z. Kunszt, C.B. Lang: Phys. Lett. B**110**, 282 (1982);
     H. Hamber, G. Parisi: Phys. Rev. D**27**, 208 (1983);
     D. Weingarten: Nucl. Phys. B**215**, 1 (1983);
     F. Fucito, G. Martinelli, C. Omero, G. Parisi, R. Petronzio, F. Rapuano: Nucl. Phys. B**210**, 407 (1982);
     C. Bernard, T. Draper, K. Olynyk: Nucl. Phys. B**220**, 508 (1983)

10. Recent Developments

K. Binder, A. Baumgärtner, J. P. Hansen, M. H. Kalos, K. W. Kehr, D. P. Landau,
D. Levesque, H. Müller-Krumbhaar, C. Rebbi, Y. Saito, K. E. Schmidt,
D. Stauffer, and J. J. Weis

This chapter adds some new material to the Topics volume originally published in 1984.

Although only less than three years have elapsed since then, the field of computer simulation has expanded so enormously that a complete account of all the recent work would require a whole book, rather than an additional chapter. Thus we shall present a very condensed guide to the literature only, emphasizing some highlights, as well as recent general trends in the community of workers applying Monte-Carlo (MC) methods, but cannot go into the details of the physics which is being studied. One reason for this growth of the field is that the usefulness of MC simulations has now been realized more widely than a couple of years ago: in fact, MC simulations belong to the main tasks of supercomputers which are becoming increasingly available for scientific research. In addition, special purpose computers have been dedicated specifically to the MC simulation of various models in statistical mechanics, and some of them have yielded very useful results. Some of these recent trends and developments have been briefly reviewed by one of the authors of this update in a chapter in the companion volume [10.1] but clearly this is a subject in flux.

The outline of this new chapter now more or less follows the plan of the book, each section corresponds to one of the first nine chapters of this book.

10.1 Introduction and Some Specialized Topics

10.1.1 Size Effects and Self-Averaging

In simulations of the percolation problem [10.2] the main limitation usually is the finite-size of the simulated lattice. It now is possible to make these finite-size effects simply very small by the brute-force approach of simulating extremely large lattices: for two-dimensional percolation, a lattice of size 160.000×160.000 was simulated [10.3], and for three-dimensional percolation, a $1000 \times 1000 \times 1000$ lattice [10.4].

In MC computer experiments on problems in statistical mechanics where thermal fluctuations are simulated by importance sampling simulations, the situation is not so simple, of course: *Both—finite-size and finite "observation time" of the MC run —are limiting factors.* Although it has been possible to simulate a three-dimensional Ising model as large as $1080 \times 1080 \times 1080$ at one temperature [10.5], we note this as a curiosity only. Here the problem rather is to use a given amount of effort (CPU time) as efficiently as possible for the problem under consideration, taking into account how the performance of the MC code depends on the system's linear dimensions (this is a non-trivial question particularly for vector processors), as well as the systematic limitations due to finite-size effects and finite "observation time" effects. For certain problems, like the sampling of the mean-square order parameter $\langle\psi^2\rangle_t$ of a system at time t starting at time t = 0 in a disordered configuration but at a temperature where the system should be ordered [10.6,7] it is clear that a single run on a very large system is rather meaningless —for an example see, e.g., [10.7] —, due to a *complete lack of self-averaging.* This occurs because the relative error $\Delta(\langle\psi^2\rangle_t)/\langle\psi^2\rangle_t$ behaves as [10.6]

$$\Delta(\langle\psi^2\rangle_t)/\langle\psi^2\rangle_t = \sqrt{2}/n \quad . \tag{10.1}$$

In (10.1), the average $\langle......\rangle$ is defined by averaging over n statistically independent runs. "Lack of self-averaging" now means in the present context that the relative error in (10.1) does not decrease when the linear dimensions L of the system increases. This property results from the fact that sampling $\langle\psi^2\rangle_t$ means *sampling of fluctuations,* we have $\langle\psi\rangle_t \equiv 0$, and these fluctuations are Gaussian distributed: $P_L(\psi,t) \propto \exp[-\psi^2 L^2/2\langle\psi^2\rangle_t]$, for a d-dimensional system. Now $\Delta(\langle\psi^2\rangle_t) \equiv n^{-\frac{1}{2}}(\langle\psi^4\rangle_t - \langle\psi^2\rangle_t^2)^{\frac{1}{2}} = n^{-\frac{1}{2}}\langle\psi^2\rangle_t\sqrt{2}$, for a Gaussian distribution. Although $\langle\psi^2\rangle_t$ itself decreases with L proportional to L^{-d}, its *relative error* does not. Thus, for a given total effort we will obtain n the larger the smaller we choose L. On the other hand, one wishes to have the linear dimension L much larger than the size of the ordered domains l(t), which can be defined as [10.7]

$$[l(t)]^d \equiv \langle\psi^2\rangle_t L^d/\psi_T^2 \quad , \tag{10.2}$$

ψ_T being the thermal equilibrium value of the order parameter at the considered temperature, T. Since one expects the law $l(t) = \Omega t^x$, for large t, where in general neither the amplitude Ω nor the exponent x are known in beforehand, it is clear that the proper choice of n and L, satisfying $l(t) \ll L$ for the time t up to which one wishes to simulate the system and which optimizes the effort, is a subtle matter [10.7]. It can often be decided only after some data on the system are already taken and preliminary estimates of Ω and x have been obtained. Simulation of the kinetics of domain growth, as described above, is now a rather popular application of MC sampling [10.8]; but the problem of the *lack of self-averaging when fluctuations are sampled* is much more general: e.g., the standard case is the

sampling of the specific heat from energy fluctuations, the sampling of the susceptibility from magnetization fluctuations [10.6], etc.; these fluctuations are
Gaussian distributed if the correlation length $\xi \ll L$, and analogous error formulae
apply if one averages over n statistically independent observations (which may come
from a single run if the observations are separated from each other by a time interval Δt exceeding the correlation time τ distinctly, rather than from separate runs
as discussed above). Of course, such simple considerations are probably nothing new
for the experienced practitioner of MC, but should be helpful for the beginner in
this field.

Finite-size effects are small in the domain-growth problem, if $l(t) \ll L$, and in
the study of thermal equilibrium states off a phase transition, if $\xi \ll L$. On the
other hand, *close enough to a phase transition finite-size effects always are important*, and have received much attention already [10.1,9-13]. Recent progress concerns
the analysis of finite-size effects for critical points for which the hyperscaling
relation fails [10.10] and the analysis of finite-size effects at temperature-driven
first-order phase transitions [10.11]. In such a phase transition one normally does
not find any symmetry between the low-temperature phase and the high-temperature
phase: As a consequence, by finite-size the transition is *both rounded and shifted*
(unlike the case of the field-driven Ising magnet below T_c discussed in Chap.1,
where by symmetry no shift can occur). It turns out that both shift and rounding
vary inversely proportional to the volume L^d of the system, and also the corresponding proportionality constants can be quantitatively predicted [10.11], as well as
the full finite-size scaling function for the specific heat at the transition
[10.11]. Only for phase transitions which seem very weakly of first-order, perhaps
some unexplained problems remain [10.12].

10.1.2 Slowing Down at Phase Transitions: Can we get Around it?

Although the simple phenomenological theory of the rounding and shifting of first-
order phase transition [10.11] has been confirmed by extensive simulations for the
10-state Potts model, its usefulness is, in general, restricted by the fact that we
expect the correlation time τ to observe the full thermal equilibrium at first-
order transitions (i.e., no hysteresis!) to scale as

$$\ln \tau \propto L^{d-1} \quad, \quad L \to \infty \quad, \tag{10.4}$$
$$\text{(first-order transition, discrete symmetry broken)} \quad.$$

This is even worse than the critical slowing down at second-order phase transitions
(see, Sect.3.6 and [10.1,13], where instead we have

$$\ln \tau = z \ln L \quad, \quad L \to \infty \quad, \tag{10.5}$$

z being the "dynamic exponent"). For phase transitions in most pure systems z is
about 2, while for spin glasses with d = 3 the situation seems much worse ($z \approx 5$

[10.14]). This slowing down at phase transitions obviously hampers their MC study extremely much.

This slowing down at phase transitions possibly could be beaten, if instead of sampling the standard local moves (in an Ising model, a single spin flip) one would be able to sample the appropriate "collective modes". Ideas in this direction have been developed by a number of workers [10.15-22]. The "multigrid method" [10.19] has been applied to a ϕ^4-model on a square lattice (this model is also an example which shows that finite-size scaling methods are not always useful for very small L [10.23], due to substantial corrections to finite-size scaling, unlike the two-dimensional Ising model where such corrections are rather small). In this method, one also defines block variables for blocks of sizes 2^d, 4^d, ..., up to L^d; one standard MC step per site is then replaced by an updating on all levels of these variables. The method is constructed in such a way that for a Gaussian model on the lattice, it would eliminate the critical slowing down completely, while for the ϕ^4-model critical slowing down is reduced by about one order of magnitude.

Another approach was taken for the plane-rotator model [10.20]: adding to the Hamiltonian a suitably invented ad-hoc kinetic energy term, one can define a molecular dynamics algorithm for the model instead of a MC algorithm! In practice, it turns out advantageously to apply a "hybrid algorithm" where one suitably mixes "molecular dynamics" and "Brownian dynamics" techniques, and a considerable reduction of critical slowing down is, in fact, achieved. A qualitatively similar method has also been used in the context of simulations of linear polymers [10.24] and star polymers [10.25]; in the latter case systems with up to 50 arms consisting of up to 50 monomers each could be simulated: due to entanglements of the arms near the center of the star it would be very difficult, if not impossible, to simulate such a system with standard MC techniques, as described in Chap.5.

The next approach which we discuss in this section has been formulated for nearest-neighbour Ising and Potts models [10.21]. It rests on the equivalence between the configurations of these models and configurations of clusters [10.26]. Suppose we consider an Ising model in a particular configuration and introduce "virtual bonds" *between parallel spins* with a probability $p_B = 1 - \exp(-2J/k_B T)$, J being the exchange energy. Clusters are then defined by requesting that all spins of a cluster must be connected by virtual bonds. Every spin then belongs to a cluster (the sizes of these clusters may range from monomers up to percolating clusters spanning the whole lattice). The next configuration of the system then is found by assigning to each cluster (i.e., all spins which it contains) a spin orientation according to a transition probability for which one can show that detailed balance holds [10.21,26]. This spin orientation may be the previous one or a new one: with these steps, correlations between subsequent configurations are destroyed much quicker than by the standard method, and a considerable reduction of critical slowing down is, in fact, observed [10.21]. Finally we draw attention to ideas ("improved MC renormalization group" IMCRG) [10.27] where spins and block spins are simulated together choosing weights such that the block spins are uncorrelated, to reduce critical slowing down.

Still another approach, useful for random systems, was proposed also by *Swendsen* [10.22] simulating in parallel many "real replicas" of the same random system at somewhat different temperatures, transferring certain information about large clusters which can be flipped with zero or small energy cost, from one replica to the other. By this method, the two-dimensional ±J Ising spin glass could be studied at considerably lower temperatures than by the standard method [10.28], although the latter work used a very efficient program at the ICL-DAP computer ("distributed array processor"). Thus we see, there are many different routes by which one can make the problem of slowing down less severe, at least for certain models. Clearly, there is both room and need for more thought on these problems!

10.1.3 Pushing MC Calculations to their Limits: Superfast MC Programs on "Supercomputers"; Special Purpose Computers for MC Methods

For every problem there does not exist a clever method to reduce slowing down, as discussed in the previous subsection; in extreme cases even though one uses such a method one still may be hampered by the limitation of unsufficient statistical accuracy. Thus often the need arises to make MC programs very fast. As an example of what can be done we mention the Ising-model simulation of *Reddaway* et al. [10.29] achieving 218 steps per microsecond on an ICL-DAP computer. This superfast code required the developement of a particular new random-number generator, which clearly is a major effort. Alternative very fast Ising algorithms have been discussed by *Williams* and *Kalos* [10.30] and *Bhanot* et al. [10.31]. In this context we also mention the "microcanonical" MC simulation of the Ising model due to *Creutz* et al. [10.32,33]: Apart from the construction of the critical configuration, this "demon algorithm" does not use random numbers.

At institutions where fast vector processors or array processors have been installed (e.g., the universities of Edinburgh, Minnesota, Georgia, and Florida State) MC programs are being developed and optimized. This line of research is well respected and part of a long-term effort (some of these institutions have formally created "supercomputer institutes" etc.). This research has also led to applications as diverse as lattice-gauge theory (Sects.10.9), spin glasses [10.28,34-38] and percolation [10.4,39], liquid crystals [10.40] and fluids (Sect.10.2), studies of complicated phase diagrams [10.41], static [10.42,43] and dynamic [10.44,45] critical phenomena, surface [10.46] and size effects [10.11,31] at phase transitions, polymers [10.25,47], see also Sect.10.5, and many other problems. For many of these applications, the availability of "supercomputers" certainly was absolutely crucial.

For some problems, such as three-dimensional Ising spin glasses, even very fast programs running for many hours on supercomputers still yield results not yet satisfactory in all respects (because of the slowing down mentioned above, the need to sample many realizations of the random-bond configuration, etc.). In such cases the use of special purpose machines is very helpful, as demonstrated by *Ogielski* and co-workers [10.14,48-50]. The first such machine dedicated to MC calculations is the Delft Ising special purpose computer [10.51] which is still operating and produces useful results [10.52], although its speed is only of the order of 1.5 MC steps per microsecond and hence much less than what can be reached on vector processors [10.29,31]. Here, we shall not review the work devoted to the construction of special purpose machines in detail, as other competent recent reviews exist [10.53]. Also we refer to the proceedings of a recent conference on Quantum MC methods [10.54] for details on new concepts for parallel computers (GF 11, Caltech Hypercube, etc.) which potentially will have a large impact for the applications of MC methods.

10.2 Simulation of Classical Fluids

The years since 1983 have witnessed a steady expansion, both in scope and in precision, of the application of MC methods to the field of classical fluids. The following brief overview is devoted to a non-exhaustive sample of some of the most significant contributions.

Among the methodological developments, the problem of the precise evaluation of free energies, in view of the location of phase transitions, has been thoroughly reviewed by *Frenkel* [10.55]. Finite-size and boundary effects in canonical ensemble simulations have been examined in detail in the case of the hard-disk fluid [10.56, 57], and the corresponding equation-of-state has been computed with an accuracy of the order of one part in 10^4 [10.58]. A novel method for the direct evaluation of the chemical potential has been put forward [10.59], while practical aspects of the use of force-biased sampling in MC simulations have been discussed in [10.60].

Recent work on simple Coulombic systems has focused on the two-dimensional Coulomb gas which undergoes pair formation and collapse in the case of oppositely charged point particles [10.61], while it exhibits a Kosterlitz-Thouless transition in the case of extended charges [10.62]; the latter work gives the complete phase diagram for charged hard disks.

Systems of molecules interacting via non-spherical hard core potentials of various shapes have been studied extensively including mixtures of rods and plates [10.63], oblate sphero-cylinders [10.64], hard dumbells with point quadrupoles [10.65], infinitely-thin hard rods [10.66], hard ellipsoids [10.67], and hard spheroids [10.68]. The main result of these studies is the clear evidence of a nematic transition [10.66,67], while the smectic transition has also been observed for hard sphero-cylinders [10.69].

Much recent effort has gone into the determination of the dielectric properties of polar liquids with significantly improved accuracy (relative errors less than 10%). The correct way of taking into account the boundary effects in the computer simulation of dielectric fluids has been discussed by *Neumann* and collaborators [10.70-72], while size effects were treated in [10.73]. Extensive calculations of the dielectric and thermo-dynamic properties of water have been carried out on the basis of several models, and comparison of the results with the experiment was used as a test of these models [10.74-77]. A systematic comparison of theoretical results and simulation data for realistic models of polar liquids has been presented in [10.78,79]. The solvation of ions has studied by MC simulations in [10.80], but most of the work on this problem was carried out by MD simulations [10.81].

One of the most active areas of MC research has recently been the study of fluid surfaces and interfaces. An analysis of the size effects on the liquid-vapour interface of Lennard-Jones droplets showed that systems of at least 2000 particles are needed to reproduce the surface tension of a planar interface [10.82]. System sizes of less than 800 molecules were found to be insufficient to allow a valid extrapolation of the surface tension of droplets of polar molecules to the planar-interface limit [10.83]. The orientation of water molecules was studied in simulations of the liquid-vapour interface of a water model [10.84].

The fluid-wall interface as a model for the liquid-solid interface has been studied for the hard-sphere fluid [10.85], a diatomic fluid [10.86], and for dipolar hard spheres [10.87]. The structure of water near a soft wall representing a hydro-phobic boundary has been determined by MD simulations [10.88,89]. Wetting of a fluid-wall interface has been considered in [10.90].

Electric double layers near planar electrodes have also been the subject of extensive simulations. *Valleau* and co-workers examined the long-range interaction of ions with the surface [10.91] and the effects of image charges [10.92], which were also investigated by *Bratko* et al. [10.93].

Adsorption of gases on solid substrates, like graphite, or on porous media has likewise been the subject of several simulation studies [10.94-97].

Another very active area is the simulation of quantum processes in classical fluids. The path-integral method has been used to represent a quantum particle in a classical fluid as a cyclic chain polymer. This method has been applied to study the behaviour of an excess electron in helium gas [10.98], in a water cluster [10.99], in ammonia [10.100], and in a hard-sphere fluid [10.101]. Trapping of a quantum particle in a disordered system has been examined in [10.102].

The present trends in computer simulations of fluids include very accurate computations on simple models, the study of orientational order in systems of very elongated molecules and the simulation of quantum impurities in classical fluids.

10.3 Critical and Multicritical Phenomena

A rather substantial amount of work on relatively simple classical spin models has revealed a wide range of interesting behaviour. Although the behaviour of simple 2-D ferromagnetic models is rather well understood, until recently virtually nothing was known about the corresponding antiferromagnets. *Lee* et al. [10.103] studied the antiferromagnetic plane rotator

$$\mathcal{H} = -J \sum_{(ij)} (S_{ix}S_{jx} + S_{iy}S_{jy}) + H \sum_i S_{ix} \quad , \tag{10.6}$$

where $|S_i| = 1$ is a two-dimensional vector, on both square and triangular lattices. The square lattice results showed a single 2nd-order phase boundary around a spin-flop phase in contrast to the theoretical prediction of *Dotsenko* and *Uimin* [10.104]. In contrast, due to the frustration which occurs naturally, the triangular lattice antiferromagnet shows a rich variety of phase transitions (and long-range order!) differing in character as a function of field and temperature. In order to observe all phase boundaries as well as to properly identify all ordered phases, it was necessary to use a two-dimensional complex order parameter. This fact came to light as a result of a group-theoretical analysis; the interplay between theory and simulation was essential for this work. *Miyashita* and *Shiba* [10.105] used a 36-state clock model to approximate this problem and carried out zero-field simulations; the results and interpretation were similar to the continuous spin work. *Miyashita* and *Kawamura* [10.106,107] examined the Heisenberg antiferromagnet on the triangular lattice first in zero field where they found Z_2 vortex unbinding and a Kosterlitz-Thouless-like specific heat peak and then in a uniform magnetic field where they found a rich phase diagram which was in some ways similar to that found for the XY antiferromagnet on a triangular lattice [10.103]. *Kawamura* investigated the phase transition of a Heisenberg antiferromagnet [10.103] on a layered triangular lattice using a symmetry analysis and MC simulations; he found a new universality class-SO(3) An XY antiferromagnet on the layered triangular lattice showed [10.109] similar critical behaviour.

The finite size behaviour of simple, nearest-neighbour Ising models became the subject of intense scrutiny. *Barber* et al. [10.110] used a special-purpose processor to study finite-size behaviour and hyperscaling on the simple cubic Ising model. *Binder* et al. [10.10] showed that this work did not actually test hyperscaling. Several different simulations using different algorithms on different computers [10.31,111,112] demonstrated that the anomalous finite-size behaviour shown in [10.110] was spurious. *Binder* [10.113] examined the five-dimensional hypercubic Ising model and showed that the appropriate length scale for finite-size behaviour was a "thermodynamic" length $\ell \propto |T - T_c|^{2/d}$ instead of the correlation length ξ. The finite-size behaviour of systems undergoing first-order transitions was studied using large scale simulations. *Binder* and *Landau* [10.114] determined the properties of the Ising Square lattice below T_c as a uniform magnetic field was swept through zero. *Challa* et al. [10.11] then carried out very long simulations (up to 3.5×10^7 MC steps/site) to study the $q = 10$ Potts model as the temperature was varied. Both studies showed a simple dependence of all thermodynamic quantities upon the "volume" L^d.

Large-scale simulations were also used to extract information about dynamic critical behaviour. *Williams* [10.44] used a "dynamic MCRG method" (Monte Carlo renormalization group) to analyze data generated on the DAP to obtain a value for the dynamic exponent $z = 2.13 \pm 0.03$ for the Ising square lattice. *Wansleben* and *Landau* [10.45] showed that biased-sampling effects could be noticeable in time-displaced correlation functions (possibly affecting other recent high-resolution work [10.5, 115] and extracted a corrected estimate for the simple cubic Ising model of $z = 2.03 \pm 0.04$. These two results together provide strong evidence to support the predictions obtained using ε-expansions.

Antiferromagnetic Potts models with competing interactions were studied by different groups. *Ono* [10.116] investigated the $q = 3$ antiferromagnet with ferromagnetic next-nearest neighbour coupling and found evidence for an intermediate Kosterlitz-Thouless phase instead of the previously conjectured "broken sublattice phase". In three-dimensions MC data suggest a Kosterlitz-Thouless transition for $q = 3$ [10.117], a long-range ordered state for $q = 4$ and 5 [10.118,119] and no transition for $q = 6$ [10.119].

Surface critical behaviour and scaling of the layer magnetization for the simple cubic Ising model were further studied by *Kikuchi* and *Okabe* [10.46]. They also determined the dynamic-surface critical behaviour and showed that it could be described solely in terms of bulk exponents [10.120]. *Binder* and *Landau* [10.121], and *Binder* et al. [10.122-124] added a surface field and studied wetting transitions in this model finding critical and tricritical transitions, in addition to first-order wetting.

A quantum MC method was used by *Jacobs* et al. [10.125] to examine a two-dimensional model for granular superconductors. Using a path integral formalism they calculated several thermodynamic quantities which were used to extract a phase diagram as a function of a parameter which measures the importance of quantum effects. They found evidence that quantum fluctuations produce a low-temperature first-order re-entrant transition.

10.4 Few- and Many-Fermion Problems

The basic problem of finding a general method to calculate the properties of fermion problems remains. Since the original article (Chap.4) was published, some progress has been made both in understanding the validity of various approximation schemes as well as developing exact fermion methods. In this short addendum we will try to mention some of this progress and give the interested reader some related references.

As in the original article, we will not cover lattice calculations in depth. However, a number of calculations have now been completed on two- and three-dimensional systems [10.126-128]. Most success has been obtained from the application of the Hubbard-Stratanovich method. Here the interaction term which is typically short range and bi-quadratic in the fermion fields is converted to terms that are linear in the fermion-number operator. This is accomplished by introducing an auxiliary boson field which mediates the fermion interaction. If these boson degrees of freedom are integrated over, the new Hamiltonian is converted back to the original Hamiltonian. However, just as in lattice-gauge calculations, the fermion fields can be integrated or summed analytically first. This leaves a sum or integral over the bose fields. Because of the antisymmetry of the fermion variables, the interaction between the boson fields is a complicated determinant. The boson fields can be integrated or summed using the MC method. For some two- or three-dimensional systems (for example, the Hubbard model in two dimensions [10.127] or the magnetic impurity interacting with an electron band in three dimensions [10.128]) these methods give good results in reasonable amounts of computer time. Unfortunately, for systems with continuous spatial degrees of freedom, the required number of boson degrees of freedom is infinite or at least very large so that the method seems not to be practical.

For continuous systems like liquid ^{3}He, the mirror-potential method [10.129] shows promise. In this method, the Schrödinger equation is rewritten as two coupled equation, viz

$$H\psi_+(R) + C(R)\psi_+(R)\psi_-(R) = E\psi_+(R) \quad ,$$

$$H\psi_-(R) + C(R)\psi_+(R)\psi_-(R) = E\psi_-(R) \quad .$$

(10.7)

By subtracting these two equations we see that $\psi_+ - \psi_-$ is a solution of the original Hamiltonian for any symmetric function $C(R)$. The method gets its name from the term $C(R)\psi_+(R)\psi_-(R)$ which acts as a potential for ψ_+ that is proportional to ψ_-, and vice versa. Since these two functions are antisymmetric images of each other this term is called the mirror potential. It acts to keep ψ_+ out of regions of large ψ_-, and vice versa, ensuring that the difference $\psi_+ - \psi_-$ has nonzero overlap with the true fermion ground state. For few particle systems these equations can be used directly much like in [10.130]. By using approximate mirror potentials various approximations to the Fermi ground state can be obtained. Currently work is being done on iteratively correcting the mirror potential and using multiple path methods to obtain improved results.

Elser [10.131] has recently published an approximate fermion algorithm that is essentially an approximation to the method devised by *Arnow* et al. [10.130] using the path-integral formalism. It offers no advantages over that method and has the great disadvantage that configurations are cancelled (rejected in his Metropolis et al. integration) whenever they are within a prescribed range in configuration space. A limiting procedure is then suggested to decrease this range to zero. This limit is likely to be difficult to carry out effectively for a many-body calculation in more than one dimension, since the probability of the overlap will become small rapidly. Extrapolation to zero range will involve an amplification of errors. The method of *Arnow* et al. cancelled configurations with a probability given by the overlap of their Green's functions which is, of course, an exact method, and no limiting procedure is necessary.

The fixed-node method has been applied to many systems with good results. Recently, calculations have been done on liquid ^{3}He, both in bulk [10.129] and for droplets both using a many-body wave function that contains Feynman-Cohen backflow. The latter improves the nodal surfaces compared with those derived from a pure Slater determinant. The fixed-node energy at the experimental equilibrium density is -2.38 ±0.01 K compared to the experimental value of -2.47 K. Similar agreement is found over the entire density range.

10.5 Simulation of Polymer Models

10.5.1 Monte-Carlo Techniques

Recently *Meirovitch* suggested a new technique, based on the concepts of the scanning method, which enables one to estimate approximately the value of the probability with which a conformation is sampled and hence to extract entropy and free energy from a relatively small smaple of chain conformations [10.132]. This method has been applied to single chains [10.132] as well as to multiple chain systems [10.133]. An interesting "grand-canonical" method for single-chain statistics has been propounded recently in [10.134]. Assigning to an N-step walk a fugacity β^N, β being the inverse temperature of an associated Euclidean lattice field theory [10.135], it is possible to estimate the grand partition function and correlation length, and hence the corresponding critical exponents γ and ν, respectively. It should be noted that this procedure is not the same as the usual constant-fugacity MC renormalisation group by *Redner* and *Reynolds* [10.136].

10.5.2 Polymer Networks

Polymer networks have been investigated by *Eichinger* and co-workers. Static properties [10.137] as, e.g., the sol-gel distribution and the population of various types of dangling ends and loop defects in the gel, as well as dynamic properties [10.138] as, for example, the longest relaxation time and its dependency on dangling ends, have been studied.

10.5.3 Polymer Blends

Critical phenomena of polymer blends have been studied by *Sariban* et al. with a novel "grand-canonical" simulation technique [10.139] where chains of type A are transformed into chains of type B, and vice versa, keeping the chemical potential difference fixed. They found that the critical exponents are those of the Ising model ($\beta = 0.32$, $\nu = 0.63$) rather than those of the Flory-Huggins mean-field theory ($\beta = \gamma = 0.5$). Simulations of two-dimensional polymer mixtures [10.140] have demonstrated that, as predicted [10.141], the classical linear theory of spinodal decomposition is not valid in $d = 2$, which is in contrast to three-dimensional blends, where the initial growth is according to the classical theory exponentially fast.

10.5.4 Polymer Melting

Progress in the problem of polymer melting has been made recently by MC simulations. It has been shown that, in contrast to predictions based on Flory's mean-field theory [10.142], semi-rigid chains in the bulk do not exhibit a phase transition to a crystalline state where all chains are stretched and oriented in one direction if excluded volume constraints are present only [10.143-147]. Inclusion of local orientation-dependent interactions of van der Waals type are inevitable in order to observe a phase transition of second order in two dimensions [10.143-145] and of first-order in three dimensions [10.145-147].

10.5.5 Dynamics of Polymers

The dynamics of polymers confined into tubes has been studied by *Kremer* and *Binder* [10.148] in order to elucidate the crossover between Rouse and reptation behaviour. Scaling laws for static and dynamic properties of mean-square displacements and structure factor parallel and perpendicular to the tube are given. Polymer dynamics of a distinctly different type, as compared to the classical Rouse and reptation models, have recently been found for polymer chains performing Brownian motion between randomly distributed impenetrable fixed obstacles (porous medium) [10.149]. Diffusion coefficient and longest relaxation time have been estimated to $D \propto N^{-2.9}$ and $\tau \propto N^{4.0}$, respectively. Dynamics of melts are studied in [10.149a].

10.6 Diffusion in Lattice Gases and Related Kinetic Phenomena

The influence of a memory between several steps of a tracer atom was investigated by *Kutner* [10.150] for lattice gases on the honeycomb lattice. In this lattice memory effects are especially important because of its low coordination number. It turned out that the correlation factor, resulting from the assumption of memory between consecutive steps only, gives a poor description of the simulation data at intermediate concentrations. Some improvement is achieved by taking correlations between four successive transitions into account. The simulation data agree well with the theories of *Nakazato* and *Kitahara* [6.16], and *Tahir-Kheli* and *Elliott* [6.17], respectively. These theories apparently include correlations over many steps, albeit in an approximate way.

An improved theoretical expression for the correlation factor of tracer diffusion in non-interacting lattice gases at arbitrary concentration c was derived by *Tahir-Kheli* [10.151,152] and by *van Beijeren* and *Kutner* [10.153]. These improvements treat the correlated tracer-vacancy motion in a self-consistent fashion. Accurate simulations by *van Beijeren* and *Kutner* [10.153] on a square lattice with 360.000 sites at two concentrations (c $\approx$ 0.501 and c $\approx$ 0.923) verified the validity of the above-mentioned expression for the correlation factor. Further, a logarithmic correction to the asymptotic mean-square displacement was observed in the simulations. This correction corresponds to a long-time tail $\propto t^{-2}$ in the velocity autocorrelation function of the tagged particle. The existence of such long-time tails is predicted by the above-mentioned theories, as well as by mode-coupling theory [10.154]. Tracer diffusion in two-dimensional isotropic and anisotropic lattice gases was investigated by *Tahir-Kheli* and *El-Meshad* [10.155] by precise MC simulations. They first showed that the previous theories (*Nakazato* and *Kitahara* [6.16], *Tahir-Kehli* and *Elliott* [6.17]) are not completely satisfactory for these lattices. With the improved version of the theory good agreement between the simulations and the theory is achieved. Further, these authors observed the logarithmic correction in the mean-square displacement of a particle in two dimensions, too.

Very recently a major extension of the diffusion studies in lattice gases was achieved by the inclusion of drift to the particles. This means that diffusion is investigated in stationary states, rather than in equilibrium. The diffusion of a tagged particle must now be referred to the mean motion of this particle under the influence of the drift field. In linear chains, the mean-square displacement of a tagged particle from its mean position becomes linear in time, for long times [10.156,157]. The simple result $D_{tr} = (1-c)(\Gamma_+ - \Gamma_-)/2$ was obtained for the tracer diffusion coefficient, $(\Gamma_+ - \Gamma_-)$ being the difference in the transition rates between transitions in the field direction and against the field direction. The numerical simulations of *Kutner* and *van Beijeren* [10.157] confirmed this result. Moreover, an approximate theory for the behaviour of the mean-square displacement at arbitrary times was given in [10.157] and verified by the simulations. The collective diffusion which describes the decay of density gradients in the stationary situation turned out to be even more interesting. A mode-coupling theory predicted [10.158] that a density disturbance in a one-dimensional lattice gas, under the influence of a drift field, will spread for long times $\propto t^{2/3}$, i.e. faster than $\propto t^{\frac{1}{2}}$ being characteristic for diffusion. The diffusive behaviour $\propto t^{\frac{1}{2}}$ was observed for short times. The predicted asymptotic $t^{4/3}$ behaviour of the squared spread of a density disturbance was confirmed by numerical simulations at c = 0.503 and several values of the drift field [10.158].

The study of lattice gases was also extended to tracer diffusion in A-B alloys where the constituents A, B have different transition rates and where a finite vacancy concentration is present. In the one-dimensional case not only the mean-square displacement of tagged A or B atoms was estimated by simulations, but also their space- and time-dependent correlation functions [10.159]. *Tahir-Kheli* and *El-Meshad* generalized theoretical expressions of Levitt for the asymptotic form of these correlation functions by scaling arguments to the two-component case. Good agreement with the simulations was found [10.159]. In d = 2 and 3 only the mean-square displacements of A and B atoms and the ensuring correlation factors were studied by the simulations. The results could be well described by theory as long as the two transition rates were not too different [10.160]. Finally, *Tahir-Kehli* simulated also the tagged particle diffusion in a one-dimensional lattice gas with statistically distributed random transition rates [10.161]. The numerical results could be well accounted for by the coherent-potential approximation.

At the time when Chap.6 was written the problem of diffusion in a partially blocked lattice at or near the percolation threshold of the vacancies found increasing attention. Some pertinent references are [10.162-167]. A very high precision in the simulations was achieved particularly by *Pandey*, *Stauffer*, and co-workers [10.168,169]. The conjectures on relations between static and dynamic critical exponents by *Alexander* and *Orbach* [10.170] and *Aharony* and *Stauffer* [10.171] raised great interest. Their validity was investigated by numerical simulations, e.g., in [10.172] and [10.173], respectively. Biased diffusion in percolating systems was studied by *Pandey* [10.173]. See *Havlin* [10.175] for a recent review of diffusion in percolating systems.

We now turn briefly to simulations dealing with the dynamics of systems out of equilibrium. *Marro* and co-workers [10.176] have continued their studies of non-equilibrium phase transitions in a system where jump rates in a particular lattice direction differ from the rates in the inverse direction so that a net drift occurs. Some additional work has also been done continuing simulations of nucleation and growth (e.g. [10.177]) and spinodal decomposition (e.g. [10.140,178]). A relatively new subject in this context, however, is the simulation of the growth of wetting layers [10.179]. There one considers a system with free surfaces and assumes conditions where the forces at all walls would require wetting layers at these walls in equilibrium. As an initial condition, however, it is assumed that the surface is nonwet. For a short range attractive wall potential it is then found that the thickness of the wetting layer grows with time according to a logarithmic law [10.179].

Simulations studying the kinetics of domain growth have been a particularly active area [10.6-8,177,180-189]. We have already discussed in Sect.10.1 the difficulty involved in these studies due to the lack of self-averaging. It turns out that the factors determining the exponent x in the law relating the average domain linear dimension l(t) to the time t elapsed since the start in an initially dis-ordered configuration, $l(t) \propto t^x$, are still controversial [10.6,8,180,182]. More work will be needed to clarify the situation. Particularly interesting seems also the problem of domain growth in the presence of randomly frozen impurities [10.183] or random fields [10.181,189].

A particularly active area are simulations relating to kinetic aspects of per-colation and other clustering processes (diffusion on percolation clusters, diffu-sion-limited aggregation, clustering of clusters, Eden's model, growing random walks, etc.). The recent literature on these subjects simply is too extensive to be listed here in detail; we rather refer to two recent reviews [10.190,191]. In these problems, one is particularly interested in the relationship between the kinetics of the process and the fractal structure of the objects under consider-ation.

10.7 Roughening and Melting in Two Dimensions

10.7.1 Two-Dimensional Melting

The problem whether the melting transition in two dimensions is a first-order transition, qualitatively similar to three-dimensional melting, or whether melting in two dimensions proceeds continuously via two consecutive Kosterlitz-Thouless transitions, with the hexatic phase in between, has still found much interest; both pertinent MC studies [10.192-197] and molecular-dynamics simulations [10.198-201] appeared, as well as very interesting experiments [10.202-205]. Although not all questions are settled (compare, e.g., the discussion in [10.200] versus [10.203] which both deal with monolayer xenon on graphite, and unexplained discrepancies between the molecular dynamics simulation and the experiment remain), most of the evidence points towards a first-order melting transition, *for realistic potentials*. Highlights in this direction are studies at the Delft molecular dynamics special purpose processor [10.198], which show that the orientational correlation length stays finite at the melting transition, and the extensive molecular dynamics work of *Abraham* et al. [10.199,200]. The simulation of krypton on graphite [10.199], using more than 10^5 atoms, certainly is a world record in molecular dynamics, as far as the size of the simulated system is concerned. Among the recent MC work we draw attention to MCRG studies [10.193] and studies of bond-angle order in Lennard-Jones and hard-disk systems [10.194], and the study of the phase diagram of the two-dimensional Coulomb gas [10.196].In a Laplacian roughening model for two-dimensional melting, on the other hand, a crossover from a single first-order transition to a hexatic phase does in fact occur [10.195].

10.7.2 Roughening Transition

Here we can report on much less recent activity with MC methods [10.206,207] but again there exist, for the first time in this field, convincing and stimulating experiments [10.208,209] demonstrating the roughening of stepped crystal surfaces. The simulation of *Selke* and *Szpilka* [10.207] is a nice example of the close interaction between experiment, analytical theory, and MC simulation which now exists for this problem. Finally, we draw attention to the process where a rough surface is brought below its roughening temperature and healing of the roughness begins [10.210]. In a pioneering paper, *Villain* [10.210] maked detailed predictions for the dynamics of this process and draws analogies to the coarsening behaviour seen in spinodal decomposition (Chap.6). Simulations of this problem clearly would be most instructive.

A new direction of research which we like to mention in this context is the study of the configurational properties of rough surfaces [10.211]. Both conformations and dynamical properties of "tethered surfaces" (i.e., two-dimensional surfaces of fixed connectivity embedded in d dimensions, as exemplified by hard spheres tethered together by strings into a triangular net) are obtained. These surfaces differ from the so-called "random surfaces", which are more ramified objects, resembling branched polymers, and were also studied by MC methods [10.212].

10.8 "Random" Systems: Spin Glasses, Percolation, etc.

10.8.1 Spin Glasses

The problem of understanding the physics of spin glasses has continued to be an area of very active research [10.213]. There has been remarkable progress in clarifying the question at which dimensionalities short-range spin glasses have phase transitions at a nonzero freezing temperature; the current consensus of most work is that only the Ising spin glass in d = 3 dimensions has a $T_f \neq 0$ [10.14,34,48, 214] while $T_f = 0$ for two-dimensional Isin spin glasses [10.22,28,216,217] and for isotropic two- and three-dimensional spin glasses [10.35,36,218-222]. Still, questions concerning the precise critical behaviour remain; also for the XY spin glass the possibility of a more subtle phase transition (involving "chirality" variables as suggested by *Villain* [10.223]) has been raised [10.224]. A problem somewhat similar to the isotropic n-vector spin glass is the ordering of quadrupolar glass models. Simulations show that in the isotropic case in three dimensions one again has $T_f = 0$ [10.225].

Due to the very slow relaxation encountered in spin glasses, it is very hard to obtain valid results by the standard Metropolis sampling method. It now is clear that a lot of the early work, as reviewed in [10.226] and Chap.8, is inconclusive due to insufficient equilibration and lack of statistical accuracy. In this situation, the use of vector processors [10.28,34-39,218,219,222,225] or special-purpose computers [10.14,48,214] was extremely valuable for achieving progress. In addition, there is still the need to find more efficient algorithms for spin glasses, avoiding the dramatic slowing down at low temperatures. A promising step in this direction was already mentioned [10.22]. Finally we refer to a recent review [10.213] for a more detailed description of recent spin glass simulations.

10.8.2 Random Fields, Random Impurities, etc.

While in the random-field Ising problem an exact solution has shown that in three dimensions the ground state for weak random fields is ferromagnetic [10.227], the behaviour near the phase transition is still controversial [10.49,228]: Some work claims that the ferromagnetic-paramagnetic transition in the presence of weak random fields is a weak first-order transition [10.228], while other work claims that it is second-order [10.49]. The problem with huge relaxation times is similar to the situation in spin glasses. Thus, it is clear that also for this problem the trend goes towards the use of vector processors [10.228] or special-purpose computer [10.49]. We have already drawn attention to the interesting problem of domain growth in the random field Ising model [10.184,229] and the related problem of domain growth in the presence of random impurities [10.183].

Randomly diluted Ising models with ferromagnetic interactions near the percolation threshold were simulated, with emphasis on their dynamics [10.230,231] to check the expected violation of dynamic scaling. Static aspects of random Ising systems have been discussed by *Labarta* et al. [10.232].

10.8.3 Percolation

We disregard here kinetic aspects of percolation (such a diffusion on percolation clusters) and rather refer to the recent review by *Herrmann* [10.190]. The elastic behaviour of percolation networks has been studied in [10.233] and seems to follow a critical exponent $\mu + 2\nu$ under certain conditions. The related problem of the fracture of disordered solids has recently also been investigated [10.234]. Noise, distribution of voltages, and various moments of the voltage distribution in random resistor networks seems to be described near the percolation threshold by an infinite number of apparently independent exponents [10.237]. Finally we draw attention to a new algorithm to produce lattice animals [10.238].

10.9 Lattice Gauge Theories

Since Chap.9 was originally written, progress has been achieved through calculations of much larger scale, enabled by the availability of supercomputers, through a better understanding of the underlying principles and algorithms, through the development of new techniques and their application to a wide class of systems. The innovative aspects of this latter reserach, however, although non-negligible, have not been as impressive as in the early stages, when the application of the MC method had represented a real break-through in the study of non-perturbative field-theoretical phenomena. To some extent the techniques currently used are the same, as described in the chapter, with most of the progress coming from consideration of systems of greater volume, use of larger statistical samples, more careful analysis of finite-size effects and other possible sources of error. The implementation of renormalization-group transformations and the introduction and testing of a variety of algorithms to include dynamical fermions and to accelerate the simulations probably constitute the most important new methodological developments.

From the point of view of physical results, however, the progress has been remarkable. While almost all of the original calculations were made on computers of the former generation, with scalar architectures, during the last few years supercomputers with larger memories, vector architectures and even multiple processors have been widely employed. This has changed some of the strategies, for instance, the use of discrete subgroups as approximation to the continuous gauge groups is no longer really advantageous, and especially has allowed the consideration of systems of greater extent. For example, lattices as large as $24^3 \times 48$, with 2,654,208 SU(3) matrices as dynamical variables, have been used in investigations of QCD ("quantum chromodynamics").

For the same reasons, as explained in Chap.9, QCD continued to be the major field of interest for lattice MC simulations. Results for many physical quantities, such as the force between static quarks, the deconfinement temperature and the lowest masses in the hadronic spectrum, have been made much more precise, at least within the approximation (commonly called quenched approximation), discussed in the text, of neglecting the dynamical effects of virtual quark-antiquark pairs. New observables such as the spin-dependent potentials, the computation of which is more demanding, have been evaluated with a reasonable accuracy. These investigations have shown, in particular, that finite-size effects in the earlier studies were more important than formerly thought and that scaling towards the continuum limit may not be as precocious as formerly believed. The use of real-space renormalization-group transformations by which the theory is simulated on a large, say 16^4, lattice and renormalized systems over smaller lattices, e.g., 8^4, 4^4, are defined by suitably blocking the gauge variables, has proved very useful.

The problems of including dynamical effects of fermions in a MC simulation remains a fundamental one. Various schemes of approximate solutions have been proposed and applied to the theory of QCD. The questions of how the presence of light quark pairs can affect the thermodynamical properties of the quark gluon plasma and the calculations of the spectrum are of particular interest. Reliable results have been obtained, but the algorithms proposed for the simulations, albeit approximate, are at least one order of magnitude slower than without dynamical quarks, and the systems which can be considered are correspondingly smaller.

Going beyond QCD, interesting applications of lattice MC methods have been made to systems involving scalar bosonic fields, i.e. Higgs fields, together with gauge fields. Since Higgs-gauge systems are the prototype for many theories for particle interactions (the electroweak theory being a notable example) the importance of finding out whether such systems have features not manifested in a perturbative analysis is obvious. The renormalization-group analysis of a system with just scalar fields shows that only a trivial continuum limit of non-interacting fields can be defined. The question is then to what extent such result carries over to the theory with gauge fields and what possible constraints the structure of critical surfaces and fixed points may imply for the continuum limit. Renormalization-group methods have proved useful in this context as well.

Finally, one expects for the near future that computers, yet more powerful, will become available. The scale of the simulations will increase again, but this will bring into evidence problems that, with the present lattice sizes, merely begin to surface. Critical slowing down, in particular, if not brought under control, may well render ineffective the consideration of larger lattices. Anticipating such problems, current research efforts are focused on developing techniques, for instance, based on Fourier acceleration, which may produce a faster evolution of the modes with long wavelength and therefore avoid critical slowing down. These forefront investigations are conceptually very interesting and may prove essential for the forthcoming MC simulations.

The very large amount of work done in the field of MC lattice simulations during the last two years makes it impossible to present here an updated list of references. Singling out just a few contributions would be unfair, while an exhaustive list of even the most important contributions would take several pages. Practitioners of lattice gauge theories have gathered regularly for meetings and conferences. Recent international symposia, sponsored by NATO, have been held at the Brookhaven National Laboratory in September 1986 and at the University of Wuppertal in November 1985. The proceedings, published (Wuppertal) [10.237] or to be published (Brookhaven) by Plenum Press in the NATO.ASI series, are offering good reviews of the most recent developments in the field.

10.10 Concluding Remarks

This chapter clearly shows that the field of MC computer simulation is very well alive and rapidly progressing. In fact, the progress is so fast that the authors hardly could catch up even during the time interval it took writing up this chapter: e.g. while the fastest algorithm mentioned in the main text, the microcanonical

Ising simulation of *Herrmann* [10.32] reaches a speed of 6.7×10^8 MC updates per second, the rumor is spreading out that this world record is broken again by *J.G. Zabolitzky* and *H. Herrmann*, who seem to have reached a speed of 4.2×10^9 MC updates per second (on a CRAY-2 4-processor machine). At this point, however, we also add a word of caution: Although the Herrmann algorithm [10.32] is so fast, it conserves energy precisely, unlike that of *Creutz* [10.32], whose "demon algorithm" conserves energy only approximately; this may be somewhat inconvenient in practical applications, and also the relaxation times needed to reach equilibrium seem to be distinctly larger than in the Metropolis algorithm (perhaps this reflects "hydrodynamical slowing down", see Chap.1). Also the convergence to equilibrium in the Herrmann algorithm is doubtful.

Also it must be emphasized that this chapter certainly could present just a selection of topics and results, *and much other very interesting work so far has not been mentioned*: e.g., work on interfacial adsorption (e.g. [10.238]) lipid layer transitions (e.g. [10.239]), simulation of alloy phase diagrams (e.g. [10.240]), frustration effects in simple cubic lattices (e.g. [10.241]) or in models for mixed ferro-antiferroelectric systems [10.242], etc. *We apologize to all colleagues who have done related interesting work but could not be mentioned in this review.*

References

10.1 K. Binder: In *Monte Carlo Methods in Statistical Physics*, 2nd ed., ed by K. Binder (Springer, Berlin, Heidelberg 1986) Chap.10
10.2 D. Stauffer: *Introduction to Percolation Theory* (Taylor and Francis, London 1985)
10.3 D.C. Rapaport: J. Phys. A18, L175 (1985)
10.4 D. Stauffer, J.G. Zabolitzky: J. Phys. A19, 3705 (1986)
10.5 C. Kalle: J. Phys. A17, L801 (1984)
10.6 A. Milchev, K. Binder, D.W. Heermann: Z. Phys. B63, 521 (1986)
10.7 A. Sadiq, K. Binder: J. Stat. Phys. 35, 617 (1984)
10.8 For a review and recent references, see: K. Binder, D.W. Heermann, A. Milchev, A. Sadiq, in *Glassy Dynamics and Optimization*, ed. by J.L. van Hemmen, I. Morgenstern (Springer, Berlin, Heidelberg 1987) in print
10.9 For a recent review, see: K. Binder: Ferroelectrics (in press, 1987)
10.10 K. Binder, M. Nauenberg, V. Privman, A.P. Young: Phys. Rev. B31, 1499 (1985)
10.11 M.S.S. Challa, D.P. Landau, K. Binder: Phys. Rev. B34, 1841 (1986)
10.12 S. Katznelson, P.G. Lauwers: A potentially dangerous finite-size phenomenon at first-order phase transitions (preprint, University of Bonn, 1986)
10.13 K. Binder: In [Ref.10.1, Chap.1]
10.14 A.T. Ogielski: Phys. Rev. B32, 7384 (1985)
10.15 M.H. Kalos: Brookhaven Conference on Monte Carlo Methods and Future Computer Architectures, May 1983 (unpublished)
10.16 K.E. Schmidt: Phys. Rev. Lett. 51, 2175 (1983)
10.17 G. Parisi: In *Progress in Gauge Field Theory*, ed. by G.'t Hooft et al., NATO Advanced Study Institute, Series B, Vol.115 (Plenum, New York 1984) p.531
10.18 G.G. Batrouni, G. Katz, A. Kronfeld, G. Lepage, S. Svetitsky, K. Wilson: Phys. Rev. D32, 2736 (1985)
10.19 J. Goodman, A. Sokal: Phys. Rev. Lett. 56, 1015 (1986)

10.20 J. Kogut: Paper presented at ICTP Conference on Perspectives in Computational Physics, Trieste, Oct. 1986 (unpublished);
E. Dagotto, J.B. Kogut: preprint; see also S. Duane: Nucl. Phys. B**257** [FS14], 652 (1985);
J. Kogut: Nucl. Phys. B**275** [FS17], 1 (1986)
10.21 R.H. Swendsen, J.W. Wang: Preprint
10.22 R.H. Swendsen, J.S. Wang: Phys. Rev. Lett. **57**, 2607 (1986)
10.23 A. Milchev, D.W. Heermann, K. Binder: J. Stat. Phys. **44**, 749 (1986)
10.24 G.S. Grest, K. Kremer: Phys. Rev. A**33**, 3628 (1986)
10.25 G.S. Grest, K. Kremer, T.A. Witten: Preprint
10.26 C.-K. Hu: Phys. Rev. B**29**, 5103, 5129 (1985); B**32**, 7325 (1985); J. Phys. A**19**, 3067 (1986)
10.27 R. Shankar, R. Gupta, G. Murphy: Phys. Rev. Lett. **55**, 1812 (1985);
R. Gupta, K.G. Wilson, C. Ummrigar: J. Stat. Phys. **43**, 1095 (1986);
see also R. Gupta, R. Cordery: Phys. Lett. A**105**, 415 (1984)
10.28 A.P. Young: Phys. Rev. Lett. **50**, 917 (1983)
10.29 S.F. Reddaway, D.M. Scott, K.A. Smith: Comp. Phys. Comm. **37**, 351 (1985)
10.30 G.O. Williams, M.H. Kalos: J. Stat. Phys. **37**, 283 (1984)
10.31 G. Bhanot, D. Duke, R. Salvador: Phys. Rev. B**33**, 7841 (1986); J. Stat. Phys. **44**, 985 (1986);
see also S. Wansleben: Comp. Phys. Comm. **43**, 315 (1987)
10.32 M. Creutz: Phys. Rev. Lett. **50**, 1411 (1983)
For a much faster alternative see H.J. Herrmann: J. Stat. Phys. **45**, 145 (1986); this code probably suffers from slow relaxation and lack of ergodicity, however
10.33 M. Creutz, P. Mitra, K.J.M. Moriarty: Comp. Phys. Commun. **33**, 361 (1984); J. Stat. Phys. **42**, 823 (1986)
10.34 R.N. Bhatt, A.P. Young: Phys. Rev. Lett. **54**, 924 (1985)
10.35 R.E. Walstedt, L.R. Walker: Phys. Rev. Lett. **47**, 1624 (1981); J. Appl. Phys. **53**, 7985 (1982);
L.R. Walker, R.E. Walstedt: J. Magn. Mat. **31-34**, 1289 (1983)
10.36 A. Chakrabarti, C. Dasgupta: Phys. Rev. Lett. **56**, 1404 (1986); and preprint
10.37 W. Kinzel: Phys. Rev. B**33**, 5086 (1986)
10.38 J.D. Reger, K. Binder: Z. Phys. B**60**, 137 (1985)
10.39 J.G. Zabolitzky, D.J. Bergman, D. Stauffer: J. Stat. Phys. **44**, 211 (1986)
10.40 C. Dasgupta: Phys. Rev. Lett. **55**, 1771 (1985); and preprint
10.41 Y.-L. Wang, F. Lee, J.D. Kimel: preprint;
J.D. Kimel, S. Black, P. Carter, Y.-L. Wang: preprint
10.42 G.S. Pawley, R.H. Swendsen, D.J. Wallace, K.G. Wilson: Phys. Rev. B**29**, 4030 (1984)
10.43 A.D. Bruce: J. Phys. A**18**, L873 (1985)
10.44 J.K. Williams: J. Phys. A**18**, 49 (1985)
10.45 S. Wansleben, D.P. Landau: J. Appl. Phys. (in press); and to be published
10.46 M. Kikuchi, Y. Okabe: Progr. Theor. Phys. **73**, 32 (1985); **74**, 458 (1985)
10.47 A. Baumgärtner: J. Phys. Lett. (Paris) **45**, L-515 (1984)
10.48 A. Ogielski, I. Morgenstern: Phys. Rev. Lett. **54**, 928 (1985)
10.49 A. Ogielski, D.A. Huse: Phys. Rev. Lett. **56**, 1298 (1986)
10.50 J.H. Condon, A.T. Ogielski: Rev. Sci. Instr. **56**, 1691 (1985)
10.51 A. Hoogland, J. Spaa, B. Selman, A. Compagner: J. Comp. Phys. **51**, 250 (1983)
10.52 A. Compagner: preprint;
A. Compagner, A. Hoogland: preprint;
A. Hoogland, H.W.J. Blöte, A. Compagner: preprint;
H.W.J. Blöte, A. Hoogland, A. Compagner: preprint
10.53 H. Herrmann, H.J. Hilhorst: Phys. Blätter **42**, 52 (1986);
H. Herrmann: Proc. Statphys. XVII, Boston, Aug. 1986 [Physica **140**A, 421 (1986)]
10.54 Articles in J. Stat. Phys. **43**, Nos. 3/4 (1986)
10.55 D. Frenkel: In "Molecular Dynamics Simulation of Statistical-Mechanical Systems", ed. by G. Cicotti and W.G. Hoover, to appear in Nuovo Cim.
10.56 W. Schreiner, K.W. Kratky: Chem. Phys. **80**, 245 (1983)
10.57 W. Schreiner, K.W. Kratky: Chem. Phys. **89**, 177 (1984)

10.58 J.J. Erpenbeck: Phys. Rev. A**32**, 2920 (1985)
10.59 K.K. Mon, R.B. Griffiths: Phys. Rev. A**31**, 956 (1986)
10.60 S. Goldman: J. Comp. Phys. **62**, 441 (1986)
10.61 J.P. Hansen, L. Viot: J. Stat. Phys. **38**, 823 (1985)
10.62 J.M. Caillol, D. Levesque: Phys. Rev. B**33**, 499 (1986)
10.63 R. Hasrun, G.R. Luckurst, S. Romano: Mol. Phys. **53**, 1535 (1984)
10.64 M. Wojcik, K.E. Gubbins: Mol. Phys. **53**, 397 (1984)
10.65 M. Wojcik, K.E. Gubbins: Mol. Phys. **51**, 951 (1984)
10.66 R. Eppenga, D. Frenkel: Mol. Phys. **52**, 1303 (1984)
10.67 D. Frenkel, B.M. Mulder: Mol. Phys. **55**, 1171 (1985)
10.68 J.N. Perram, M.S. Wertheim, J.L. Lebowitz, G.O. Williams: Chem. Phys. Lett. **105**, 277 (1984)
10.69 A. Stroobanks, H.N.W. Lekkerkerker, D. Frenkel: Phys. Rev. Lett. **57**, 1452 (1986)
10.70 M. Neumann: Mol. Phys. **50**, 841 (1983)
10.71 M. Neumann, O. Steinhauser, G.S. Pawley: Mol. Phys. **52**, 97 (1984)
10.72 M. Neumann: Mol. Phys. **57**, 97 (1986)
10.73 C.G. Gray, Y.S. Singh, C.G. Joslin, P.T. Cummings, S. Goldman: J. Chem. Phys. **85**, 1592 (1986)
10.74 M. Neumann: J. Chem. Phys. **82**, 5663 (1985)
10.75 M. Neumann: J. Chem. Phys. **85**, 1567 (1986)
10.76 W. Jorgensen, J.D. Madura: Mol. Phys. **56**, 1381 (1985)
10.77 J. Dietrich, G. Corongiu, E. Clementi: Chem. Phys. Lett. **112**, 426 (1984)
10.78 D. Levesque, J.J. Weis, G.N. Patey: Mol. Phys. **51**, 333 (1984)
10.79 J.M. Caillol, D. Levesque, J.J. Weis, P.G. Kusalik, G.N. Patey: Mol. Phys. **55**, 65 (1985)
10.80 G. Alagona, G. Ghio, P. Kollman: J. Am. Chem. Soc. **188**, 185 (1986)
10.81 D.G. Bounds: Mol. Phys. **54**, 1335 (1985)
10.82 S.M. Thompson, K.E. Gubbins, J.P.R.B. Walton, R.A.R. Chantry, J.S. Rowlinson: J. Chem. Phys. **81**, 530 (1984)
10.83 A.P. Shreve, J.P.R.B. Walton, K.E. Gubbins; J. Chem. Phys. **85**, 2178 (1986)
10.84 R.M. Toownsend, J. Gyko, S.A. Rice: J. Chem. Phys. **82**, 4391 (1985)
10.85 J.R. Henderson, F. Van Swol: Mol. Phys. **51**, 921 (1984)
10.86 S.M. Thomson, K.E. Gubbins, D.E. Sullivan, C.G. Gray: Mol. Phys. **51**, 21 (1984)
10.87 D. Levesque, J.J. Weis: J. Stat. Phys. **40**, 29 (1985)
10.88 C.Y. Lee, J.A. McCammon, P.J. Rossky: J. Chem. Phys. **80**, 4448 (1984)
10.89 A.C. Belch, M. Berkowitz: Chem. Phys. Lett. **113**, 278 (1985)
10.90 J.R. Henderson, F. Van Swol: Mol. Phys. **56**, 1313 (1985)
10.91 J.P. Valleau, G.M. Torrie: J. Chem. Phys. **81**, 6291 (1984)
10.92 G.M. Torrie, J.P. Valleau, C.W. Outhwaite: J. Chem. Phys. **81**, 6296 (1984)
10.93 D. Bratko, B. Jönsson, H. Wennerström: Chem. Phys. Lett. **128**, 449 (1986)
10.94 J.C. Talbot, D.J. Tildesley, W.A. Steele: Mol. Phys. **51**, 1331 (1984)
10.95 R.M. Lynden-Bell, J.C. Talbot, D.J. Tildesley, W.A. Steele: Mol. Phys. **54**, 183 (1985)
10.96 J.F. Knight, P.A. Monson: J. Chem. Phys. **84**, 1909 (1986)
10.97 W. Van Megen, I.K. Snook: Mol. Phys. **54**, 741 (1985)
10.98 J. Barrtholomew, R. Hall, B.J. Berne: Phys. Rev. B**32**, 548 (1985)
10.99 A. Wallquist, D. Thirumalai, B.J. Berne: J. Chem. Phys. **85**, 1533 (1986)
10.100 M. Sprik, R.W. Impey, M.L. Klein: J. Chem. Phys. **83**, 5802 (1985)
10.101 M. Sprik, M.L. Klein: J. Chem. Phys. **83**, 3042 (1985)
10.102 M. Sprik, M.L. Klein, D. Chandler: Phys. Rev. B**32**, 545 (1985)
10.103 D.H. Lee, J.D. Joannopoulos, J.W. Negele, D.P. Landau: Phys. Rev. Lett. **52**, 433 (1984); Phys. Rev. B**33**, 450 (1986); and to be published
10.104 V.S. Dotsenko, G.V. Uimin: Zh. Eksp. Teor. Fiz. **40**, 236 (1984)
10.105 S. Miyashita, H. Shiba: J. Phys. Soc. Japan **53**, 1145 (1984)
10.106 H. Kawamura, S. Miyashita: J. Phys. Soc. Japan **53**, 9 (1984)
10.107 H. Kawamura, S. Miyashita: J. Phys. Soc. Japan **54**, 3220 (1985)
10.108 H. Kawamura: J. Phys. Sco. Japan **54**, 3220 (1985)
10.109 H. Kawamura: to be published

10.110 M.N. Barber, R.B. Pearson, D. Toussaint, J.L. Richardon: Phys. Rev. B32, 1720 (1985)
10.111 G. Parisi, F. Rapuano: Phys. Lett. 157B, 301 (1985)
10.112 A. Hoogland, A. Compagner, H.W.J. Blöte: Physica 132A, 457 (1985)
10.113 K. Binder: Z. Phys. B61, 13 (1985)
10.114 K. Binder, D.P. Landau: Phys. Rev. B30, 1477 (1984)
10.115 R.B. Pearson, J.L. Richardson, D. Toussaint: Phys. Rev. B31, 4472 (1985)
10.116 I. Ono: J. Phys. Soc. Japan 53, 4102 (1984)
10.117 I. Ono: to be published
10.118 J.R. Banavar, F.Y. Wu: Phys. Rev. B29, 1511 (1984)
10.119 B. Hoppe, L.L. Hirst: J. Phys. A18, 3375 (1985)
10.120 M. Kikuchi, Y. Okabe: Phys. Rev. Lett. 55, 1220 (1985)
10.121 K. Binder, D.P. Landau: J. Appl. Phys. 57, 3306 (1985)
10.122 K. Binder, D.P. Landau, D.M. Kroll: Phys. Rev. Lett. 56, 2272 (1986)
10.123 K. Binder, D.P. Landau, D.M. Kroll: J. Magn. Magn. Mat. 54-57, 669 (1986)
10.124 K. Binder, D.P. Landau: to be published
10.125 L. Jacobs, J.V. José, M.A. Novotny: Phys. Rev. Lett. 53, 2177 (1984), to be published
10.126 J. Hirsch: Rev. B28, 4059 (1983)
10.127 J.E. Gubernatis, D.J. Scalapino, R.L. Sugar, D.W. Toussaint: Phys. Rev. B32, 103 (1985)
10.128 J. Hirsch, R.M. Frye: Phys. Rev. Lett. 56, 2521 (1986)
10.129 J. Carson, M.H. Kalos: Phys. Rev. C32, 1735 (1985)
10.130 D.M. Arnow, M.H. Kalos, M.A. Lee, K.E. Schmidt: J. Chem. Phys. 77, 5562 (1982)
10.131 V. Elser: Phys. Rev. A34, 2293 (1986)
10.132 H. Meirovitch: Macromolecules 16, 249; 1628 (1983)
10.133 H.C. Öttinger: Macromolecules 18, 92 (1985)
10.134 C. Aragao de Carvalho, S. Caracciolo: J. Physique 44, 323 (1983); B. Berg, W. Förster: Phys. Lett. 106B, 323 (1981)
10.135 C. Aragao de Carvalho, S. Caracciolo, J. Fröhlich: Nucl. Phys. B215, 209 (1983)
10.136 S. Redner, P.J. Reynolds: J. Phys. A14, 2679 (1981)
10.137 Y.-K. Leung, B.E. Eichinger: J. Chem. Phys. 80, 3877 (1984); J. Chem. Phys. 80, 3885 (1984)
10.138 N.A. Neuburger, B.E. Eichinger: J. Chem. Phys. 83, 884 (1985)
10.139 A. Sariban, K. Binder, D.W. Heermann: Phys. Rev. B (1987, in press)
10.140 A. Baumgärtner, D.W. Heermann: Polymer 27, 1777 (1986)
10.141 K. Binder: Phys. Rev. A29, 341 (1984)
10.142 P.J. Flory: Proc. Roy. Soc. A234, 60 (1956)
10.143 A. Baumgärtner: J. Chem. Phys. 81, 484 (1984)
10.145 A. Baumgärtner: J. Physique Lett. 46, 659 (1985)
10.146 A. Kolinski, J. Skolnick, R. Yaris: Macromolecules 19, 2550 (1986)
10.147 R. Dickman, C.K. Hall: J. Chem. Phys. 85, 3023 (1986)
10.148 K. Kremer, K. Binder: J. Chem. Phys. 81, 6381 (1984)
10.149 A. Baumgärtner, M. Muthukumar: (preprint) 1986
10.149a A. Kolinski, J. Skolnick, R. Yaris: J. Chem. Phys. 84, 1922 (1986), and preprints
10.150 R. Kutner: J. Phys. C18, 6323 (1985)
10.151 R.A. Tahir-Kheli: Phys. Rev. B28, 3049 (1983)
10.152 R.A. Tahir-Kheli: Philos. Mag. (to be published)
10.153 H. van Beijeren, R. Kutner: Phys. Rev. Lett. 55, 238 (1985)
10.154 H. van Beijeren: J. Stat. Phys. 35, 399 (1984)
10.155 R.A. Tahir-Kheli, N. El-Meshad: Phys. Rev. B32, 6166 (1985)
10.156 A. De Masi, P.A. Ferrari: J. Stat. Phys. 38, 603 (1985)
10.157 R. Kutner, H. van Beijeren: J. Stat. Phys. 39, 317 (1985)
10.158 H. van Beijeren, R. Kutner, H. Spohn: Phys. Rev. Lett. 54, 2026 (1985)
10.159 R.A. Tahir-Kheli, N. El-Meshad: Phys. Rev. B32, 6184 (1985)
10.160 N. El-Meshad, R.A. Tahir-Kheli: Phys. Rev. B32, 6176 (1985)
10.161 R.A. Tahir-Kheli: Phys. Rev. B31, 644 (1985)
10.162 D. Ben-Avraham, S. Havlin: J. Phys. A15, L691 (1982)

10.163 S. Havlin, D. Ben-Avraham, H. Sompolinsky: Phys. Rev. A**27**, 1730 (1983)
10.164 T. Vicsek: J. Phys. A**16**, 1215 (1983)
10.165 I. Majid, D. Ben-Avraham, S. Havlin, H.E. Stanley: Phys. Rev. B**30**, 1626 (1984)
10.166 P. Argyrakis, R. Kopelman: J. Chem. Phys. **81**, 1015 (1984); **83**, 3099 (1985); Phys. Rev. B**29**, 511 (1984)
10.167 A. Bunde, W. Dieterich, E. Roman: Phys. Rev. Lett. **55**, 5 (1985)
10.168 R.B. Pandey, D. Stauffer: Phys. Rev. Lett. **51**, 527 (1983)
10.169 R.B. Pandey, D. Stauffer, A. Margolina, J.G. Zabolitzky: J. Stat. Phys. **34**, 427 (1984)
10.170 S. Alexander, R. Orbach: J. Physique (Paris) **43**, L625 (1982)
10.171 A. Aharony, D. Stauffer: Phys. Rev. Lett. **52**, 2368 (1984)
10.172 J.G. Zabolitzky: Phys. Rev. B**30**, 4077 (1984)
10.173 H.E. Stanley, I. Majid, A. Margolina, A. Bunde: Phys. Rev. Lett. (Comments) **53**, 1706 (1984)
10.174 R.B. Pandey: Phys. Rev. B**30**, 489 (1984)
10.175 S. Havlin: Preprint
10.176 J. Marro, J.L. Lebowitz, H. Spohn, M.H. Kalos: J. Stat. Phys. **38**, 725 (1985);
J.L. Vallés, J. Marro: J. Stat. Phys. **43**, 441 (1986);
S. Katz, J.L. Lebowitz, H. Spohn: J. Stat. Phys. **34**, 497 (1984);
H. Spohn: In Proc. Enrico Fermi School on "Molecular Dynamics" Simulations of Mechanical-Statistical Systems (unpublished)
10.177 K. Binder, D.W. Heermann: In *Sacling Phenomena in Disordered Systems*, ed. by R. Pynn and A. Skjeltrop (Plenum, New York 1985) p.207
10.178 A. Milchev, D.W. Heermann, K. Binder: preprint;
J. Marro, R. Toral: preprint for Physica B;
R. Toral, J. Marro: Phys. Rev. Lett. **54**, 1424 (1985)
10.179 K.K. Mon, K. Binder, D.P. Landau: preprint
10.180 K. Binder: Ber. Bunsenges. Phys. Chem. **90**, 257 (1986)
10.181 J. Vinals, J.D. Gunton: Surf. Sci. **157**, 473 (1985)
10.182 O.G. Mouritsen: Phys. Rev. B**31**, 2613 (1985); B**32**, 1632 (1985); Phys. Rev. Lett. **56**, 850 (1986)
10.183 G.S. Grest, D.J. Srolovitz: Phys. Rev. B**32**, 3014 (1985);
D.J. Srolovitz, G.S. Grest: Phys. Rev. B**32**, 3021 (1985)
10.184 M. Grant, D.J. Gunton: Phys. Rev. B**29**, 6266 (1984);
E.T. Gawlinski, K. Kaski, M. Grant, J.D. Gunton: Phys. Rev. Lett. **53**, 2266 (1984);
E.T. Gawlinski, S. Kumar, M. Grant, J.D. Gunton, K. Kaski: Phys. Rev. B**32**, 1575 (1985)
10.185 E.T. Gawlinski, M. Grant, J.D. Gunton, K. Kaski: Phys. Rev. B**31**, 281 (1985);
C. Dasgupta, R. Pandit: Phys. Rev. B**33**, 4752 (1986)
10.186 J. Vinals, M. Grant, M. San Miguel, J.D. Gunton, E.T. Gawlinski: Phys. Rev. Lett. **54**, 1264 (1985);
K. Kaski, T. Ala-Nissila, J.D. Gunton: Phys. Rev. B**31**, 310 (1985);
K. Kaski, J. Nieminen, J.D. Gunton: Phys. Rev. B**31**, 2998 (1985)
10.187 G.S. Grest, P.S. Sahni: Phys. Rev. B**30**, 2261 (1984)
10.188 G.S. Grest, S.A. Safran, P.S. Sahni: J. Appl. Phys. **55**, 2432 (1986)
10.189 E. Pytte, J.F. Fernandez: Phys. Rev. B**31**, 616 (1985);
D. Chowdhury, D. Stauffer: Z. Physik B**60**, 249 (1985);
see also D. Stauffer, C. Hartzstein, K. Binder, A. Aharony: Z. Physik B**55**, 325 (1984)
10.190 H.J. Herrmann: Phys. Rep. **136**, 143 (1986);
T.A. Witten: In Proceedings of the 1986 Les Houches Summer School, to be published
10.191 L. Pietroneri, E. Tosatti (eds.): *Fractals in Physics* (North-Holland, Amsterdam 1986);
N. Boccara, M. Daoud (eds.): *Physics of Finely Divided Matter*, Springer Proc. Phys., Vol.5 (Springer, Berlin 1985);
H.E. Stanley, N. Ostrowsky (eds.): *On Growth and Form* (Martinus Nijhoff, The Hague 1985)

10.192 M.P. Allen, D. Frenkel, W. Gignac, J.P. MacTague: J. Chem. Phys. **78**, 4206 (1983)
10.193 D. Nicolaides: Phys. Rev. B**30**, 3824 (1984)
10.194 K.J. Strandburg, J.A. Zollweg, G.V. Chester: Phys. Rev. B**30**, 2755 (1984)
10.195 K.J. Strandburg: Phys. Rev. B**34**, 3536 (1986)
10.196 J.M. Caillol, D. Levesque: Phys. Rev. B**33**, 499 (1986)
10.197 B. Joos, M.S. Duesberry: Phys. Rev. B**33**, 8632 (1986)
10.198 A.F. Bakker, C. Bruin, H.J. Hilhorst: Phys. Rev. Lett. **52**, 449 (1984)
10.199 F.F. Abraham, W.E. Rudge, D.J. Auerbach, S.W. Koch: Phys. Rev. Lett. **52**, 445 (1984)
10.200 F.F. Abraham: Phys. Rev. B**29**, 2606 (1984)
10.201 J.D. Weeks, J.Q. Broughton: J. Chem. Phys. **78**, 4197 (1983)
10.202 M.D. Migone, Z.R. Li, M.H.W. Chan: Phys. Rev. Lett. **53**, 810 (1984)
10.203 P. Dimon, P.M. Horn, M. Sutton, R.J. Birgeneau, D.E. Moncton: Phys. Rev. A**31**, 437 (1985)
10.204 Q.M. Zhang, H.K. Kim, M.H.W. Chan: Phys. Rev. B**33**, 5149 (1986)
10.205 R. Marx: Phys. Repts. **125**, 1 (1985)
10.206 Y. Saito, M. Uchiyame: In *Materials Science of Minerals and Rocks*, ed. by I. Sunagawa (Terra, Tokyo/Reidel, Dordrecht) to be published
10.207 W. Selke, A.M. Szpilka: Z. Phys. B**62**, 381 (1986)
10.208 E.H. Conrad, R.M. Aten, D.S. Kaufman, L.R. Allen, M. den Nijs, E.K. Riedel: J. Chem. Phys. **84**, 1015 (1986)
10.209 M. den Nijs, E.K. Riedel, E.H. Conrad, T. Engel: Phys. Rev. Lett. **55**, 1689 (1985)
10.210 J. Villain: Europhysics Lett. **2**, 531 (1986)
10.211 Y. Kantor, M. Kardar, D.R. Nelson: Phys. Rev. Lett. **57**, 791 (1986); and preprint
10.212 B. Berg, A. Billoire: Phys. Lett. **139**B, 297 (1984); B. Baumann, B. Berg: Phys. Lett. **164**B, 131 (1985)
10.213 K. Binder, A.P. Young: Rev. Mod. Phys. **58**, 801 (1986)
10.214 A.T. Ogielski: preprint; I. Morgenstern: preprint
10.215 W.L. McMillan: Phys. Rev. B**29**, 4026 (1984); B**30**, 476 (1984)
10.216 D. Huse, I. Morgenstern: Phys. Rev. B**32**, 3032 (1985)
10.217 W. Kinzel, K. Binder: Phys. Rev. B**29**, 1300 (1984)
10.218 S. Jain, A.P. Young: J. Phys. C**19**, 3913 (1986)
10.219 B.W. Morris, S.G. Colborne, M.A. Moore, A.J. Bray, J. Canisius: J. Phys. C**19**, 1157 (1986)
10.220 M. Cieplak, J.R. Banavar: Phys. Rev. B**29**, 469 (1984)
10.221 W.L. McMillan: Phys. Rev. B**31**, 342 (1985)
10.222 A. Olive, D. Sherrington, A.P. Young: Phys. Rev. B**34**, 6341 (1986)
10.223 J. Villain: In *Ill-Condensed Matter*, ed. by R. Balian, R. Maynard, G. Toulouse (North-Holland, Amsterdam 1979) p.521
10.224 H. Kawamura, M. Tanemura: J. Phys. Soc. Japan **54**, 4479 (1985)
10.225 H.-O. Carmesin, K. Binder: unpublished
10.226 K. Binder, D. Stauffer: In *Monte Carlo Methods in Statistical Physics*, ed. by K. Binder, Topics Curr. Physics, Vol.7 (Springer, Berlin, Heidelberg 1979) Chap.8
10.227 J.Z. Imbrie: Phys. Rev. Lett. **53**, 1747 (1984)
10.228 A.P. Young, M. Nauenberg: Phys. Rev. Lett. **54**, 2429 (1985)
10.229 E. Pytte, J.F. Fernandez: Phys. Rev. B**31**, 616 (1985)
10.230 D. Chowdhury, D. Stauffer: J. Phys. A**19**, L19 (1986); J. Stat. Phys. **44**, 203 (1986)
10.231 S. Jain: J. Phys. A**19**, L57 (1986)
10.232 A. Labarta, J. Marro, T. Tejada, J. Marro, A. Labarta, J. Tejada: Phys. Rev. B**34**, 347 (1986); B. Derrida, B. Southern, D. Stauffer: J. Physique, in press (1987)
10.233 S. Roux: J. Phys. A**19**, L351 (1986); M. Sahimi: J. Phys. C**19**, L79 (1986); and [Ref.10.39]
10.234 L. de Arcangelis, H.J. Herrmann, S. Redner: J. Phys. (Paris) **46**, L585 (1985); B.K. Chakrabarti, D. Chowdhury, D. Stauffer: Z. Phys. B**62**, 343 (1986)

10.235 L. de Arcangelis, S. Redner, A. Coniglio: Phys. Rev. B31, 4725 (1985);
       R. Blumenfeld, A. Aharony: J. Phys. A18, L443 (1985);
       A.B. Harris: preprint
10.236 D. Dhar, P.M. Lam: J. Phys. A19, L1057 (1986) and preprints
10.237 B. Bunk, K.H. Mütter, K. Schilling (eds.): *Lattice Gauge Theory, A Challenge in Large-Scale Computing* (Plenum, New York 1986)
10.238 P. Rujan, W. Selke, G. Uimin: Z. Physik B65, 235 (1986);
       W. Selke: Ber. Bunsenges. Phys. Chem. **90**, 232 (1986);
       T. Ala-Nissila, J. Amar, J.D. Gunton: J. Phys. A19, L41 (1986);
       T. Ala-Nissila, J.D. Gunton, K. Kaski: Phys. Rev. B33, 7583 (1986);
       references to less recent work can be found in the quoted papers
10.239 O.G. Mouritsen, M.J. Zuckermann: Eur. Biophys. J. **12**, 75 (1985), and preprints;
       O.G. Mouritsen, M. Bloom: Biophys. J. **46**, 141 (1984), and references therein
10.240 U. Gahn: Phys. Chem. Solids **47**, 1153 (1986);
       K. Binder: Festkörperprobleme (Advances in Solid State Physics) **26**, 133, ed. by P. Grosse (Vieweg, Braunschweig 1986);
       H.T. Diep, A. Ghazali, B. Berge, P. Lallemand: Europhys. Letters **2**, 603 (1986);;
       H. Ackermann, S. Crusius, G. Inden: Acta Met. **34**, 2311 (1986)
       B. Dünweg, K. Binder: preprint

Additional References with Titles

Chapter 1

M.N. Barber, R.B. Pearson, D. Toussaint, J.L. Richardson: Finite size scaling in the three-dimensional Ising model. Phys. Rev. B**32**, 1720 (1985)

N.H. Christ, E.A. Terrano: A very fast parallel processor. Unpublished

G. Ciccotti, G. Jacucci, I.R. McDonald: 'Thought-Experiments' by Molecular Dynamics. J. Stat. Phys. **21**, 1 (1979)

M. Creutz: Microcanonical Monte Carlo Simulation. Phys. Rev. Lett. **50**, 1411 (1983)

V. Gerold, J. Kern: The determination of atomic interaction energies in solid solutions from short range order coefficients - an inverse Monte Carlo method. in *Atomic Transport and Defects in Metals by Neutron Scattering*, ed. by C. Janot, W. Petry, D. Richter, T. Springer (Springer, Berlin 1986) p.17

H.J. Hilhorst, A.F. Bakker, C. Bruin, A. Compagner, A. Hooghland: Special Purpose Computers in Physics, J. Stat. Phys. **34**, 987 (1984)

H. Meirovitch: Methods for estimating entropy with computer simulation: the simple cubic Ising lattice. J. Phys. A**16**, 839 (1983)

K.E. Schmidt: Using renormalization group ideas in Monte Carlo Sampling. Phys. Rev. Lett. **51**, 2175 (1983)

Chapter 2

D.J. Adams: On the use of the Ewald summation in computer simulation. J. Chem. Phys. **78**, 2585 (1983)

M.P. Allen, D. Frenkel, W. Gignac, J-P. McTague: A Monte Carlo simulation study of the two-dimensional melting mechanism. J. Chem. Phys. **78**, 4206 (1983)

R. Bacquet, P.J. Rossky: Corrections to the HNC equation for associating electrolytes. J. Chem. Phys. **79**, 1419 (1983)

A.F. Bakker, C. Bruin, F. van Dieren, H.J. Hilhorst: Molecular dynamics of 17000 Lennard-Jones particles. Phys. Lett. **93A**, 67 (1982)

H.J. Böhm, I.R. McDonald, P.A. Madden: An effective pair potential for liquid acetonitrile. Mol. Phys. **49**, 347 (1983)

P. Bopp, G. Jancsó, K. Heinzinger: An improved potential for non-rigid water molecules in the liquid phase. Chem. Phys. Lett. **98**, 129 (1983)

D.G. Bounds: On the MCY potential for liquid water. Chem. Phys. Lett. **96**, 604 (1983)

H. Breitenfelder-Manske: The use of the superposition approximation in perturbation theories of liquids. Mol. Phys. **48**, 209 (1983)

Ph. Choquard, J. Clerouin: Cooperative phenomena below melting of the one-component two-dimensional plasma. Phys. Rev. Lett. **50**, 2086 (1983)

M. Claessens, M. Ferrario, J.P. Ryckaert: The structure of liquid benzene. Mol. Phys. **50**, 217 (1983)

E. Clementi, P. Habitz: A new two-body water-water potential. J. Phys. Chem. **87**, 2815 (1983)

D.F. Coker, J.R. Reimers, R.O. Watts: The infrared absorption spectrum of water. Aust. J. Phys. **35**, 623 (1982)

M. Creutz: Microcanonical Monte Carlo simulation. Phys. Rev. Lett. **50**, 1411 (1983)

T. Croxton, D.A. McQuarrie, G.N. Patey, G.M. Torrie, J.P. Valleau: Ionic solution near an uncharged surface with image forces. Can. J. Chem. **59**, 1998 (1981)

M.P. d'Evelyn, S.A. Rice: Comment on the configuration space diffusion criterion for optimization of the force bias Monte Carlo method. Chem. Phys. Lett. **77**, 630 (1981)

M.P. d'Evelyn, S.A. Rice: A study of the liquid-vapor interface of mercury: Computer simulation results. J. Chem. Phys. **78**, 5081 (1983)

M.P. d'Evelyn, S.A. Rice: A pseudoatom theory for the liquid-vapor interface of simple metals: Computer simulation studies of sodium and cesium. J. Chem. Phys. **78**, 5225 (1983)

S.W. De Leeuw, J.W. Perram, E.R. Smith: Simulation of electrostatic systems in periodic boundary conditions. III. Further theory and applications. Proc. Roy. Soc. London A**388**, 177 (1983)

P.A. Egelstaff, J.H. Root: The temperature dependence of the structure of water. Chem. Phys. **76**, 405 (1983)

M.W. Evans, M. Ferrario: Computer simulation of the molecular dynamics of liquid dichloro-methane. Adv. Mol. Rel. Int. Proc. **24**, 75 (1982)

M. Ferrario, M.W. Evans: Molecular dynamics computer simulation of liquid dichloro-methane. Chem. Phys. **72**, 141 (1982)

M. Fixman: Direct simulation of the chemical potential. J. Chem. Phys. **78**, 4223 (1983)

A. Fukumoto, A. Ueda, Y. Hiwatari: Molecular dynamics studies of superionic conductors. Interionic potential and ion-conducting phase. J. Phys. Soc. Japan **51**, 3966 (1982)

G.A. Gaballa, G.W. Neilson: The effect of pressure on the structure of light and heavy water. Mol. Phys. **50**, 97 (1983)

I.P. Gibson, J.C. Dore: Neutron diffractive studies of water. III Structural change as a function of temperature. Mol. Phys. **48**, 1019 (1983)

S. Goldman: A simple new way to help speed up Monte Carlo convergence rates: Energy-scaled displacement Monte Carlo. J. Chem. Phys. **79**, 3938 (1983)

J.M. Gonzáles Miranda, V. Torra: A molecular dynamics study of liquid sodium at 373 K. J. Phys. F: Met. Phys. **13**, 281 (1983)

J. Gryko, S.A. Rice: The structure of the liquid-vapour interface of sodium caesium alloys. J. Phys. F: Met. Phys. **12**, L245 (1982)

S. Gupta, J.M. Haile, W.A. Steele: Use of computer simulation to determine the triplet distribution function in dense fluids. Chem. Phys. **72**, 425 (1982)

D.M. Heyes: Molecular dynamics simulations of restricted primitive model 1: 1 electrolytes. Chem. Phys. **69**, 155 (1982)

C. Hoheisel, U. Deiters, K. Lucas: The extension of pure fluid thermodynamic properties to supercritical mixtures. A comparison of current theories with computer data over a large region of states. Mol. Phys. **49**, 159 (1983)

W.L. Jorgensen, J. Chandrasekhar, J.D. Madura, R.W. Impey, M.L. Klein: Comparison of simple potential functions for simulating liquid water. J. Chem. Phys. **79**, 926 (1983)

W.L. Jorgensen, J.D. Madura: Solvation and conformation of methanol in water. J. Am. Chem. Soc. **105**, 1407 (1983)

R.H. Kincaid, H.A. Scheraga: Acceleration of converge in Monte Carlo simulations of aqueous solutions using the Metropolis algorithm. Hydrophobic hydration of methane. J. Comp. Chem. **3**, 525 (1982)

P.N. Kusalik, S.F. O'Shea: Hard discs with embedded three dimensional quadrupoles. Mol. Phys. **49**, 33 (1983)

S. Labik, I. Nezbeda: Fluid of general hard triatomic molecules II. Monte Carlo simulation results for a non-linear molecule model. Mol. Phys. **48**, 97 (1983)

D. Levesque, J.J. Weis: Surface properties of the three-dimensional one-component plasma. J. Stat. Phys. **33**, 549 (1983)

D. Levesque, J.J. Weis, G.N. Patey: Fluids of Lennard-Jones spheres with dipoles and tetrahedral quadrupoles. A comparison between computer simulation and theoretical results. Mol. Phys. **51**, 333 (1984)

P. Linse, B. Jönsson: A Monte-Carlo study of the electrostatic interaction between highly charged aggregates. A test of the cell model applied to micellar systems. J. Chem. Phys. **78**, 3167 (1983)

M. Marchesi: Molecular dynamics simulation of liquid water between two walls. Chem. Phys. Lett. **97**, 224 (1983)

W.J. McNeil, W.G. Madden, A.D.J. Haymet, S.A. Rice: Triplet correlation functions in the Lennard-Jones fluid: Tests against molecular dynamics simulations. J. Chem. Phys. **78**, 388 (1983)

P.A. Monson, W.A. Steele, W.B. Streett: Equilibrium properties of molecular fluids with charge distributions of quadrupolar symmetry. J. Chem. Phys. **78**, 4126 (1983)

P.A. Monson, K.E. Gubbins: Equilibrium properties of the Gaussian overlap fluid. Monte Carlo simulation and thermodynamic perturbation theory. J. Phys. Chem. **87**, 2852 (1983)

C.S. Murthy, K. Singer, R. Vallauri: Computer simulation of liquid chlorine. Mol. Phys. **49**, 803 (1983)

M. Neumann, O. Steinhauser: On the calculation of the dielectric constant using the Ewald-Kornfeld tensor. Chem. Phys. Lett. **95**, 417 (1983)

M. Neumann: Dipole moment fluctuation formulas in computer simulations of polar systems. Mol. Phys. **50**, 841 (1983)

S. Okazaki, K. Nakanishi, H. Touhara: Computer experiments on aqueous solution. I. Monte Carlo calculation on the hydration of methanol in an infinitely dilute aqueous solution with a new water-methanol pair potential. J. Chem. Phys. **78**, 454 (1983)

P. Perez, W.K. Lee, E.W. Prohofsky: Study of hydration of the Na^+ ion using a polarizable water model. J. Chem. Phys. **79**, 388 (1983)

G. Ravishanker, M. Mezei, D.L. Beveridge: Monte Carlo computer simulation study of the hydrophobic effect. Faraday Symposium 17 (1983)

F.H. Ree: Simple mixing rule for mixtures with exp-6 interactions. J. Chem. Phys. **78**, 409 (1983)

F.H. Ree: Solubility of H_2-He mixtures in fluid phases to 1 Gpa. J. Phys. Chem. **87**, 2846 (1983)

R.O. Rosenberg, R. Mikkilineni, B.J. Berne: Hydrophobic effect on chain folding. The trans to gauche isomerization of n-butane in water. J. Am. Chem. Soc. **104**, 7647 (1982)

L.F. Rull, S. Toxvaerd: The structure and thermodynamics of a solid-fluid interface II. J. Chem. Phys. **78**, 3273 (1983)

W. Schreiner, K.W. Kratky: Finiteness effects in computer simulation of fluids with spherical boundary conditions. Mol. Phys. **50**, 435 (1983)

K.S. Shing, K.E. Gubbins: The chemical potential in non-ideal liquid mixtures. Computer simulation and theory. Mol. Phys. **49**, 1121 (1983)

O. Steinhauser: On the structure and dynamics of liquid benzene. Chem. Phys. **73**, 155 (1982)

O. Steinhauser: On the orientational structure and dielectric properties of water. A comparison of ST2 and MCY potential. Ber. Bunsenges. Phys. Chem. **87**, 128 (1983)

B. Svensson, B. Jönsson: On the mean spherical approximation (MSA) for colloidal systems. A comparison with results from Monte Carlo simulations. Mol. Phys. **50**, 489 (1983)

J. Sys, S. Labik, A. Malijevsky: Parametrization of the radial distribution function and thermodynamic properties of the restricted primitive model of electrolyte solutions. Czech. J. Phys. **B33**, 763 (1983)

H. Tanaka, K. Nakanishi, N. Watanabe: Constant temperature molecular dynamics calculation on Lennard-Jones fluid and its application to water. J. Chem. Phys. **78**, 2626 (1983)

W.E. Thiessen, A.H. Narten: Neutron diffraction study of light and heavy water mixtures at 25° C. J. Chem. Phys. **77**, 2656 (1982)

S.M. Thompson, K.E. Gubbins, D.E. Sullivan, C.G. Gray: Structure of a diatomic fluid near a wall. II. Lennard-Jones fluid. Mol. Phys. **51**, 21 (1984)

D.J. Tildesley, E. Enciso, P. Sevilla: Monte Carlo simulation and thermodynamic perturbation theory for mixtures of diatomic molecules. Chem. Phys. Lett. **100**, 508 (1983)

Chr. Votava, R. Ahlrichs, A. Geiger: The HCl-HCl interaction: From quantum mechanical calculations to properties of the liquid. J. Chem. Phys. **78**, 6841 (1983)

J.S. Whitehouse, D. Nicholson, N.G. Parsonage: A grand ensemble Monte Carlo study of krypton adsorbed on graphite. Mol. Phys. **49**, 829 (1983)

M. Wojcik, K.E. Gubbins: Thermodynamics of hard dumbbell mixtures. Mol. Phys. **49**, 1401 (1983)

J. Yao, R.A. Greenkorn, K.C. Chao: Thermodynamic properties of Stockmayer molecules by Monte Carlo simulation. J. Chem. Phys. **76**, 4657 (1982)

Chapter 5

A. Baumgärtner, K. Kremer, K. Binder: Dynamics of entangled flexible polymers: Monte Carlo simulations and their interpretation. Discuss. Faraday Chem. Soc. **18**, 37 (1983)

A. Baumgärtner: Diffusion and Brownian motion of polymeric liquids. J. Poly. Sci. C. Symp. **73**, 181 (1985)

A. Baumgärtner, D.Y. Yoon: Phase transition of dense square lattice polymer systems. J. Chem. Phys. **79**, 521 (1983)

M. Bishop, M.H. Kalos, H.L. Frisch: The influence of attraction on the static and dynamic properties of simulated single and multichain systems. J. Chem. Phys. **79**, 3500 (1983)

M. Bishop, M.H. Kalos, A.D. Sokal, H.L. Frisch: Scaling in multichain polymer systems in two and three dimensions. J. Chem. Phys. **79**, 3496 (1983)

T.M. Birshtein, A.M. Skvortsov, A.A. Sariban: Structure of polymer solutions: scaling and modelling on an electronic computer. Polymer **24**, 1145 (1983)

V. Buscio, M. Vacatello: Lipid bilayers in the 'fluid' state: Computer simulations. Mol. Cryst. Liq. Cryst. **97**, 195 (1983)

J.M. Deutsch: Dynamic Monte Carlo simulation of an entangled many-polymer system. Phys. Rev. Lett. **27**, 926 (1982)

Y.-K. Leung, B.E. Eichinger: Computer simulations of end-linked elastomers, J. Chem. Phys. **80**, 3877 (1984)

P.G. Khalatur, S.G. Pletneva, Yu.G. Papulov: Monte Carlo simulation of multiple chain systems: the concentration dependence of osmotic pressure. Chem. Phys. **83**, 97 (1984)

D.E. Kranbuehl, P.H. Verdier: Simulation of the dynamic and equilibrium properties of many-chain polymer systems. Macromolecules **17**, 749 (1984)

K. Kremer: Statics and dynamics of polymeric melts. Macromolecules **16**, 1632 (1983)

M. Muthukumar: Monte Carlo renormalization group calculation for polymers. J. Stat. Phys. **30**, 457 (1983)

R.J. Needs, S.F. Edwards: Computer simulations of the dynamics of star molecules. Macromolecules **16**, 1492 (1983)

R.J. Needs: A computer simulation of the effect of primitive path length fluctuations in the reptation model. Macromolecules **17**, 437 (1984)

D.Y. Yoon, A. Baumgärtner: Phase transition of cubic lattice polymer systems. Macromolecules **17**, 2864 (1984)

Chapter 6

J.C. Angles d'Auriac, A. Benoit, R. Rammal: Random walk on fractals: numerical studies in two dimensions. J. Phys. A**16**, 4039 (1983)

J.C. Angles d'Auriac, R. Rammal: Scaling analysis for random-walk properties on percolation clusters. J. Phys. C**16**, L825 (1983)

D. Ben-Avraham, S. Havlin: Diffusion on percolation clusters at criticality. J. Phys. A**15**, L691 (1982)

D. Bensimon, E. Domany, A. Aharony: Crossover of Fractal Dimension in Diffusion-Limited Aggregates. Phys. Rev. Lett. **51**, 1394 (1983)

S. Havlin, D. Ben-Avraham: Diffusion and fraction dimensionality on fractals and on percolation clusters. J. Phys. A**16**, L483 (1983)

S. Havlin, D. Ben-Avraham, H. Sompolinsky: Scaling behavior of diffusion on percolation clusters. Phys. Rev. A (Rapid Commun.) **27**, 1730 (1983)

D.W. Heermann: Metastability and Spinodal Nucleation. Preprint

D.W. Heermann: Test of the Validity of the Classical Theory of Spinodal Decomposition. Phys. Rev. Lett. **52**, 1126 (1984)

D.W. Heermann: Mean-Field-Like Behaviour in the Metastable and Spinodal Region of Binary Systems. Z. Phys. B**61**, 311 (1985)

D.W. Heermann, A. Coniglio, W. Klein, D. Stauffer: Nucleation and Metastability in Three-Dimensional Ising Models. J. Stat. Phys. **36**, 447 (1984)

H.J. Herrmann: The moles' labyrinth: a growth model. J. Phys. A**16**, L611 (1983)

K. Kaski, K. Binder, J.D. Gunton: A study of cell distribution functions of the three-dimensional Ising model. Phys. Rev. B**29**, 3996 (1984)

K. Kaski, J.D. Gunton: Universal dynamical scaling in the clock model. Phys. Rev. B**28**, 5371 (1983)

K. Kaski, M.C. Yalabik, J.D. Gunton, P.S. Sahni: Dynamics of random interfaces in an order-disorder transition. Phys. Rev. B**28**, 5263 (1983)

S.L. Katz, J.L. Lebowitz, H. Spohn: Phase transitions in stationary nonequilibrium states of model lattice systems. Phys. Rev. B**28**, 1655 (1983)

K. Kawasaki, T. Nagai: Statistical dynamics of Interacting Kinks I. Physica **121**A, 175 (1983)

W. Klein, C. Unger: Pseudospinodals, spinodals and nucleation. Phys. Rev. B**28**, 445 (1983)

S.W. Koch, R.C. Desai, F.F. Abraham: Dynamics of phase separation in two-dimensional fluids: spinodal decomposition. Phys. Rev. A**27**, 2152 (1983)

S.W. Koch, R. Liebmann: Comparison of molecular dynamics and Monte-Carlo computer simulations of spinodal decomposition. J. Stat. Phys. **33**, 31 (1983)

M. Kolb, R. Botet, R. Jullien: Scaling of Kinetically Growing Clusters. Phys. Rev. Lett. **51**, 1127 (1983)

R. Kutner: Tracer diffusion in honeycomb lattice. Correlations over several consecutive jumps. Solid State Ionics **9-10**, 1409 (1983)

R. Kutner: Hopping in concentrated two-dimensional lattice-gases. Correlation factor for tracer diffusion of noninteracting particles in a honeycomb lattice. J. Phys. C**18**, 6323 (1983)

P. Meakin: Effects of particle drift on diffusion-limited aggregation. Phys. Rev. B**28**, 5221 (1983)

P. Meakin: Formation of Fractal Clusters and Networks by Irreversible Diffusion-Limited Aggregation. Phys. Rev. **51**, 1119 (1983)

P. Meakin, H.E. Stanley: Spectral Dimension for the Diffusion-Limited Aggregation Model of Colloid Growth. Phys. Rev. Lett. **51**, 1457 (1983)

P. Meakin, T.A. Witten: Growing interface in diffusion-limited aggregation. Phys. Rev. A**28**, 2985 (1983)

O.G. Mouritsen: Domain growth kinetics of herringbone phases. Phys. Rev. B**28**, 3150 (1983)

G.E. Murch: Simulation of Diffusion Kinetics with the Monte Carlo method, in *Diffusion in Solids II*, ed. by G.E. Murch, N.S. Nowick (Academic, New York), to be published

T. Nagai, K. Kawasaki: Molecular Dynamics of Interacting Kinks I. Physica **120**A, 587 (1983)

M. Nauenberg: Critical growth velocity in diffusion-controlled aggregation. Phys. Rev. B**28**, 449 (1983)

M. Nauenberg, R. Richter, L.M. Sander: Crossover in diffusion-limited aggregation. Phys. Rev. B**28**, 1649 (1983)

R.B. Pandey, D. Stauffer: Confirmation of dynamical scaling at the percolation threshold. Phys. Rev. Lett. **51**, 527 (1983)

R.B. Pandey, D. Stauffer: Diffusion on random systems above, below, and at their percolation threshold in two and three dimensions. J. Stat. Phys. **34**, 427 (1984)

A. Sadiq, K. Binder: Dynamics of the formation of two-dimensional ordered structures. J. Stat. Phys. **35**, 517 (1984)

S.A. Safran, P.S. Sahni, G.S. Grest: Kinetics of ordering in two dimensions: I Model Systems. Phys. Rev. B**28**, 2686 (1983)

P.S. Sahni, D.J. Srolovitz, G.S. Grest, M.P. Anderson, S.A. Safran: Kinetics of ordering in two dimensions: II Quenched systems. Phys. Rev. B**28**, 2705 (1983)

R.A. Tahir-Kheli: Correlated diffusion in quasi-low-dimensional systems. Phys. Rev. B**27**, 6072 (1983)

T. Vicsek: Fractal models for diffusion-controlled aggregation. J. Phys. A**16**, L647 (1983)

T.A. Witten Jr., P. Meakin: Diffusion limited aggregation at multiple growth sites. Phys. Rev. B**28**, 5632 (1983)

Chapter 7

On Roughening

C. Jayaprakash, W.F. Saam, S. Teitel: Roughening and Facet Formation in Crystals. Phys. Rev. Lett. **50**, 2017 (1983)

R. Miranda, E.V. Albano, S. Daiser, G. Ertl, K. Wandelt: Experimental Evidence of a Roughening Transition in Adsorbed Xenon Multilayers. Phys. Rev. Lett. **51**, 782 (1983)

P.E. Wolf, S. Balibar, F. Gallet: Experimental Observation of a Third Roughening Transition on hcp ^{4}He Crystals. Phys. Rev. Lett. **51**, 1366 (1983)

R.K.P. Zia, J.E. Avron: Total Surface Energy and Equilibrium Shapes: Exact Results for the d=2 Ising Crystal. Phys. Rev. B**25**, 2042 (1982)

On Melting

P. Choquard, J. Clerouin: Cooperative Phenomena below Melting of the One-Component Two-Dimensional Plasma. Phys. Rev. Lett. **50**, 2086 (1983)

S.W. Koch, F.F. Abraham: Freezing Transition of Xenon on Graphite: A Computer-Simulation Study. Phys. Rev. B**27**, 2964 (1983)

T.F. Rosenbaum, S.E. Nagler, P.M. Horn, R. Clarke: Experimental Observation of Continuous Melting into a Hexatic Phase. Phys. Rev. Lett. **50**, 1791 (1983)

K.J. Strandburg, S.A. Solla, G.V. Chester: Monte Carlo Studies of a Laplacian Roughening Model for Two Dimensional Melting. Phys. Rev. B**28**, 2717 (1983)

Chapter 8

D. Andelman, H. Orland, L.C.R. Wijwardhana: The lower critical dimension of the random field Ising model: A Monte Carlo study. Phys. Rev. Lett. **52**, 145 (1984)

K. Binder, A.P. Young: Logarithmic dynamic scaling in spin glasses. Phys. Rev. B**29**, 2864 (1984)

H.M.J. Boots, R.B. Pandey: Qualitative analysis of free-radical cross-linking polymerization by computer simulation. Unpublished

R.B. Grzonka, M.A. Moore: Computer studies of two-level systems of the three-dimensional planar spin glass. J. Phys. C**17**, 2785 (1984)

C.L. Henley: Computer search for defects in a d=3 Heisenberg spin glass. Ann. Phys. (N.Y.) **156**, 324 (1984)

H.J. Herrmann, D. Stauffer: Corrections to scaling and finite size effects. Phys. Letters **100**A, 366 (1984); Erratum **102**A, 446 (1984)

N. Jan, T. Lookman, D.L. Hunter: Phase diagram, critical properties and dimensional effects of the kinetic gelation model. J. Phys. A**16**, L757 (1983)

T. Kawasaki: Concentration dependence of modulated phases in diluted ANNNI models — A Monte Carlo simulation. Progr. Theor. Phys. **71**, 246 (1984)

T. Lookman et al.: Real space renormalization group for kinetic gelation. Phys. Rev. B**29**, 2805 (1984)

A. Margolina, H. Nakanishi, D. Stauffer, H.E. Stanley: Monte Carlo and series study of corrections to scaling in 2D percolation. J. Phys. A**17**, 1683 (1984)

A. Margolina, H.J. Herrmann: One finite-size scaling of the order parameter in the disordered phase. Phys. Lett. **104**A, 295 (1984)

D. Matthews-Morgan, D.P. Landau, H.J. Herrmann: Effects of solvent in a kinetic gelation model. Phys. Rev. B**29**, 6328 (1984)

W.L. McMillan: Monte Carlo simulation of the two-dimension random (±J) model. Phys. Rev. B**28**, 5216 (1983)

B.W. Morris, A.J. Bray: Monte Carlo studies of spin glasses with uniaxial anisotropy. J. Phys. C**17**, 1717 (1984)

R.W. Walstedt: Spin glass behavior in finite numerical samples. In *Heidelberg Colloquium on Spin Glasses*, ed. by J.L. van Hemmen, I. Morgenstern (Springer, Berlin, Heidelberg, New York 1983) p.177

S. Wansleben, J.G. Zablitzky, C. Kalle: Monte Carlo simulation of Ising models by multi spin coding technique on a vector computer. J. Stat. Phys. **37**, 271 (1984)

S. Wilke, E. Guyon, G. de Marsily: Water penetration through rocks: A percolation description. Unpublished

D. Wilkinson, J.F. Willemsen: Invasion percolation: a new form of percolation theory. J. Phys. A**16**, 3365 (1983)

Subject Index

Domain coarsening 8,181,209,214,215, 300,301,312,313,314,315

Domain wall 113,117,204,214,254,263

Drift 311,312

Droplets 210-212,305,309

Dual representative 223

Duality 117,224,227-231

Dumbbells 55,56,59,304

Dynamic critical phenomena 93,114,115, 301,307,315

Dynamic structure factor 173,174

Dynamical matrix 243

Dzyaloshinskii-Moriya (DM) interaction 252

Edwards-Anderson (EA) model 242,244, 248,249

Edwards-Anderson order parameter 245, 246-249

Electrolytes 46,50,81

Electron gas 139,218

Electron glass 244

Electron layers 48,305

Encounter model 183

End-to-end distance 151,153,154,160, 161,168

Entanglements 168,170-172,302

Entropy 1,99,107,148,247,252,253,309

Enumeration techniques 151,160-162,243

Equation-of-motion technique 243

Essential singularity 224,227,232,233, 278

Ewald summation 31,46-50,57,58,81

Exact partition function calculations 243-250,253,255

Exchange interactions 241,242

Excitation spectrum 11,243,244

Exciton transport 195

Excluded volume interaction 147,149, 152-154,160,175

ε-Expansion 155,158,161,163

Facets 231

Fermi ground state 134,136,157

Fermions 294,307,308,316

Feynman diagrams 282

Filtering method 141

Finite size effects 13,14,17,20,22,29, 31,34,103,106,115,120,227,229,236,247, 249,253,256,262,289,299,300,303

Finite size scaling 5,14,17,19,21,23, 29,31,94,97,115,117,151,159,160,175, 262-267,301

First-order phase transition 1,14,16, 22,23,24,28,48-50,95,97,104,105,109, 116-119,166,224,232-236,301,310,315

Fixed node approximation 126,137,139, 141,142,309

Fixed point 19,98,99,109,265,317

Flory-Huggins approximation 167,370

Fluctuation-dissipation theorem 1,18, 30,245

Fluid-solid transition 48

Force bias sampling 38,39,73,74

Fractal dimension 5,262,313

Free boundary conditions 21,48,101

Free draining limit 168,169

Free energy 1,7,16,20,23,24,27-29,73, 78,79,103,107,166,210,212,231,242, 249,292,309

Freezing transition 243-245,251,252, 314

Frequency-dependent diffusion coefficient 188,189

Friedel oscillations 49

Frustration 106,242,248,251,253,306,318

Fully frustrated systems 251

Gauge fields 277-281,284,294,295,316

Gauge transformations 250,289

Gaussian model 242,245,247,249

Gelation 209,216-218,268

Gels 153

Glass transition 44,165,173

Glasses 163,173,243,244,270

Glauber kinetic Ising model 7,115,209, 212,253

Globule state 157,158

Goldstone modes 15